Methods in Molecular Biology

Series Editor
John M. Walker
School of Life and Medical Sciences
University of Hertfordshire
Hatfield, Hertfordshire, UK

For further volumes:
http://www.springer.com/series/7651

For over 35 years, biological scientists have come to rely on the research protocols and methodologies in the critically acclaimed *Methods in Molecular Biology* series. The series was the first to introduce the step-by-step protocols approach that has become the standard in all biomedical protocol publishing. Each protocol is provided in readily-reproducible step-by-step fashion, opening with an introductory overview, a list of the materials and reagents needed to complete the experiment, and followed by a detailed procedure that is supported with a helpful notes section offering tips and tricks of the trade as well as troubleshooting advice. These hallmark features were introduced by series editor Dr. John Walker and constitute the key ingredient in each and every volume of the *Methods in Molecular Biology* series. Tested and trusted, comprehensive and reliable, all protocols from the series are indexed in PubMed.

Plant Meiosis

Methods and Protocols

Edited by

Mónica Pradillo

Departamento de Genética, Fisiología y Microbiología, Facultad de Biología, Universidad Complutense de Madrid, Madrid, Spain

Stefan Heckmann

Leibniz Institute of Plant Genetics and Crop Plant Research (IPK) OT Gatersleben, Stadt Seeland, Germany

Humana Press

Editors
Mónica Pradillo
Departamento de Genética
Fisiología y Microbiología
Facultad de Biología
Universidad Complutense
de Madrid
Madrid, Spain

Stefan Heckmann
Leibniz Institute of Plant Genetics and Crop
Plant Research (IPK) OT Gatersleben
Stadt Seeland, Germany

ISSN 1064-3745 ISSN 1940-6029 (electronic)
Methods in Molecular Biology
ISBN 978-1-4939-9817-3 ISBN 978-1-4939-9818-0 (eBook)
https://doi.org/10.1007/978-1-4939-9818-0

This Humana imprint is published by the registered company Springer Science+Business Media, LLC, part of Springer Nature.
The registered company address is: 233 Spring Street, New York, NY 10013, U.S.A.

Preface

Meiosis is one of the key processes of heredity in sexually reproducing eukaryotes, assuring genetic variation and accurate transmission of the genetic material from generation to generation. Despite being a highly evolutionarily conserved process, specific features of meiosis within kingdoms and even within species are found. Historically, plants provided a good model for studies of meiosis due to their rather large chromosomes or accessibility of meiotic material. In recent years, advanced cytological studies, facilitated by improvements in microscopy, combined with genetic, bioinformatic, and biochemical approaches, have greatly contributed to further elucidate the genetic control of plant meiosis while providing valuable knowledge for plant breeders to improve and accelerate breeding programs.

Despite the small size of its chromosomes and without agro-economical interest, *Arabidopsis thaliana*, a species within the Brassicaceae family, has brought forward our knowledge of meiosis in plants. It was the first plant genome to be sequenced, and meiotic studies allowed an in-depth dissection of meiotic progression and recombination in plants leading to the identification of several key genes and factors involved in these processes. More recently, much progress has been made in various crop species.

In addition to getting a deeper insight into the genetic control of plant meiosis, some of the current challenges for researchers in this field are to understand this division in a situation where more than two chromosome sets are present (polyploid genome organization) or in species with much larger and complex genomes than Arabidopsis. A comprehensive understanding of how polypoidization and genome size/organization affect meiosis is essential for developing breeding strategies and innovative crop improvement programs since a large number of crops are polyploids and/or have large and complex genomes. By taking advantage of the enhanced chromosome visualization protocols (e.g., by super-resolution microscopy), as well as of the advances in molecular procedures and -omics techniques, specifically conducted in meiocytes and applied to both diploid and polyploid plants (both model and crop species), it will be possible to have a holistic view of how meiosis works in plants and will allow to improve and accelerate plant breeding programs.

In this book, expert researchers provide effective and innovative tools that will serve the meiosis research community in achieving these goals. Part I (Chapters 1–6) includes cytological and imaging approaches to study meiosis and chromosome dynamics in Arabidopsis, in both diploid and polyploid backgrounds. Part II (Chapters 7–15) focuses on cytological techniques to study meiosis in species other than Arabidopsis. Part III (Chapters 16–20) presents protocols related to molecular and biochemical approaches to study plant meiosis. Finally, Part IV (Chapters 21–27) compiles further procedures beneficial for plant meiosis research.

We are highly grateful to all the contributors for their effort in turning complex methods into practical resources, and we hope that this book will be useful for both experienced and new researchers in the field of plant meiosis and chromosome biology.

Madrid, Spain — *Mónica Pradillo*
Stadt Seeland, Germany — *Stefan Heckmann*

Contents

Contributors

YUN-JAE AHN • *Leibniz Institute of Plant Genetics and Crop Plant Research (IPK), Stadt Seeland, Germany*

LORINDA K. ANDERSON • *Department of Biology, Colorado State University, Fort Collins, CO, USA*

SUSAN J. ARMSTRONG • *School of Biosciences, University of Birmingham, Birmingham, UK*

MIKEL ARRIETA • *Cell and Molecular Sciences, The James Hutton Institute, Dundee, Scotland, UK*

MOHAMMAD A. AYOUB • *Leibniz Institute of Plant Genetics and Crop Plant Research (IPK), Stadt Seeland, Germany*

AURÉLIE BERARD • *Etude du Polymorphisme des Génomes Végétaux (EPGV), INRA, Université Paris-Saclay, Evry, France*

NADIA BESSOLTANE • *Institut Jean-Pierre Bourgin, INRA, AgroParisTech, CNRS, Université Paris-Saclay, Versailles, France*

ALEXANDER R. BLACKWELL • *Department of Plant Sciences, University of Cambridge, Cambridge, UK*

AURÉLIEN BLARY • *Institut Jean-Pierre Bourgin, INRA, AgroParisTech, CNRS, Université Paris-Saclay, Versailles, France*

PABLO BOLAÑOS-VILLEGAS • *Laboratory of Cell and Molecular Biology, Fabio Baudrit Agricultural Research Station, University of Costa Rica, Alajuela, Costa Rica*

AQIN CAO • *College of Life Sciences, Wuhan University, Wuhan, China*

DELPHINE CHARIF • *Institut Jean-Pierre Bourgin, INRA, AgroParisTech, CNRS, Université Paris-Saclay, Versailles, France*

CHANGBIN CHEN • *Department of Horticultural Science, University of Minnesota, Saint Paul, MN, USA*

KYUHA CHOI • *Department of Plant Sciences, University of Cambridge, Cambridge, UK; Department of Life Sciences, Pohang University of Science and Technology, Pohang, Gyeongbuk, Republic of Korea*

JIHED CHOUAREF • *Swammerdam Institute for Life Sciences, Universiteit van Amsterdam, Amsterdam, The Netherlands*

NICOLAS CHRISTOPHOROU • *Institut Jean-Pierre Bourgin, INRA, AgroParisTech, CNRS, Université Paris-Saclay, Versailles, France*

ISABELLE COLAS • *Cell and Molecular Sciences, The James Hutton Institute, Dundee, Scotland, UK*

OLIVIER CORITON • *Institut de Génétique, Environnement et Protection des Plantes, INRA, UMR 1349 IGEPP, Molecular cytogenetics Platform, Le Rheu Cedex, France*

MARIA CUACOS • *Leibniz Institute of Plant Genetics and Crop Plant Research (IPK), Stadt Seeland, Germany*

NIEVES CUÑADO • *Departamento de Genética, Fisiología y Microbiología, Facultad de Biología, Universidad Complutense de Madrid, Madrid, Spain*

BENOIT DARRIER • *INRA UMR 1095, Génétique, Diversité & Ecophysiologie des Céréales, Clermont-Ferrand, France; School of Agriculture, Food and Wine, Faculty of Science, The University of Adelaide, Urrbrae, SA, Australia*

Elena V. Deineko • *Institute of Cytology and Genetics, Russian Academy of Sciences, Siberian Branch, Novosibirsk, Russian Federation*

Stuart Desjardins • *Department of Genetics and Genome Biology, University of Leicester, Leicester, UK*

Nico De Storme • *Horticell Lab, Department of Plants and Crops, Faculty of Bioscience Engineering, Ghent University (UGent), Ghent, Belgium*

Annika Dorn • *Botanical Institute, Karlsruhe Institute of Technology, Karlsruhe, Germany*

Steven Dreissig • *Leibniz Institute of Plant Genetics and Crop Plant Research (IPK), Stadt Seeland, Germany*

Stefanie Dukowic-Schulze • *Department of Horticultural Science, University of Minnesota, Saint Paul, MN, USA*

Rocio Escobar-Guzman • *Grupo de Desarrollo Reproductivo y Apomixis, UGA Langebio, Cinvestav Irapuato, Mexico*

Xiaobing Feng • *College of Life Science, Xinyang Normal University, Xinyang, Henan, China*

Jörg Fuchs • *Leibniz Institute of Plant Genetics and Crop Plant Research (IPK), Stadt Seeland, Germany*

Nelson Garcia • *Department of Horticultural Science, University of Minnesota, Saint Paul, MN, USA*

Danny Geelen • *Horticell Lab, Department of Plants and Crops, Faculty of Bioscience Engineering, Ghent University (UGent), Ghent, Belgium*

Adrián Gonzalo • *Institut Jean-Pierre Bourgin, INRA, AgroParisTech, CNRS, Université Paris-Saclay, Versailles, France*

Santiago Valentín Galvan Gordillo • *Instituto de Biotecnología/UNAM, Cuernavaca, Morelos, Mexico*

Yuki Hamamura • *Department of Developmental Biology, University of Hamburg, Hamburg, Germany*

Stefan Heckmann • *Leibniz Institute of Plant Genetics and Crop Plant Research (IPK) OT Gatersleben, Stadt Seeland, Germany*

Ian R. Henderson • *Department of Plant Sciences, University of Cambridge, Cambridge, UK*

James D. Higgins • *Department of Genetics and Genome Biology, University of Leicester, Leicester, UK*

Axel Himmelbach • *Leibniz Institute of Plant Genetics and Crop Plant Research (IPK), Stadt Seeland, Germany*

Andreas Houben • *Leibniz Institute of Plant Genetics and Crop Plant Research (IPK), Stadt Seeland, Germany*

Menghui Hu • *College of Life Science, Xinyang Normal University, Xinyang, Henan, China*

Virginie Huteau • *Institut de Génétique, Environnement et Protection des Plantes, INRA, UMR 1349 IGEPP, Molecular cytogenetics Platform, Le Rheu Cedex, France*

Lydia Jaffrelo • *GDEC, INRA, Université Clermont Auvergne, Clermont-Ferrand, France*

Eric Jenczewski • *Institut Jean-Pierre Bourgin, INRA, AgroParisTech, CNRS, Université Paris-Saclay, Versailles, France*

Sateesh Kagale • *National Research Council Canada, Saskatoon, SK, Canada*

Kostya Kanyuka • *Department of Biointeractions and Crop Protection, Rothamsted Research, Harpenden, UK*

Judith Kappermann • *Leibniz Institute of Plant Genetics and Crop Plant Research (IPK), Stadt Seeland, Germany*
Marie-Therese Kurzbauer • *Max Perutz Labs, Department of Chromosome Biology, Vienna BioCenter (VBC), University of Vienna, Vienna, Austria*
Christophe Lambing • *Department of Plant Sciences, University of Cambridge, Cambridge, UK*
Marie-Christine le Paslier • *Etude du Polymorphisme des Génomes Végétaux (EPGV), INRA, Université Paris-Saclay, Evry, France*
Helian Liu • *College of Life Sciences, Wuhan University, Wuhan, China*
Leslie D. Lohmiller • *Department of Biology, Colorado State University, Fort Collins, CO, USA*
Malcolm Macaulay • *Cell and Molecular Sciences, The James Hutton Institute, Dundee, Scotland, UK*
Martin Mascher • *Leibniz Institute of Plant Genetics and Crop Plant Research (IPK), Stadt Seeland, Germany*
Sybille U. Mittmann • *Cell and Molecular Sciences, The James Hutton Institute, Dundee, Scotland, UK; Division of Plant Sciences, University of Dundee at The James Hutton Institute, Dundee, Scotland, UK*
Emilie Montes • *IGEPP, INRA, Agrocampus Ouest, Université de Rennes, Le Rheu, France*
José Mora-Calderón • *Laboratory of Cell and Molecular Biology, Fabio Baudrit Agricultural Research Station, University of Costa Rica, Alajuela, Costa Rica*
Chris Morgan • *John Innes Centre, Norwich, UK*
Sergey R. Mursalimov • *Institute of Cytology and Genetics, Russian Academy of Sciences, Siberian Branch, Novosibirsk, Russian Federation*
Tomás Naranjo • *Departamento de Genética, Fisiología y Microbiología, Facultad de Biología, Universidad Complutense de Madrid, Madrid, Spain*
Nathalie Nesi • *IGEPP, INRA, Agrocampus Ouest, Université de Rennes, Le Rheu, France*
Pablo Parra-Nunez • *Departamento de Genética, Fisiología y Microbiología, Facultad de Biología, Universidad Complutense de Madrid, Madrid, Spain*
Marina Pfalz • *Institut Jean-Pierre Bourgin, INRA, AgroParisTech, CNRS, Université Paris-Saclay, Versailles, France*
Dylan W. Phillips • *Institute of Biological, Environmental and Rural Sciences, Aberystwyth University, Aberystwyth, UK*
Charles Poncet • *GDEC, INRA, Université Clermont Auvergne, Clermont-Ferrand, France*
Mónica Pradillo • *Departamento de Genética, Fisiología y Microbiología, Facultad de Biología, Universidad Complutense de Madrid, Madrid, Spain*
Pilar Prieto • *Departamento de Mejora Genética, Instituto de Agricultura Sostenible (IAS), Consejo Superior de Investigaciones Científicas (CSIC), Córdoba, Spain*
Maria Ada Prusicki • *Department of Developmental Biology, University of Hamburg, Hamburg, Germany*
Holger Puchta • *Botanical Institute, Karlsruhe Institute of Technology, Karlsruhe, Germany*
Luke Ramsay • *Cell and Molecular Sciences, The James Hutton Institute, Dundee, Scotland, UK*
Daniel Rodriguez-Leal • *Grupo de Desarrollo Reproductivo y Apomixis, UGA Langebio, Cinvestav Irapuato, Mexico; Inari Agriculture, Cambridge, MA, USA*
Arnaud Ronceret • *Instituto de Biotecnología/UNAM, Cuernavaca, Morelos, Mexico*

ALEVTINA RUBAN • *Leibniz Institute of Plant Genetics and Crop Plant Research (IPK), Stadt Seeland, Germany*

TWAN RUTTEN • *Leibniz Institute of Plant Genetics and Crop Plant Research (IPK), Stadt Seeland, Germany*

JUAN LUIS SANTOS • *Departamento de Genética, Fisiología y Microbiología, Facultad de Biología, Universidad Complutense de Madrid, Madrid, Spain*

PETER SCHLÖGELHOFER • *Max Perutz Labs, Department of Chromosome Biology, Vienna BioCenter (VBC), University of Vienna, Vienna, Austria*

ARP SCHNITTGER • *Department of Developmental Biology, University of Hamburg, Hamburg, Germany*

KALANI SCOTT-MORAGA • *Laboratory of Cell and Molecular Biology, Fabio Baudrit Agricultural Research Station, University of Costa Rica, Alajuela, Costa Rica*

ZEESHAN SHAMIM • *School of Biosciences, University of Birmingham, Birmingham, UK; Department of Biotechnology, Mirpur University of Science and Technology (MUST), Mirpur Azad Kashmir, Pakistan*

LINDSAY A. SHEARER • *Department of Biology, Colorado State University, Fort Collins, CO, USA*

ARUN S. K. SHUNMUGAM • *National Research Council Canada, Saskatoon, SK, Canada*

YURIY V. SIDORCHUK • *Institute of Cytology and Genetics, Russian Academy of Sciences, Siberian Branch, Novosibirsk, Russian Federation*

JASON SIMS • *Max Perutz Labs, Department of Chromosome Biology, Vienna BioCenter (VBC), University of Vienna, Vienna, Austria*

PIERRE SOURDILLE • *INRA UMR 1095, Génétique, Diversité & Ecophysiologie des Céréales, Clermont-Ferrand, France; School of Agriculture, Food and Wine, Faculty of Science, The University of Adelaide, Urrbrae, SA, Australia*

STEPHEN M. STACK • *Department of Biology, Colorado State University, Fort Collins, CO, USA*

JEAN-PHILIPPE VIELLE-CALZADA • *Grupo de Desarrollo Reproductivo y Apomixis, UGA Langebio, Cinvestav Irapuato, Mexico*

JIANBO WANG • *College of Life Sciences, Wuhan University, Wuhan, China*

ROBBIE WAUGH • *Cell and Molecular Sciences, The James Hutton Institute, Dundee, Scotland, UK; Division of Plant Sciences, University of Dundee at The James Hutton Institute, Dundee, Scotland, UK*

EVA WEGEL • *John Innes Centre, Norwich, UK*

DAN D. WU • *Leibniz Institute of Plant Genetics and Crop Plant Research (IPK), Stadt Seeland, Germany; Triticeae Research Institute, Sichuan Agricultural University, Wenjiang, China*

LIYU YANG • *College of Life Sciences, Wuhan University, Wuhan, China*

CHI ZHANG • *College of Life Science, Xinyang Normal University, Xinyang, Henan, China*

ZAIBAO ZHANG • *College of Life Science, Xinyang Normal University, Xinyang, Henan, China; Institute for Conservation and Utilization of Agro-Bioresources in Dabie Mountains, Xinyang, Henan, China*

YONG H. ZHOU • *Triticeae Research Institute, Sichuan Agricultural University, Wenjiang, China*

Part I

Cytological and Imaging Approaches to Study Arabidopsis Meiosis

Chapter 1

A Practical Guide to Live-Cell Imaging of Meiosis in Arabidopsis

Maria Ada Prusicki, Yuki Hamamura, and Arp Schnittger

Abstract

Plants are powerful model systems to study meiosis. Our knowledge about the cytology of plant meiosis is mainly based on the analysis of fixed material. Although highly informative, this approach is limited in understanding the dynamics of meiosis. Here, we present a step-by-step instruction for a newly developed method to follow meiosis in male meiocytes of Arabidopsis in real time by confocal laser scanning microscopy. We envision that this method can be easily translated to other plant species and especially crops (e.g., *Brassica*, maize, and potato).

Key words Meiosis, Arabidopsis, Confocal laser scanning microscopy, Live cell imaging, Time lapse

1 Introduction

Over the last decades, live-cell imaging has become an increasingly important methodology to study biological processes spanning from the observation of organs and tissues to the analysis of cellular behavior and subcellular structures. This also holds true for the analysis of meiosis. In yeast and animals, this is exemplified by studies addressing questions about meiotic progression and chromosome behavior, for example, in *S. pombe* and *S. cerevisiae* [1, 2], in *C. elegans* [3–5], and in mammals [6–9]. In contrast, live-cell imaging of meiosis in plants has so far been limited to a handful of studies in maize [10–12] and Arabidopsis [13]. At the same time, plants are a powerful model system to study meiosis due to several advantages, for example the finding that plants appear to have relaxed checkpoints in meiosis that allow following meiotic progression in mutants that are arrested or even lethal in animals [14]. Moreover, in many plant species, flowers, and with that meiocytes, are generated over the course of several days or weeks providing samples for a long time and at least male meiocytes are relatively easily accessible.

Mónica Pradillo and Stefan Heckmann (eds.), *Plant Meiosis: Methods and Protocols*, Methods in Molecular Biology, vol. 2061, https://doi.org/10.1007/978-1-4939-9818-0_1, © Springer Science+Business Media, LLC, part of Springer Nature 2020

To close the gap in live-cell analysis of meiosis in plants, we have developed a robust imaging setup that relies on the observation of male meiosis in anthers of living flower buds [15]. Our method relies on a standard confocal microscopy system, which can be found in many universities and research institutions. This imaging system has been applied to Arabidopsis, since this model plant has many advantages for an immediate application of our method (e.g., the ease by which transgenic plants that contain meiotic reporters can be generated and the large collection of mutants affecting meiosis).

So far, the technique has been used to quantitatively dissect progression of male meiosis with high temporal and spatial resolution in diploid Arabidopsis plants as well as for the analysis of meiotic mutants and the monitoring of cohesin dynamics [15, 16].

We envision that this method can in future be adjusted with small amendments for the study of female meiosis in Arabidopsis as well as the analysis of meiosis in many other plant species (e.g., *Brassica*, maize, and potato).

2 Materials

2.1 Plant Material and Growth Conditions

1. Seeds of Arabidopsis plants carrying transgenes that encode fluorescent reporters (*see* **Note 1**) are surface-sterilized, for example, with chlorine gas (6.5% NaClO, 37% HCl) under vacuum, sown on 1% agar plates (half-strength Murashige and Skoog (MS) salts and 1% sucrose, pH 5.8) and stored 3 days at 4 °C in the dark for stratification. Herbicides may be added to the medium to select for seedlings containing the reporter genes (*see* **Note 2**). Plants are germinated at long-day conditions (16 h day/8 h night regime at 22 °C/18 °C) for 10 days. After germination, plants are transferred on soil and grown under short-day conditions for 2 weeks (12 h day–12 h night regime at 21 °C/18 °C). Subsequently, they are transferred to long-day conditions to induce flowering (*see* **Note 3**). Inflorescences in the second and third week after flower initiation are in the optimal state for imaging. In general, six plants are often enough to carry out an imaging experiment for over 2 weeks. It is, however, good practice to plan the sowing well ahead to have material available continuously.

2.2 Flower Dissection and Sample Mounting

1. Petri dishes (60 × 15 mm) filled with autoclaved 1% agarose in distilled water as dissection support. They can be sealed and stored at 4 °C until needed.
2. Petri dishes (35 × 10 mm) filled with Apex Culture Medium (ACM) supplemented with vitamins as described in [17]:

MS is dissolved in distilled water to obtain a concentration of 0.22% (2.2 g/l) together with 1% sucrose (10 g/l), pH is adjusted to 5.8 using KOH. After adjusting the pH to 5.8 with KOH, 0.8% agarose (8 g/l) is added and the final solution is sterilized by autoclaving. Vitamins from the 1000× stock are added to the hand-warm solution. The 1000× vitamin stock is prepared as follows: 10% myoinositol, 0.1% nicotinic acid, 0.1% pyridoxine hydrochloride, 0.1% thiamine hydrochloride, 0.2% glycine in distilled water. The solution is filter-sterilized (22 μm filters) under laminar hood, aliquoted, and store at −20 °C).

The petri dishes can be sealed and stored at 4 °C.

3. Autoclaved 2% agarose in distilled water or in ACM.
4. Dissecting microscope with magnification power of approximately 5× (e.g., Zeiss Stemi2000 and Stemi508).
5. Fine tweezers: suggested thickness: 0.02/0.01 mm.
6. Micropipette for 40–50 μl volumes with tips.

2.3 Image Acquisition

1. Upright Confocal Microscope (CLSM) equipped with water dipping objective (e.g., Zeiss LSM880 Airyscan® and Zeiss W Plan-Apochromat 40×/1.0 DIC objective).
2. Double-sided adhesive tape and glass slides or microscope-stage adaptor for petri dish (35 × 10 mm).
3. Syringe pump and disposable syringe 10 ml.
4. Autoclaved distilled water (ca. 15–20 ml of which about 5 ml for the petri dish at first, and slightly more than 10 ml for the syringe).

3 Methods

3.1 Flower Dissection

1. Grow plants to an age of 5 or 6 weeks (Fig. 1a).
2. Clip one entire inflorescence off using fine tweezers, right above the last opened flower leaving 2 or 3 mm of the stem attached. Place the inflorescence on the 1% agarose dissection plate (Fig. 1b, c).
3. Remove one by one all flower buds that have an elongated shape by cutting them at the lower part of the pedicel (Fig. 1d). In the wild type (Col-0), male meiosis takes place in flower buds that are 0.3–0.6 mm long and have a round shape (Fig. 1d, e and *see* **Note 4**). Healthy wild-type inflorescences have approximately four flower buds that proceed through meiosis in a slightly shifted manner with respect to each other. When the flower primordium of interest is reached, anchor the entire stem of the inflorescences into the medium, and, with the help of the medium as support, gently remove the

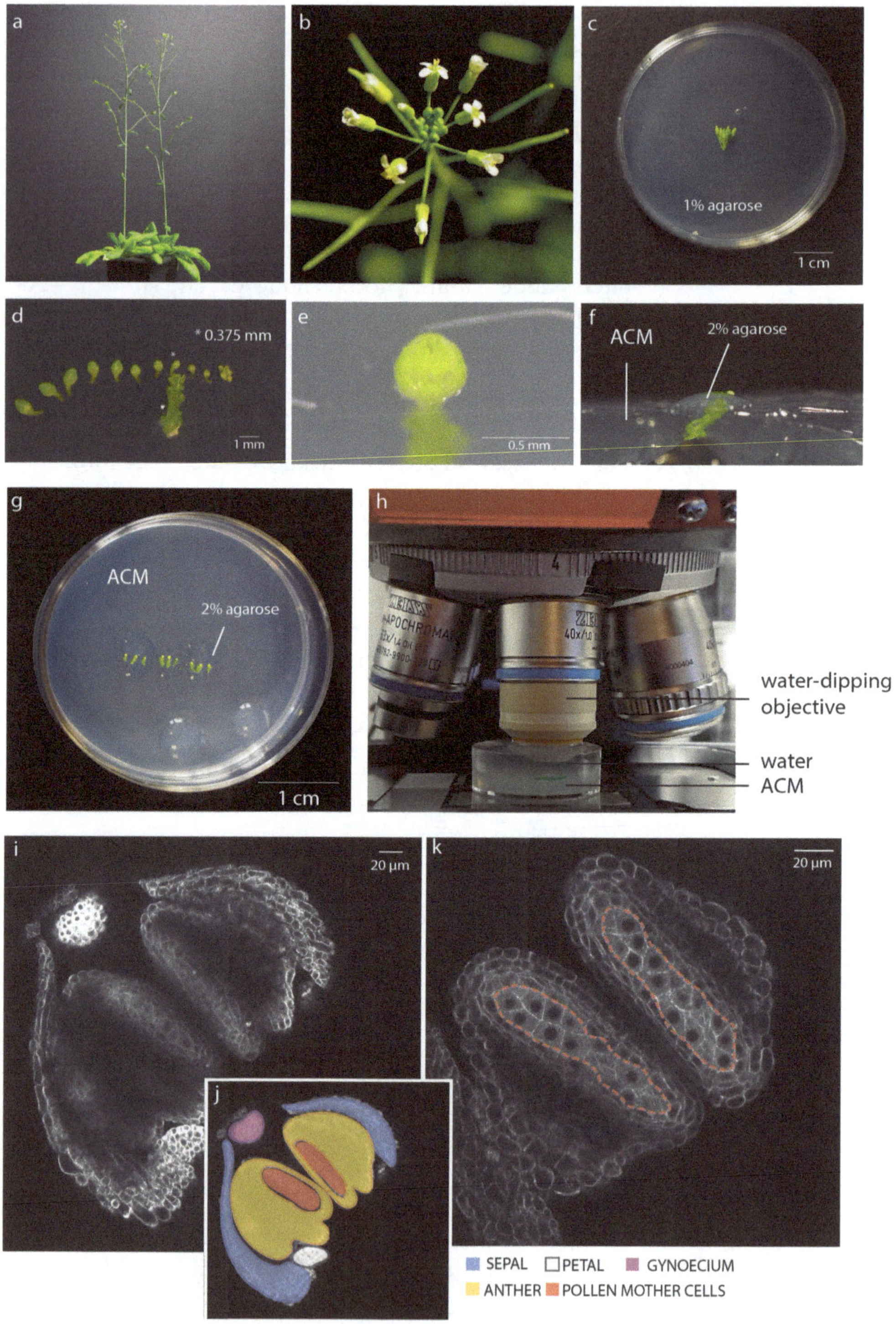

Fig. 1 This figure illustrates the main steps to perform live-cell imaging of male meiocytes in Arabidopsis. (**a**) Examples of two plants at the right stage to perform live-cell imaging of meiosis. The main inflorescences have already produced a few, yet not too many siliques, and multiple secondary inflorescences have been initiated. (**b**) Close-up of an inflorescence ready for microscopy observation. (**c**) The closed flowers of the

uppermost sepal. It is crucially important to not touch the anther to avoid damaging the reproductive tissues.

4. After one sepal has been removed, two of the six anthers are exposed since petals are smaller than anthers at this stage (Fig. 1e, j). If the petals are not smaller than the anthers, the flower has already completed meiosis (wild-type background). In this case, move to the next smaller flower bud and proceed with the removal of the upper sepal.
5. Once the right flower bud in meiosis has been found, remove all the smaller buds in the inner part of the inflorescence, leaving only the flower of interest attached to the stem (Fig. 1d).
6. Transfer the flower to the small petri dish containing ACM. Use the stem to anchor the flower bud to the medium in such a way that the exposed anthers point upward (Fig. 1e, f). The flower bud should be placed in the center of the petri dish to facilitate the image acquisition (Fig. 1g).
7. Melt the 2% agarose and stabilize the sample by covering the pedicel and its lower half with a drop of it. (Fig. 1f). To cover only the lower half of the flower bud, point the pipette tip a few millimeters away from the flower bud to release the agarose. If the agarose gel is warm enough, it will expand around the sample without covering the anthers. However, it is possible to image and obtain good quality pictures even when the flower is fully covered by agarose.
8. It is possible to mount more than one flower bud on the same petri dish (Fig. 1g and *see* **Note 5**) and acquire images from several specimens in parallel. This can be useful when comparing different genotypes or to accelerate the speed of data acquisition. Male meiosis in wild-type Arabidopsis plants lasts

Fig. 1 (continued) inflorescence are clipped off and laid on a 1% agarose support, which will help to hold the flowers while preparing the samples for microscopy. (**d**) Selection of a flower bud containing meiocytes at the right stage for live-cell imaging. Flowers of the clipped inflorescence are removed one by one, from the largest to the smallest until the flower bud is reached in which meiosis takes place (asterisk). The flower of interest is left attached to the main stem of the inflorescence while the smaller flowers are cut off as well. (**e**) Close up of a flower bud with meiocytes. The upper sepal of the selected flower bud is removed using fine forceps to expose the two uppermost anthers. (**f**) Cross section through the support medium with a sample mounted. The flower is anchored by placing the stem into the ACM medium and further stabilized by a drop of 2% agarose. (**g**) Top view of the samples mounted and ready to be observed. (**h**) Microscope setup. The images are acquired with a water-dipping lens that is immersed into autoclaved water, which covers the entire support medium. (**i–k**) Optical cross sections through a complete flower bud (**i**, **j**) and of two pollen sacs (**k**) as they appear in the microscope. The meiocyte outlines and the nuclear positions within the meiocytes are detected by excitation of the tubulin marker TagRFP:TUB4. In panel (**j**) the main visible floral organs are highlighted, and in panel (**k**) the meiocytes of two different pollen sacs of two neighboring anthers are recognizable (red dotted lines) for around 30–35 h [15, 18–20]. Therefore, following more samples at the same time can greatly reduce the time needed at the microscope.

3.2 Sample Positioning

1. Image acquisition conditions protected from vibrations and motions are fundamental to obtain a successful time series. Hence, it is of great help to turn on the microscope and lasers at least 30 min before the start of the imaging.
2. Position the sample on the microscope stage. For this, a petri dish holder can be used. Alternatively, two standard microscopy glass slides can be used as a support by fixing the petri dish onto them using double-sided adhesive tape to limit unwanted movements (Fig. 1h).
3. Fill up the petri dish with autoclaved distilled water and let the sample rest for at least 1 h. In this way the medium can absorb the water until saturated and will not swell during image acquisition, preventing a Z-drift.
4. Using fast scanning/preview settings, localize the sample on the plate, and check their developmental stage. An outline of meiotic time course of male Arabidopsis obtained with the same technique is published in [15] and can be used as a guideline for stage duration and the decision about when the acquisition should be started. If the dissected flower is too young, it is possible to let the sample grow on medium until the right stage is reached. It is sufficient to seal the petri dish with Micropore tape and store it in the growth chamber for the necessary time.
5. Usually, there is a high level of water evaporation during long time observations. Water evaporation from the mounting medium causes movements of the sample resulting in the loss of the focal plane. Hence, to further stabilize the sample, especially in the case of long-term image acquisition (more than 12 h, or in case of image acquisition without autofocus, *see* Subheading 3.3), the use of a syringe pump is advised to counterbalance water evaporation and, with that, reduce sample movement. For this, a syringe of 10 ml volume is filled with autoclaved distilled water and positioned in the syringe pump. The water is delivered to the sample via a capillary extended with a pipette tip, which is anchored with double-sided adhesive tape to the edge of the petri dish (*see* **Note 6**). The speed and volume of the water supply has to be adjusted for the specific room conditions; in our case, for a room at 20 °C and 62% of humidity, we dispense 2.0–2.5 μl/min.

3.3 Image Acquisition

1. Images are acquired with an upright confocal microscope (e.g., Zeiss LSM 780 or Zeiss LSM 880) equipped with a water-dipping objective. The objective is immersed directly into the

autoclaved water covering the sample (Fig. 1f, g). In case of Arabidopsis, we found that a 40× objective allows following meiotic progression in anthers while at the same time providing enough spatial resolution to observe chromosomes. Other objectives (20×/63×) can be considered for different applications.

2. The actual settings for image acquisition depend on many factors, such as objectives, lasers used for sample excitation, detector sensitivity, the expression level and bleaching of the reporter used and, above all, on the purpose of the experiment. Hence, some time is needed for their optimization. It is of key importance to consider the spatial and temporal resolution needed. Depending on the laser intensity, acquisition time and interval times need to be adjusted to minimize phototoxicity and photobleaching without compromising on the resolution of the experiment. With this respect, it is crucial to take into account that different meiotic phases proceed at different velocities. Below, an example is provided:

   - To obtain a time course of an entire meiotic division we carried out a time-lapse experiment with four samples per plate. We performed our experiments on a Zeiss 880 Airyscan® equipped with one GaAsP detector and controlled by the ZEN 2.3 SP1 FP1 black software. Imaging was performed with the *KINGBRID* line (*Kleisin IN Green microtuBules In ReD*), which expresses a reporter for beta TUBULIN (TUB4) fused with RFP, in combination with the alpha kleisin subunit of the cohesin complex, called MEIOTIC RECOMBINATION PROTEIN 8 (REC8), also known in Arabidopsis as SYN1 for SYNAPTIC1 or DIF1 for DETERMINATE INFERTILE 1 [15]. Plants were grown at the conditions presented in Subheading 2.1. After their removal, the flowers were placed in a row, and their positions were saved using the multi positions function of the acquisition software. Images were acquired every 10 min for all four samples sequentially, over a period of ca. 30 h. At each time point and position, a *Z*-stack of six planes with a step size of 50 μm was recorded. The samples were automatically refocused at each time point and at each position by the acquisition software. A 488 nm Argon laser and a DPSS 561–10 laser were used to detect the fluorescent signal. The emitted signals were firstly filtered through the Beam splitter MBS 488/561. Green and red fluorescence were recorded in two channels by sequential line mode filtered for wavelengths between 498–550 nm and 578–649 nm, respectively. An additional third channel was used to record the autofluorescence of chloroplasts and filtered for wavelengths of 680–750 nm. The pinhole was set at 1 Airy Unit, and scan time did not exceed 4.10 μs pixel dwell. The bidirectional scan function was on and averaging was performed on 2 lines. Images were acquired in 208 nm/pixel size.

3.4 Image Processing and Analysis

The image processing and analysis has to be adapted to the biological question to be addressed (centromere tracking, time calculation, etc.). However, a few operations can be performed on all the acquired images, independently of the later steps of the analysis:

1. For images obtained by autofocus, open the .czv file with ImageJ (Fiji version 1.52b, https://imagej.net/Fiji) and save them as sequential images in a new folder. To select a focal plane from all recorded *Z*-stacks, the software MetaMorph Version 7.8.0.0 is useful. Open the sequential images via the App > Review Multi Dimensional Data. Then, select the desired planes, export, and save the file as a .tiff stack file.
2. To correct for XY-drift in the two channels, open the .tiff stack file with ImageJ and apply the plugin named "HyperStack-Reg v05" (https://sites.google.com/site/vedsharma/imagej-plugins-macros/hyperstackreg). Select "Solid Body" for the transformation function and save it as a .tiff file.
3. Adjust the brightness and contrast and stamp the time in ImageJ.

4 Notes

1. The choice of fluorescent reporters depends on the biological questions addressed. In general, a reporter for microtubules coupled to a reporter that allows visualization of chromosomes is a good option to monitor cell divisions. An example is the KINGBIRD line [15], which combines a reporter for microtubules fused with RFP, with a reporter for REC8 fused to GFP, which highlight chromosomes dynamics during the first meiotic division. In addition, REC8 is meiosis-specific and hence allows a straightforward identification of meiocytes in contrast to proteins that in general bind DNA in many cells, like (most) histones. Moreover, the localization of REC8 changes during meiosis and hence its accumulation pattern allows the identification of meiotic stages [15, 21].
2. Germination on plates offers two easy ways to select for plants carrying the reporter construct if the line is hemizygous for the transgene: selection can be done by adding herbicides to the medium, or by observing the fluorescence of a ubiquitously expressed reporter such as TUB4 in roots or shoots with a

dissecting microscope equipped with a fluorescence detection system (e.g., Olympus MVX10, coupled with UV light source X-cite series 120 Q).

3. The exact growth conditions depend on the biological question to be addressed and the genetic background used. Importantly, this imaging setup allows researchers to assess the effect of different environmental conditions on meiosis (e.g., different temperatures).
4. The size and shape of the flower buds with meiocytes in the meiotic stages of interest depend on many parameters, foremost the stage itself (e.g., early meiosis versus late meiosis), growth conditions, genetic background, and age of the plant. Thus, when establishing the technique or using a new genetic background, it is advisable to evaluate the optimal size and shape via some fast histological staining procedures such as acetocarmine staining prior to the application of the live-cell imaging.
5. The maximum number of flowers depends on the aim of the experiment and on the microscope properties (e.g., acquisition speed and time resolution). With our setup and aim to perform a time course of meiosis, we could observe up to nine flower bud simultaneously (Fig. 1g).
6. The samples on the plate have to be identified and focused individually (if available, save position function is advisable). Also, the objective has to be submerged before connecting the tip of the water supplier to the petri dish. Once it is attached, check that it stays in place when the stage is moving between positions and that the water does not spill.

Acknowledgments

This work was supported by the EU-ITN grant COMREC, the BMBF grant TarBreed, and core funding of the University of Hamburg to A.S.

References

1. Klutstein M, Fennell A, Fernández-Álvarez A et al (2015) The telomere bouquet regulates meiotic centromere assembly. Nat Cell Biol 17:458–469
2. Koszul R, Kleckner N (2009) Dynamic chromosome movements during meiosis: a way to eliminate unwanted connections? Trends Cell Biol 19:716–724
3. Mullen TJ, Wignall SM (2017) Interplay between microtubule bundling and sorting factors ensures acentriolar spindle stability during C. elegans oocyte meiosis. PLoS Genet 13: e1006986
4. Rosu S, Cohen-Fix O (2017) Live-imaging analysis of germ cell proliferation in the C. elegans adult supports a stochastic model for stem cell proliferation. Dev Biol 423:93–100
5. Mlynarczyk-Evans S, Villeneuve AM (2017) Time-course analysis of early meiotic prophase

events informs mechanisms of homolog pairing and synapsis in Caenorhabditis elegans. Genetics 207:103–114

6. Enguita-Marruedo A, Cappellen WAV, Hoogerbrugge JW et al (2018) Live cell analyses of synaptonemal complex dynamics and chromosome movements in cultured mouse testis tubules and embryonic ovaries. Chromosoma 127(3):341–359
7. Holubcová Z, Howard G, Schuh M (2013) Vesicles modulate an actin network for asymmetric spindle positioning. Nat Cell Biol 15:937
8. Clift D, Schuh M (2013) Restarting life: fertilization and the transition from meiosis to mitosis. Nat Rev Mol Cell Biol 14:549
9. Kyogoku H, Kitajima TS (2017) Large cytoplasm is linked to the error-prone nature of oocytes. Dev Cell 41:287–298.e4
10. Yu H-G, Hiatt EN, Chan A et al (1997) Neocentromere-mediated chromosome movement in maize. J Cell Biol 139:831–840
11. Sheehan MJ, Pawlowski WP (2009) Live imaging of rapid chromosome movements in meiotic prophase I in maize. PNAS 106:20989–20994
12. Nannas NJ, Higgins DM, Dawe RK (2016) Anaphase asymmetry and dynamic repositioning of the division plane during maize meiosis. J Cell Sci 129:4014–4024
13. Ingouff M, Selles B, Michaud C et al (2017) Live-cell analysis of DNA methylation during sexual reproduction in Arabidopsis reveals context and sex-specific dynamics controlled by noncanonical RdDM. Genes Dev 31:72–83
14. Wijnker E, Schnittger A (2013) Control of the meiotic cell division program in plants. Plant Reprod 26:143–158
15. Prusicki MA, Keizer EM, Rosmalen RP van et al (2019) Live cell imaging of meiosis in Arabidopsis thaliana—a landmark system. eLife. 8:e42834. https://doi.org/10.7554/eLife.42834
16. Yang C, Hamamura Y, Sofroni K et al (2019) SWITCH 1/DYAD is a novel WINGS APART-LIKE antagonist that maintains sister chromatid cohesion in meiosis. Nat Commun. 10:1755. https://doi.org/10.1038/s41467-019-09759-w
17. Hamant O, Das P, Burian A (2014) Time-lapse imaging of developing meristems using confocal laser scanning microscope. In: Plant cell morphogenesis. Humana Press, Totowa, NJ, pp 111–119
18. Armstrong SJ, Franklin FCH, Jones GH (2003) A meiotic time-course for Arabidopsis thaliana. Sex Plant Reprod 16:141–149
19. Stronghill PE, Azimi W, Hasenkampf CA (2014) A novel method to follow meiotic progression in Arabidopsis using confocal microscopy and 5-ethynyl-2′-deoxyuridine labeling. Plant Methods 10:33
20. Sanchez-Moran E, Santos J-L, Jones GH et al (2007) ASY1 mediates AtDMC1-dependent interhomolog recombination during meiosis in Arabidopsis. Genes Dev 21:2220–2233
21. Cai X, Dong F, Edelmann RE et al (2003) The Arabidopsis SYN1 cohesin protein is required for sister chromatid arm cohesion and homologous chromosome pairing. J Cell Sci 116:2999–3007

Chapter 2

Whole-Mount Immunolocalization Procedure for Plant Female Meiocytes

Santiago Valentín Galvan Gordillo, Rocio Escobar-Guzman, Daniel Rodriguez-Leal, Jean-Philippe Vielle-Calzada, and Arnaud Ronceret

Abstract

Here we present an optimized protocol for immunolocalization of meiotic proteins during female meiosis in whole mount tissues. It ensures ovule morphology integrity and homogeneous reagent penetration. The method relies on paraformaldehyde tissue fixation, polyacrylamide embedding, tissue permeabilization, antibody incubation, counterstaining, and confocal microscopy analysis. This protocol has been used in diverse Arabidopsis ecotypes and in the legume *Vigna unguiculata*.

Key words Confocal microscopy, Female meiosis, Immunolocalization, Plant, Whole-mount

1 Introduction

Meiosis is a critical biological process responsible of reshuffling and transmission of genetic information in sexual organisms. In plants, meiosis pinpoints the diploid to haploid transition state within flowers with male meiosis happening in anthers and female meiosis in ovules, (for reviews *see* [1, 2]). Plant meiosis genetic regulation has shown gender variations [3, 4]. Moreover, different recombination rates in male and female meiocytes have been reported [5, 6]. However, meiotic research in plants has been performed mainly in anthers leaving aside female meiosis study.

Plant female meiosis research limitation is mainly owed to intrinsic and technical constraints. Most plant species harbors more male meiocytes than female meiocytes (for Arabidopsis, the difference is of ~20-fold). Moreover, plant female meiocytes are not readily available since they are surrounded by several layers of sporophytic tissue. Common methodologies to study female

Mónica Pradillo and Stefan Heckmann (eds.), *Plant Meiosis: Methods and Protocols*, Methods in Molecular Biology, vol. 2061, https://doi.org/10.1007/978-1-4939-9818-0_2,

meiosis in plants disrupt the 3D structure of gynoecium in order to expose the meiocytes [6–8], restricting the scope of posterior analyzes. Novel methodologies to visualize female meiosis in plants, using the advantages of confocal microscopy in whole-mount ovules, currently exist for maize [9] and Arabidopsis [10, 11]. Unfortunately, the protocol for maize is not designed for protein immunolocalization analysis. For Arabidopsis, the protocols require extensive tissue manipulation and buffering and certain technical expertise for dissecting the tissue [10], or expensive specialized equipment for tissue dissection such as tungsten needles [11]. This protocol intend to be simple, robust, and of easy applicability using ordinary laboratory equipment. Exploiting a gentle step of ovule exposure, ensuring 3D-structure integrity in whole-mount organs or tissues, and allowing protein immunolocalization analysis by confocal microscopy. The methodology consists in fixing the ovules in paraformaldehyde and posterior embedding in a polyacrylamide matrix directly on the slides. Samples are then treated with enzymes and detergents for cell wall degradation that allow homogeneous antibody penetration in a subsequent step. Finally, the samples are counterstained and observed using confocal microscopy (Fig. 1).

The present technique is based on previous whole mount procedures: in Arabidopsis ovules [10], maize meiocytes [12] and isolated polytene nuclei of Drosophila [13]. This protocol has been used to immunolocalize ARGONAUTE proteins members during Arabidopsis female meiosis and megasporogenesis [14–16]. This method was also used by two independent groups for the elucidation of the molecular mechanisms limiting megaspore mother cell determination in Arabidopsis ovule primordia. Since it preserves the meiocyte surroundings cells intact, it was possible to uncover that MMC formation is repressed in epidermal neighboring sporophytic cells by trans-acting siRNAs (tasiRNAs) cell-to-cell transmission [17]. Similarly, a KLU unknown mobile signal is proposed to mediate non–cell autonomously WRKY28 repression of MMC formation in hypodermal somatic surrounding cells [18]. It has also been proposed as a complementary methodology to further explore meiotic COs sex bias, identified initially by SNP markers in Brassica species [19]. Finally, this technique has proved its robustness by allowing the immunolocalization of ASY1 in the crassinucellate ovules of *Vigna unguiculata* (cowpea), by just extending time of enzymatic digestion [20].

In conclusion; this methodology could allow, with minor modifications, the assessment of protein dynamics during female meiosis in different plant species, including apomictic species.

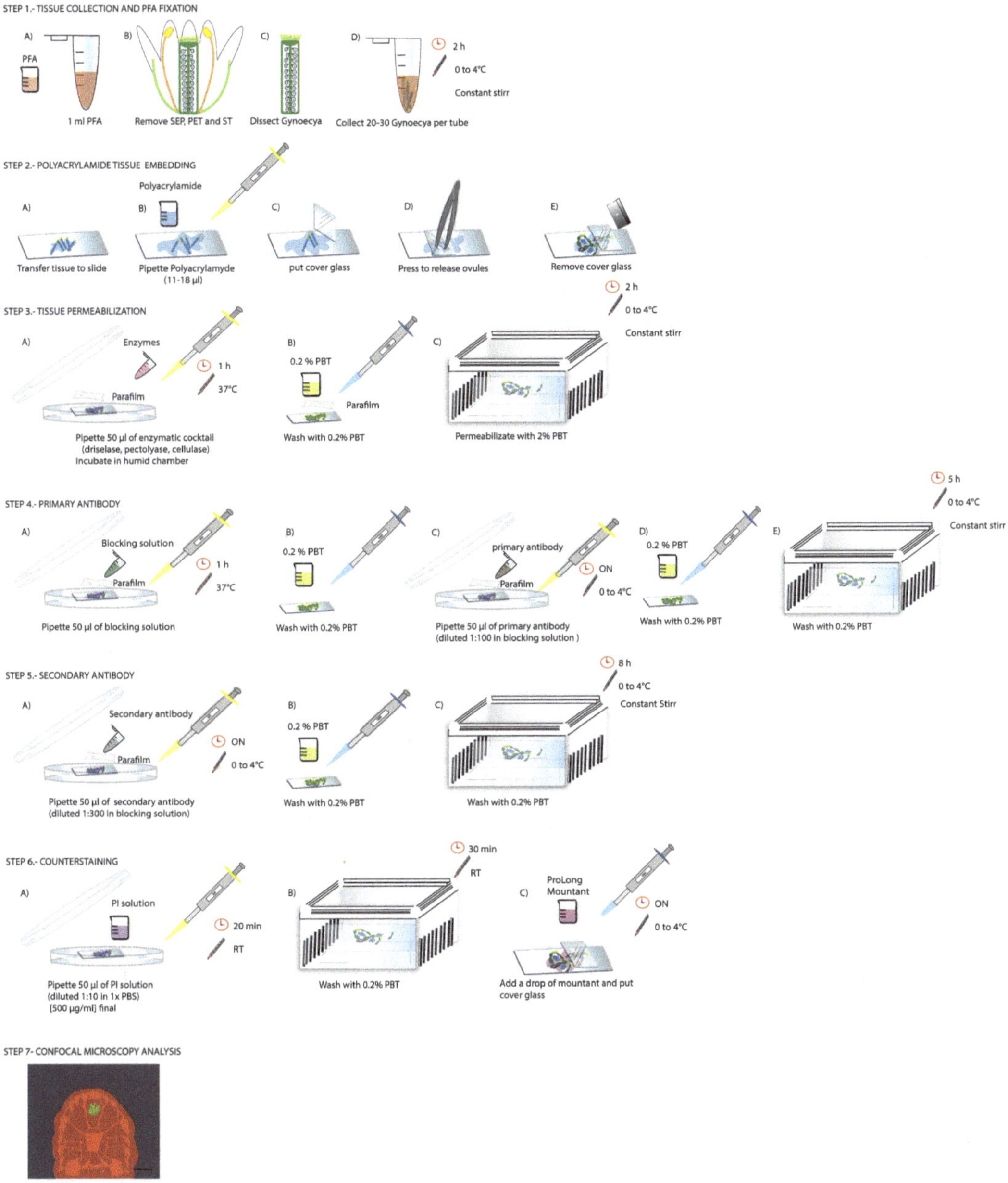

Fig. 1 Summary drawing of the steps used for the immunolocalization protocol in native plant female meiocytes. Step 1: *Tissue collection and fixation.* Prepare microcentrifuge tubes with 1 ml of PFA, collect gynoecia and keep them at 4 °C for 2 h in the magnetic stirrer. Step 2: *Tissue embedding.* Put the tissue on the slides, add embedding solution, manipulate the gynoecia quickly and put the cover glasses on the microscope slides, to release the ovules from gynoecia apply adequate pressure to the cover glasses using tweezers, let polymerization occur for at least 45 min, remove the cover glasses using a razor blade. Step 3: *Tissue permeabilization.* In humid chamber add 50 µl of enzymatic cocktail and cover the samples using a piece of Parafilm, incubate for 1 h at 37 °C, remove Parafilm, wash with 0.2% PBT, transfer the slides to a Coplin jar containing 2% PBT and apply constant agitation for 2 h at 4 °C. Step 4: *Primary antibody.* In humid chamber, add 50 µl of blocking solution, cover the samples with Parafilm, incubate for 1 h at 37 °C, wash the slides with

2 Materials

2.1 Plant Material

Sterilize Arabidopsis seeds surface with consecutive washes of pure (100%) ethanol and sodium hypochlorite (10%) and at least five rinses with distilled, deionized, water (MQ water). Vernalize the seeds for 2–4 days in the dark at 0–4 °C. Germinate in Murashige–Skoog medium (MS medium) with long day photoperiod (16 h light/8 h dark) at 22 °C. Transplant 12 days post germination seedlings into soil and keep them in green house conditions at 24 °C until use.

2.2 Solutions

All solutions are prepared using analytic grade reagents and MQ water. Autoclave all buffers and verify the pH again before each experiment. Some solutions must be freshly prepared before each experiment. For hazardous reagents follow all disposal and security guidelines.

1. *Fixative solution*: 4% paraformaldehyde.
   Dissolve the paraformaldehyde (PFA) in 2% PBT to obtain a 4% paraformaldehyde solution. For 1 ml of fixative solution dissolve 0.04 g of PFA in 2% PBT. Enough solution should be prepared before each experiment and kept on ice. To prepare working fixative solution filter 1 ml of 4% paraformaldehyde solution per sample, with 0.45 μm syringe filters, in 1.5 microcentrifuge tubes. Keep at 0–4 °C until use.
2. *30% acrylamide–bisacrylamide (29:1) embedding stock solution*: 29% acrylamide, 1% *bis*acrylamide.
   First dissolve 2.9 g acrylamide and 0.1 g *bis*acrylamide in 10 ml MQ water, then filter with 0.45 μm syringe filters. Protect from light. Keep at 0–4 °C. Prepare a 15% embedding solution before each experiment.

Fig. 1 (continued) 0.2% PBT, transfer them again to the humid chamber, add 50 μl of diluted 1:100 primary antibody, cover the samples with Parafilm and incubate overnight at 4 °C, remove Parafilm from samples, wash the slides with 0.2% PBT, and wash again with 0.2% PBT inside a Coplin jar for 5 h in constant agitation at 4 °C, changing the PBT solution every hour. Step 5: *Secondary antibody*. In humid chamber add 50 μl of diluted 1:300 secondary antibody, cover the samples using a Parafilm piece and incubate overnight at 4 °C, remove Parafilm from samples, and wash the slides with 0.2% PBT; rewash the samples inside a Coplin jar with more 0.2% PBT for 8 h on constant agitation changing washing solution every hour. Step 6: *Counterstaining with PI*. Add 50 μl of a 1:10 dilution of PI solution (final concentration 500 μg/ml), transfer the samples to the humid chamber and incubate for 20 min at room temperature, wash the slides in a Coplin jar containing 1× PBS for 30 min, add a drop of ProLong and put cover glasses, keep samples overnight at 4 °C. Step 7: *Confocal microscopy imaging*. Finally, take samples to analysis by confocal microscopy, selecting appropriate laser and filter configurations for detection

3. *15% embedding solution*: Dissolve 200 μl of 30% embedding stock solution, 2 μl APS 25% and 1 μl TEMED in MQ water up to 400 μl.
4. *Enzymatic solution*: 1% driselase, 1% pectolyase, 0.5% cellulase. Dissolve 0.1 g driselase, 0.1 g pectolyase, and 0.05 g cellulase in 1× PBS up to 10 ml and mix gently. Aliquot in microcentrifuge tubes and keep at −20 °C until use.
5. *Blocking solution*: 1% BSA.
   Dissolve 0.1 g BSA in 1× PBS up to 10 ml, filter with 0.45 μm syringe filters. Aliquot in microcentrifuge tubes and keep at −20 °C.
6. *Propidium Iodide (PI) solution*: 0.5% propidium Iodide.
   For stock PI solution add 5 mg of PI in 1 ml 1× PBS. Protect from light. Keeping solution from 4 °C to −20 °C is adequate. Prepare fresh 50 μl of a dilution 1:10 from the PI stock solution with 1× PBS before each experiment.
7. *APS solution*: Ammonium persulfate 25%.
   Dissolve 2.5 g of ammonium persulfate (APS) in MQ water up to 10 ml. Aliquot in microcentrifuge tubes and keep at 0–4 °C.

   Auxiliary buffers and washing steps:
8. *10× PBS Stock Buffer*: 1.37 M NaCl, 26.8 mM KCl, 43 mM Na_2HPO_4, 14.7 mM KH_2PO_4, pH 7.4.
   Add 80 g NaCl, 2 g KCl, 2 g KH_2PO_4 and 6.1 g Na_2HPO_4, to 1 L MQ water and dissolve. Adjust the pH to 7.4 with KOH. Autoclave and store at room temperature.
9. *PBT solutions*: 0.2% PBT, 2% PBT.
   Add 20 or 2 ml of Triton X-100, to 100 ml 10× PBS Stock Buffer (for PBT 2% and 0.2% respectively) add MQ water up to 1 L. Adjust pH to 7.4 with KOH. Autoclave and store at room temperature.

2.3 Supplies and Equipment

Except for the confocal microscope imaging, the rest of the steps can be performed in the bench easily with ordinary laboratory equipment.

1. Microscope glass slides, poly-L-lysine coated microscope glass slides.
2. Cover glasses for microscope slides (thickness 0.13–0.17 mm; L × W, 22 × 22 mm, square or smaller 18 × 18 mm, square).
3. Insulin syringes (U-100).
4. Fine tweezers.
5. Filters (0.45 μm).
6. 1.5 Microcentrifuge tubes.
7. Stereomicroscope.

8. Coplin jars.
9. Slide box.
10. Magnetic stirrer.
11. Humid chamber: a recipient, well capped, filled with PBS moistened papers and grating over to place the slides.
12. Parafilm pieces (L × W, 1 × 1 cm, square).
13. Laboratory vacuum pump (optional).
14. Incubator.
15. Confocal microscope.

3 Methods

3.1 Tissue Collection and Fixation (Fig. 1, Step 1)

1. Add 1 ml of fixative solution (*see* **Note 1**) into1.5 microcentrifuge tubes (Fig. 1, step 1a). Keep on ice.
2. Under stereomicroscope, dissect 20–30 Arabidopsis gynoecia from flower buds per sample (Fig. 1, step 1b, c) (*see* **Note 2**) and put them inside, previously prepared, 1.5 microcentrifuge tubes (Fig. 1, step 1d). Maintain tubes on ice during tissue collection.
3. (Optional) a vacuum step of 20 min could be carried out for tissue fixation and homogenization (*see* **Note 3**).
4. Keep tubes at 0–4 °C for 2 h in the magnetic stirrer (Fig. 1, step 1d); stir gently to avoid formation of bubbles caused by Triton X-100 on the fixative. Constant agitation facilitates fixative to penetrate homogeneously into the tissue (*see* **Note 4**).

3.2 Slide Preparation

1. Cover all the slides with poly-L-lysine (*see* **Note 5**) with a brush, taking care not to leave air bubbles on the slides surface, let them dry away from dust into a horizontal slide box, this ensures acrylamide attachment to the slides. This step is carried out during fixation and before embedding.
2. Clear away fixative solution and rinse the samples three times with 1 ml 1× PBS.
3. Put fixed tissue on prepared slides (*see* **Note 6**) with fine tweezers. It is critical maintain gynoecia hydrated the whole time before acrylamide embedding.

3.3 Tissue Embedding (Fig. 1, Step 2)

1. Transfer tissue to slides; remove all PBS as possible from samples with syringe needles, taking care not to dry the tissue before embedding (Fig. 1, step 2a). Prepare embedding solution (*see* **Note 7**) without TEMED. Add 1 μl of TEMED once samples are ready for the embedding step and not earlier (*see* **Note 8**).

2. Add activated embedding solution (11–18 μl per slide, according to number and size of gynoecia) to the slides (Fig. 1, step 2b), then manipulate the gynoecia quickly (*see* **Note 9**) and put the cover glasses on the microscope slides containing the samples (Fig. 1, step 2c) (*see* **Note 10**).
3. Under a stereoscope apply pressure to the cover glasses using fine tweezers to release the ovules from gynoecia (*see* **Note 11**) (Fig. 1, step 2d).
4. Let polymerization occur for at least 45 min. Save a little solution to trace polymerization. Once the solution polymerized completely, carefully remove the cover glasses from all the microscope slides using a razor blade (Fig. 1, step 2e), and continue.

3.4 Enzyme Digestion and Tissue Permeabilization (Fig. 1, Step 3)

1. Put the slides in the humid chamber and pipette 50 μl of enzymatic cocktail (*see* **Note 12**) per slide and immediately covers the samples using a piece of Parafilm (*see* **Note 13**), incubate inside the humid chamber for 1 h at 37 °C (*see* **Note 14**) (Fig. 1, step 3a).
2. Carefully remove Parafilm from samples (*see* **Note 15**) and wash the slides with 0.2% PBT (Fig. 1, step 3b).
3. To finish with sample permeabilization, put the slides in a prechilled Coplin jar containing 2% PBT in constant agitation for 2 h at 0–4 °C (*see* **Note 16**) (Fig. 1, step 3c).

3.5 Primary Antibody (Fig. 1, Step 4)

1. For sample blocking, transfers the slides back to the humid chamber and add 50 μl of blocking solution per sample (*see* **Note 17**) and cover them with a Parafilm piece, incubate for 1 h at 37 °C (Fig. 1, step 4a).
2. Wash the slides with 0.2% PBT (Fig. 1, step 4b), transfer the slides again to the humid chamber pipette 50 μl of primary antibody (*see* **Note 18**) diluted 1:100 in blocking solution, cover the samples using a piece of Parafilm and incubate overnight at 0–4 °C (Fig. 1, step 4c).
3. Carefully remove Parafilm from samples, and wash the slides with 0.2% PBT (Fig. 1, step 4d), and wash again with 0.2% PBT inside a prechilled Coplin jar for 5 h in constant agitation at 0–4 °C (Fig. 1, step 4e), taking care of changing the PBT solution every hour.

3.6 Secondary Antibody (Fig. 1, Step 5)

1. Transfer the slides to the humid chamber and pipette 50 μl per slide of diluted 1:300 secondary antibody (*see* **Note 19**). Cover the samples using a Parafilm piece and incubate overnight at 0–4 °C, protected from light to avoid fluorescence decreasing (Fig. 1, step 5a).

2. Carefully remove Parafilm from samples, and wash the slides with 0.2% PBT (Fig. 1, step 5b); rewash the samples inside a prechilled Coplin jar with more 0.2% PBT for 8 h on constant agitation (*see* **Note 20**) (Fig. 1, step 5c). Change washing solution every hour, making sure of protect slides from light to avoid florescence decay.

3.7 Counterstaining with PI (Fig. 1, Step 6)

1. Add 50 μl per slide of a 1:10 dilution of PI solution in 1× PBS (final concentration 500 μg/ml), transfer the samples into the humid chamber and incubate for 20 min at room temperature (Fig. 1, step 6a).
2. Wash the slides in a Coplin jar containing 1× PBS for 30 min (Fig. 1, step 6b).
3. Add a drop of ProLong antifade reagent (*see* **Note 21**) to all the samples and put over cover glasses. Keep samples overnight at 0–4 °C (Fig. 1, step 6c).

3.8 Confocal Microscopy Imaging (Fig. 1, Step 7)

1. Make sure to have adequate controls before image acquisition by confocal microscopy (*see* **Note 22**).
2. Select appropriate laser and filter configurations for detection of secondary antibody along with counterstain dye in this case PI (*see* **Note 23**), to perform this step is necessary to know excitation and emission wavelengths of both secondary antibody and counterstaining reagent, and that these do not overlap (*see* **Note 24**).
3. Adjust configuration (laser intensity and gain) and use the same configuration for all the samples, including controls. For analysis in different days, store the slides at −20 °C protected from light, although we do not recommend it as fluorescence decreases.

4 Notes

Notes are based in the work done in Arabidopsis (Escobar-Guzman 2015), so special emphasis is made for protocol steps adaptations in order to perform well in other plant species.

1. Paraformaldehyde is hazardous so wear appropriate protective equipment when handle it and manipulate in fume. We recommend weight the paraformaldehyde in a plastic tube (for example, Corning 50 ml centrifuge tubes), to dissolve PFA heat from 65° and mix gently, wait until solution cooled down to add Triton X-100. For plant species with larger gynoecium (i.e., maize), change the 1.5 microcentrifuge tubes for 50 ml centrifuge tubes.

2. Young plants should be used to study female meiosis as older senescent plant show flowers with declining viability affecting this analysis. During sample mounting dissect the ovules under stereoscope using fine tweezers and a fine needle to allow proper ovule exposure. For Arabidopsis ovules, detach basal gynophore and apical stigma from gynoecia using fine tweezers and cut longitudinally the septum fibers with a fine syringe needles to expose the ovules. For Arabidopsis, meiotic MMC correspond to 0.5–1 mm length gynoecia.
3. Vacuum treatment ensures tissue infiltration and could be helpful for protocol adaptations to plant species other than Arabidopsis.
4. Correct Arabidopsis ovule fixations are verified by tissue sinking at the bottom of the microcentrifuge tube. Overfixation leads to antibody epitope masking. Insufficient fixation leads to poor tissue conservation. For our purpose we determined 2 h is adequate. Adaptations in this step may be required for other plant species.
5. Ready to use, precoated poly-L-lysine slides are commercially available are a good choice. Alternatively Superfrost Plus microscope slides are also a good choice.
6. Spin microcentrifuge tubes to collect gynoecium from the bottom. Pipetting gynoecium generally provokes gynoecia loss by adhesion inside the pipette tips, wide bore tips may help with this issue. Dispose PFA in a special PFA residues container.
7. Acrylamide and bisacrylamide both affect the nervous system and may be carcinogens so wear appropriate protective equipment when handle it and manipulate in fume. To dissolve, heat the mix at 37 °C. The stock solution could be used leastwise for 1 year without compromising its polymerization efficiency. This is a critical step. Incorrect embedding could lead to tissue integrity loss. Tissue drying could also compromise its integrity, so keeping the samples inside humid chamber to avoid dehydration is advisable.
8. Ovule release and setting should be performed in 5–8 min after activation of the mix with TEMED and APS, doing it in more extended periods of time would lead to improper ovule exposure in posterior steps. Handle as much as four samples at a time when mounting on slides. Since TEMED along with APS activate the mix, is important not to add TEMED until samples are ready to mounting.
9. We recommend distributing different sizes of gynoecia in different samples and thus enable a proper (sagittal) ovule positioning during image acquisition. Do not try to use less

embedding solution than recommended since it could lead to tissue damage and inconsistent polymerization times.

10. Smaller coverslips (18 × 18 mm) could be used to perform sample replicates within the same slide; we recommend using at least two replicates per sample.
11. It is critical to apply an adequate pressure, too much pressure could affect ovule morphology and too little pressure could not be enough to release the ovules from gynoecia.
12. We recommend use fresh enzymatic cocktails as a maximum of 3 weeks old, after this time, variations between samples due to enzymatic digestion were observed. During preparation vortex vigorously, spin centrifuge and use only supernatant for aliquots. This is a critical step, shorter times than required could lead to reduced or lack of fluorescent signal, and longer times than required could lead to tissue morphology damage. Incomplete digestion can be verified by direct observation of the cell walls under a light microscope.
13. When covering the samples with Parafilm make sure of not form air bubbles as they can produce inconsistent fluorescent signal in the same sample.
14. For Arabidopsis female meiocytes, 1 h of enzymatic digestion is adequate, which means not tissue damage is observed while we verify strong fluorescent signal. Optimal digestion time should be experimentally determined for plant species different from Arabidopsis.
15. Every time removing Parafilm pieces take care to avoid detachment of polyacrylamide, otherwise tissue damage during posterior steps can occur.
16. For cooling down the slides into the Coplin jar, you can either put it in a bigger container with ice or use a cold room.
17. When different signal intensity between ovules from the same slide are observed, increase (100 μl is recommended) all solutions used during incubation steps (enzymatic solution, blocking solution, primary antibody dilution, and secondary antibody dilution).
18. A dilution of primary antibody of 1:100 was fine for analyzing meiotic proteins in our test but optimal antibody dilution should always be determined experimentally, and variations could certainly be required for other plant species or proteins. If a weak fluorescent signal is observed increase primary antibody concentration. Reduce primary antibody concentration if unspecific signal is observed.
19. For the fluorescent dye-conjugated secondary antibody used in our essays a dilution of 1:300 was optimal, but experimental optimization is required for other plant species or proteins.

20. When high background signal is observed, increase washing time and washing steps number, and/or increase secondary antibody dilution.
21. ProLong antifade reagent and variations thereof are glycerol based curing mountants with low background that protect the fluorescently labeled samples from fainting all through the visible spectrum (400–700 nm) for long periods of time. ProLong reagent is viscous so to apply to the slides use 1 ml micropipette with bore tip cut with sterile razor or scissors. Different Vectashield mounting media are also a good choice.
22. Validation and correct interpretation of data rely in correct control use. For immunohistochemical analysis, in parallel preimmune sera incubation is commonly used as negative control, a better option is using mutants for the protein of interest, where such protein is absent. For positive controls, a primary antibody tested positive in previous assays that show not cross-reaction with other related molecules is used. PCR based negative and positive controls are also valid. Finally, we recommend the use of at least three replicates.
23. In our tests we used PI absorbing at 535 nm and emitting at 575–615 nm coupled with secondary antibodies absorbing at 488 nm and emitting at 500–550 nm (green spectrum), DAPI is also a good choice to use in the green spectrum with absorption maxima at ~350 nm and emission maxima at ~450 nm.
24. Different counterstain dyes could be used (e.g., DAPI) and an array of fluorescently labeled secondary antibodies are available in the market, offering enough versatility that allows optimal experimental design.

Acknowledgments

We thank Angelica Martinez-Navarro for schematic drawings and Patricia Rueda for technical assistance; the work was supported by DGAPA PAPIIT IA201217 and IBT-UNAM Grant: p228.

References

1. Wilson ZA, Yang C (2004) Plant gametogenesis: conservation and contrasts in development. Reproduction 128(5):483–492. https://doi.org/10.1530/rep.1.00306
2. Schmidt A, Schmid MW, Grossniklaus U (2015) Plant germline formation: common concepts and developmental flexibility in sexual and asexual reproduction. Development 142 (2):229–241. https://doi.org/10.1242/dev.102103
3. Motamayor JC, Vezon D, Bajon C, Sauvanet A, Grandjean O, Marchand M, Bechtold N, Pelletier G, Horlow C (2000) Switch (swi1) an Arabidopsis thaliana mutant affected in the female meiotic switch. Sex Plant Reproduction 12((4)):209–218. https://doi.org/10.1007/s004970050002
4. Chen C, Marcus A, Li W, Hu Y, Vielle-Calzada JP, Grossniklaus U, Cyr RJ, Ma H (2002) The *Arabidopsis ATK1* gene is required for spindle

morphogenesis in male meiosis. Development 129:2401–2409

5. Lenormand T, Dutheil J (2005) Recombination difference between sexes: a role for haploid selection. PLoS Biol 3(3):e63. https://doi.org/10.1371/journal.pbio.0030063
6. Drouaud J, Mercier R, Chelysheva L, Berard A, Falque M, Martin O, Zanni V, Brunel D, Mezard C (2007) Sex-specific crossover distributions and variations in interference level along Arabidopsis thaliana chromosome 4. PLoS Genet 3(6):e106. https://doi.org/10.1371/journal.pgen.0030106
7. Golubovskaya I, Nadezhda AA, Sheridan WF (1992) Effects of several meiotic mutations on female meiosis in maize. Dev Genet 13:411–424
8. Armstrong SJ, Jones GH (2001) Female meiosis in wild-type Arabidopsis thaliana and in two meiotic mutants. Sex Plant Reprod 13:177–183
9. Barrel P, Grossniklaus U (eds) (2013) Examining female Meiocytes of maize by confocal microscopy. Plant meiosis: methods and protocols. Humana Press, New York, p 45
10. She W, Grimanelli D, Baroux C (2014) An efficient method for quantitative, single-cell analysis of chromatin modification and nuclear architecture in whole-mount ovules in Arabidopsis. J Vis Exp 88:e51530. https://doi.org/10.3791/51530
11. She W, Baroux C, Grossniklaus U (eds) (2017) Cell-Type specific chromatin analysis in whole-mount plant tissues by Immunostaining. Plant chromatin dynamics: methods and protocols. Humana Press, New York, p 651
12. Bass HW, Marshall WF, Sedat JW, Agard DA, Cande WZ (1997) Telomeres cluster De novo before the initiation of synapsis: a three-dimensional spatial analysis of telomere positions before and during meiotic prophase. J Cell Biol 137(1):5–18. https://doi.org/10.1083/jcb.137.1.5
13. Urata Y (1995) A three-dimensional structural dissection of Drosophila polytene chromosomes. J Cell Biol 131(2):279–295. https://doi.org/10.1083/jcb.131.2.279
14. Olmedo-Monfil V, Durán-Figueroa N, Arteaga-Vázquez M, Demesa-Arévalo E, Autran D, Grimanelli D, Slotkin RK, Martienssen RA, Vielle-Calzada JP (2010) Control of female gamete formation by a small RNA pathway in Arabidopsis. Nature 464(7288):628–632. https://doi.org/10.1038/nature08828
15. Rodriguez-Leal D, Leon-Martinez G, Abad-Vivero U, Vielle-Calzada JP (2015) Natural variation in epigenetic pathways affects the specification of female gamete precursors in Arabidopsis. Plant Cell 27(4):1034–1045. https://doi.org/10.1105/tpc.114.133009
16. Hernández-Lagana E, Rodríguez-Leal D, Lúa J, Vielle-Calzada JP (2016) A multigenic network of ARGONAUTE4 clade members controls early megaspore formation in Arabidopsis. Genetics 204(3):1045–1056. https://doi.org/10.1534/genetics.116.188151
17. Su Z, Zhao L, Zhao Y, Li S, Won S, Cai H, Wang L, Li Z, Chen P, Qin Y, Chen X (2017) The THO complex non-cell-autonomously represses female Germline specification through the TAS3-ARF3 module. Curr Biol 27(11):1597–1609. e1592. https://doi.org/10.1016/j.cub.2017.05.021
18. Zhao L, Cai H, Su Z, Wang L, Huang X, Zhang M, Chen P, Dai X, Zhao H, Palanivelu R, Chen X, Qin Y (2017) KLU suppresses megasporocyte cell fate through SWR1-mediated activation of WRKY28 expression in Arabidopsis. Proc Natl Acad Sci U S A 115(3):526–535. https://doi.org/10.1073/pnas.1716054115
19. Pelé A, Falque M, Trotoux G, Eber F, Negre S, Gilet M, Huteau V, Lodé M, Jousseaume T, Dechaumet S, Morice J, Poncet C, Coriton O, Martin OC, Rousseau-Gueutin M, Chevre AM (2017) Amplifying recombination genome-wide and reshaping crossover landscapes in brassicas. PLoS Genet 13(5):e1006794. https://doi.org/10.1371/journal.pgen.1006794
20. Salinas-Gamboa R, Johnson SD, Sánchez-León N, Koltunow AMG, Vielle-Calzada JP (2016) New observations on gametogenic development and reproductive experimental tools to support seed yield improvement in cowpea (*Vigna unguiculata* L. Walp.). Plant Reprod 29:165–177. https://doi.org/10.1007/s00497-015-0273-3

Chapter 3

How to Perform an Accurate Analysis of Metaphase I Chromosome Configurations in Autopolyploids of *Arabidopsis thaliana*

Pablo Parra-Nunez, Mónica Pradillo, and Juan Luis Santos

Abstract

During meiosis, accurate segregation of chromosomes requires the formation of bivalents at metaphase I. In autopolyploids, there are more than two copies of each chromosome with the same chance to form chiasmata at meiosis. This leads to the formation of multivalent configurations in which chiasma quantification is rather complicated. Here, we present an improved cytological protocol, including fluorescence in situ hybridization, to obtain high quality spreads of metaphase I chromosomes from *Arabidopsis thaliana* autotetraploids. This method allows an accurate analysis of the different meiotic configurations and enables the assessment of the number of chiasmata formed by each tetrasome (group of four homologs).

Key words Arabidopsis, Chiasma, FISH, Meiosis, Polyploidy

1 Introduction

Polyploidy is a key feature in the evolution of eukaryotes, especially in plants. Estimates suggest that between 47% and 70% of flowering plants have polyploid ancestors [1]. Polyploidy can confer several advantages such as drought and salinity tolerance, but the presence of more than two sets of chromosomes can also generate unbalanced gametes [2–4]. During polyploid meiosis, each chromosome has more than one potential partner to pair, recombine, and synapse. This can result in the presence of multivalents, homologous chromosome missegregations and reduced fertility [5]. Hence, the analysis of meiotic chromosome behavior in polyploids has a considerable practical value in terms of understanding the undesirable effects of chromosome duplication on fertility. To date, fluorescence in situ hybridization (FISH) is probably the best tool to perform this analysis. This technique was previously applied to study the partial diploidization of meiosis in successive generations of different autotetraploid lines of *Arabidopsis thaliana* [6].

Mónica Pradillo and Stefan Heckmann (eds.), *Plant Meiosis: Methods and Protocols*, Methods in Molecular Biology, vol. 2061,
https://doi.org/10.1007/978-1-4939-9818-0_3,

As early as 1907, Laibach [7] determined the correct chromosome number of *A. thaliana* as $2n = 10$. He found 10 chromocenters in interphase nuclei of somatic cells, corresponding to the number of chromosomes observed during meiosis. However, detailed cytological studies on chromosome morphology were hampered during most of the twentieth century. This was due to the small size of the chromosomes and the difficulty of getting good spreads that allow accurate light microscopic observations of both mitotic and meiotic cells [8]. This situation changed drastically with the improvement of cytological techniques, including the application of FISH [9, 10]. 45S and 5S rDNA FISH probes, combined with chromosome morphology, uniquely identify each of the Arabidopsis chromosomes and chromosome arms. The short acrocentric chromosomes 2 and 4 bear 45S rDNA sequences on their short arms in all accessions that have been examined to date, coinciding with the locations of the nucleolus organizing regions (NORs). In addition, a 5S rDNA site on the short arm of chromosome 4 is always present. This sequence helps to distinguish the two acrocentric chromosomes. A further invariant large 5S rDNA site occurs on the shorter arm of the submetacentric chromosome 5, which serves to distinguish this chromosome from chromosome 1. Chromosome 3 is the smallest of the submetacentric/metacentric group of chromosomes (1, 3, and 5) and is variable with respect to possession, location, and size of a third 5S rDNA site. Application of FISH with these DNA probes allows to identify and record the chiasma frequency of each chromosome and chromosome arm in pollen mother cells (PMCs) at metaphase I [11, 12].

In *Arabidopsis*, there are natural autopolyploid accessions as well as synthetic autotetraploid plants obtained after colchicine treatment [13–15]. In addition, polyploidy may be induced during tissue culture or may even occur spontaneously in certain accessions [16, 17]. Bivalent configurations at metaphase I can fall into two categories, rods and rings. Rods are bound by chiasmata only in one arm, whereas rings have both arms bound by chiasmata (Fig. 1a). However, as mentioned above, this situation turns out to be more complicated in autopolyploids due to the presence of multivalents [18]. Here, we describe an improved cytological protocol to obtain proper spreads of metaphase I chromosomes in autotetraploid PMCs (*see* Fig. 1b as a representative example of a conventional protocol, and Fig. 1c, d as examples of improved chromosome spreads). This protocol allows an accurate analysis of the different meiotic configurations and, with the required expertise, the assessment of the number of chiasmata formed in each tetrasome (group of four homologs). We also provide several examples of the different quadrivalent configurations that can be observed with the number of chiasmata necessary for their formation.

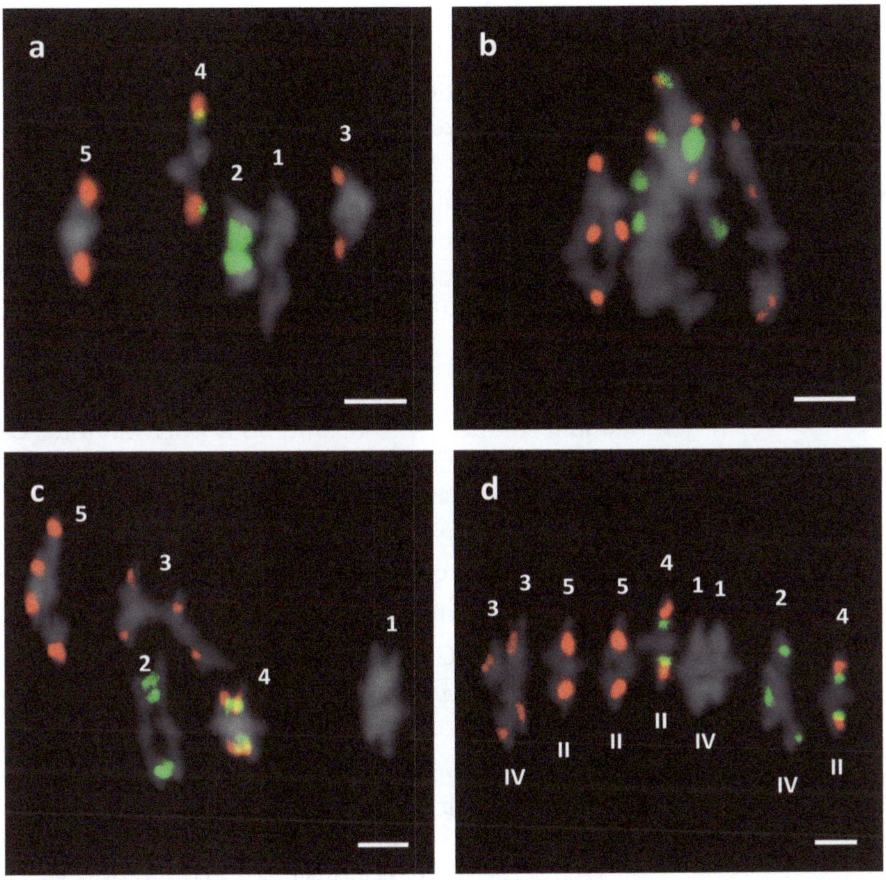

Fig. 1 Representative examples of FISH analysis of metaphase I cells. Arabic numerals (1–5) identify the different chromosomes according to their morphology and distribution of rDNA signals. Green signals identify 45S rDNA loci and red signals identify 5S rDNA loci. (**a**) Diploid cell. Individual bivalents are indicated. This cell has three ring bivalents (1, 3, and 5) and two rod bivalents (2 and 4). (**b**) Tetraploid cell obtained with the same protocol applied in (**a**). (**c**, **d**) Tetraploid cells obtained with the protocol described in this chapter. Five quadrivalents are observed in (**c**), whereas in (**d**) there are three quadrivalents (IV) and four bivalents (II). Bars represent 5 μm

2 Materials

2.1 Plants

Sow Arabidopsis Columbia-0 accession seeds in sterilized substrate with commercial soil and vermiculite (3:1), and grow plants until flowering in a constant environment chamber under the following conditions: 19 °C; photoperiod of 16 h light–8 h dark.

2.2 Colchicine

Prepare a colchicine solution of 0.25% w/v by diluting the colchicine stock with sterilized deionized water (SDW). Store solution at 4 °C.

2.3 Reagents and Equipment Required for Chromosome Spreads

1. Carnoy's fixative solution: 6 volumes of absolute ethanol, 3 volumes of chloroform, and 1 volume of glacial acetic acid.
2. 3:1 fixative solution: 3 volumes of absolute ethanol and 1 volume of glacial acetic acid. Store at 4 °C.
3. 1× citrate buffer. Prepare from a 10× stock solution: 40 mL of 0.1 M sodium citrate and 60 mL of 0.1 M citric acid, pH 4.6. Store at 4 °C.
4. Digestion enzyme mixture: 0.3% (w/v) cellulase, 0.3% (w/v) pectolyase, and 0.3% (w/v) cytohelicase in 1× citrate buffer. Store at −20 °C (*see* **Note 1**).
5. 60% glacial acetic solution diluted with SDW. Store at 4 °C.
6. Staining solution: 4′,6-diamidino-2-phenylindole (DAPI) at a concentration of 1 μg/mL in antifade mounting medium Vectashield®. Store at 4 °C.
7. Watch glasses, Pasteur pipettes, dissecting needles, and fine forceps. Microscope slides and coverslips. Diamond pen.
8. Binocular magnifier.
9. Incubator (37 °C).
10. Heating block (42 °C).
11. Hair dryer.

2.4 Preparation of DNA Probes

1. DNA probes: Clone pTa71 containing a 9-kb *Eco*RI fragment from *Triticum aestivum* consisting of the 18S–5.8S-25S rDNA genes and the spacer regions [19]. Plasmid pCT4.2 containing the 5S rDNA gene from *A. thaliana* as a 500 bp insert cloned in pBlu [20].
2. Label probes with digoxigenin-dUTP or biotin-dUTP (*see* **Note 2**) using a nick translation labeling kit and follow the manufacturer's instructions. Store the labelled probes at −20 °C until required.
3. Thermocycler (for nick translation reaction).

2.5 Hybridization and Washing

1. 20× SSC buffer (saline sodium citrate): 0.3 M NaCl, 0.03 M sodium citrate, pH 7. Store at room temperature (RT). Prepare a working dilution (2×) with SDW (*see* **Note 3**).
2. Pepsin solution: mix 90 mL of SDW and 0.01 g of pepsin powder (0.01% w/v). Later add 1 mL of 1 M HCl and SDW up to 100 mL.
3. Paraformaldehyde: solution of 4% (w/v), pH 8 (adjust with 1 M NaOH). Shake a minimum of 20 min at 60 °C. Store at 4 °C (*see* **Note 4**).

4. Alcohol series: Prepare 70% and 90% solutions with absolute ethanol and SDW. For the 100% solution use absolute ethanol. Store at 4 °C (*see* **Note 5**).
5. Hybridization mixture: 5 mL of deionized formamide, 1 mL of 20× SSC buffer, and 1 g of dextran sulfate. Dissolve at 65 °C and adjust pH to 7. Store at −20 °C.
6. 50% Formamide–2× SSC: Deionize formamide for a minimum of 1 h using ion-exchange resin beads (Amberlite). For 200 mL of solution: 100 mL of deionized formamide, 20 mL of 20× SSC, make up to 200 mL with SDW. Store at RT (*see* **Note 6**).
7. 4T buffer: Obtain 4× SSC buffer from the 20× SSC stock, 0.05% (v/v) of Tween 20 detergent, pH 7. Store at RT (*see* **Note 7**).
8. TNB: 100 mM Tris–HCl (pH 7.5), 150 mM NaCl, 0.5% (w/v) of a blocking reagent (e.g., Boehringer). Store at −20 °C.
9. Antibodies: Cy3-streptavidin and FITC-anti-digoxigenin. Dilute to the concentration of 5 ng/μL in TNB buffer (*see* **Note 8**).
10. Staining solution (*see* **step 6** in Subheading 2.3).
11. Coplin jars, coverslips and Parafilm, Eppendorf tubes.
12. Incubator (37 °C, 45 °C).
13. Heating block (72 °C, 80 °C).
14. Fluorescence microscope equipped with optical filters for DAPI, FITC, and Cy3 fluorochromes and image acquisition software.

3 Methods

3.1 Polyploidy Induction

Synthetic autotetraploids are obtained by a colchicine treatment applied to diploid plants. For that purpose, treat young plants at the preflowering rosette stage by placing 1 drop (about 7–10 μL) of 0.25% colchicine solution on the apical meristem (*see* **Note 9**).

3.2 Fixation

Fix closed flower buds in fresh Carnoy's fixative solution at RT. Exchange this solution after 24 h as many times as needed until the fixative solution remains clear. Store fixed samples at 4 °C. In polyploids, use material fixed for no more than 15 days as a crucial step for obtaining metaphase I cells in which chromosomes are not clumped (Fig. 1c, d) (*see* **Note 10**).

3.3 Flower Bud Digestion

1. Place several inflorescences on a glass well and add enough volume of 3:1 fixative to cover the samples (at least 1 mL).
2. Individualize flower buds and discard those that are not at the appropriate meiotic stage (*see* **Note 11**).

3. Wash flower buds three times (at least 5 min each) in 3:1 fixative solution.
4. Wash three times in 1× citrate buffer (at least 5 min each) to prepare flower buds for the enzyme digestion.
5. Remove the citrate buffer completely and add the enzyme mix (1 mL). Beware that the buds are submerged.
6. Cover the glass well and place it in a moist chamber to avoid evaporation. Digest flower buds in the enzyme mixture for 2.5 h at 37 °C (*see* **Note 12**).
7. Stop enzymatic digestion by replacing the enzyme mix with ice-cold SDW (also ice-cold 1× citrate buffer can be used) (*see* **Note 13**).

3.4 Chromosome Spreads

1. Place a digested flower bud on a slide with a small volume of buffer using a Pasteur pipette (*see* **Note 14**).
2. Squeeze the flower bud with a needle until the anthers are completely shattered obtaining a cell suspension (*see* **Note 15**).
3. Add 15 μL of ice-cold 60% acetic acid to the cell suspension and place the slide on a hot plate for 1 min at 42 °C (*see* **Note 16**). To fix more cells on the slide, especially in polyploids, move the cell suspension making circles with a needle without touching the surface of the slide. Then, add another 15 μL of ice-cold acetic acid to the suspension.
4. Add 200 μL of ice-cold 3:1 fixative solution making a circle around the cell suspension. Next, add further 200 μL of ice-cold 3:1 fixative solution to the center of the circle. Discard the fixative by tilting the slide. Wash again with 200 μL of ice-cold 3:1 fixative solution and dry the slide using a hair dryer.
5. Mount the slides adding 10 μL of staining solution per slide and covering with a 24 × 24 mm cover slip (*see* **Note 17**).
6. Check the slides under a microscope to select those containing metaphase I meiocytes to perform the FISH protocol (*see* **Note 18**).

3.5 DNA Hybridization

1. Remove the coverslips from the slides by washing them in a Coplin jar with 4T buffer at RT until the coverslip slips off. If the coverslip has immersion oil, remove it by washing the slides in a Coplin jar with absolute ethanol at RT. Afterward, in both cases, wash the slides in 4T buffer for a minimum of 1 h, but overnight (ON) is recommended. Skip this step when the slides were not stained before the FISH protocol.
2. Preheat 100 mL of the pepsin solution in a Coplin jar at 37 °C.
3. Wash the slides in 2× SSC buffer for 10 min at RT.

4. Incubate the slides in the pepsin solution for 90 s at 37 °C (*see* **Note 19**).
5. Wash the slides in 2× SSC buffer for 10 min at RT.
6. Fix the material by incubating the slides in 4% paraformaldehyde for 10 min at RT.
7. Rinse the slides in SDW twice for 5 min at RT.
8. Dehydrate the material in an ethanol series of 70%, 90% and 100% of absolute ethanol. Wash the slides for 2 min in each solution at RT. After the last wash, let the slides dry for a minimum of 30 min at RT in darkness.
9. To prepare the probe mixture, mix for each slide 3 μL of each probe and 14 μL of the hybridization mixture. Denature the probe mixture for 10 min at 80 °C and place the mixture on ice for 5 min.
10. Add 20 μL of the probe mixture on a coverslip. Invert the slide and touch gently the liquid to mount the coverslip on the slide. Place the slides on a hot plate for 4 min at 72 °C.
11. Incubate the slides ON at 37 °C (*see* **Note 20**).
12. Preheat four Coplin jars at 45 °C: two with 50% formamide–2× SSC, one with 2× SSC buffer and one with 4T buffer.
13. Remove the coverslip from the slides (*see* **Note 21**).
14. Wash the slides twice in 50% formamide–2× SSC buffer for 7 min at 45 °C.
15. Wash the slides for 5 min in 2× SSC buffer and for 5 min in 4T buffer at 45 °C.
16. Wash in 4T buffer for 5 min at RT.
17. Prepare the solution containing fluorescent antibodies (50 μL per slide).
18. Add 50 μL of this solution to each slide and cover with Parafilm (*see* **Note 22**).
19. Incubate the slides for 1 h in a humidified atmosphere at 37 °C in darkness.
20. Remove carefully the Parafilm using forceps and wash the slides three times in 4T buffer for 5 min at RT (*see* **Note 23**).
21. Finally, mount the slides by adding 10 μL of the staining solution on coverslips. Invert the slides and mount the coverslips upside down. Carefully, apply gentle pressure. Store slides for up to 2 weeks at 4 °C in darkness.

3.6 Microscope Captures of Cells

1. Use an epifluorescence microscope equipped with a camera and the appropriate filters to detect DAPI, FITC, and Cy3 signals.

2. Merge the three pictures obtained from each metaphase I cell (corresponding to DAPI, FITC, and Cy3). Adjust contrast and brightness using a photo editing software such as Adobe Photoshop®.

3.7 Interpretation of Meiotic Configurations

After applying the protocol mentioned above, the most difficult part to be conducted is the interpretation of the different meiotic configurations that can be observed. They are mainly conditioned by the morphology of the chromosomes involved: the appearance is different depending on whether the chromosome is acrocentric (chromosomes 2 and 4) or submetacentric (chromosomes 1, 3, and 5). The number and location of chiasmata formed also influence the configurations. FISH signals and the observation of the chromatin in grey, without the blue color of the DAPI, can help to identify the chromatids bond by chiasmata. Representative examples of the most common configurations for quadrivalents with three (Fig. 2) or more than three chiasmata (Fig. 3) are included. These images can serve as a first approximation in studies that require a record of chiasmata in tetraploids of *Arabidopsis* and species with similar chromosomal morphologies.

4 Notes

1. Alternatively, prepare the digestion enzyme mixture with 1% (w/v) of each enzyme (3×). Freshly prepare working dilution (1×) with citrate buffer (1×).
2. We traditionally apply digoxigenin-dUTP for 45S rDNA sequences and biotin-dUTP for 5S rDNA sequences, but it works just as well if it is done the other way.
3. To avoid bacterial growth in 2× SSC buffer prepare it fresh each time the protocol is conducted.
4. Paraformaldehyde solution can be reused several times.
5. Ethanol dilutions (70%, 90% and 100%) can be reused sealing the Coplin jar to avoid evaporation.
6. Deionized formamide can be stored at RT until needed.
7. 4T buffer is contaminated when the solution starts to look cloudy (in a couple of weeks). Check it before its use.
8. Usually red (Cy3) is used for detecting 5S rDNA because these signals are smaller and red is usually more clearly distinguished than green (FITC).
9. Approximately 10–20% of plants survive the colchicine treatment. The appearance of the first flowering depends on the accession used. It is typical to obtain mixoploid individuals. As a first indication for plants producing polyploid tissues, look for abnormal leaf trichomes with more than three branches [21].

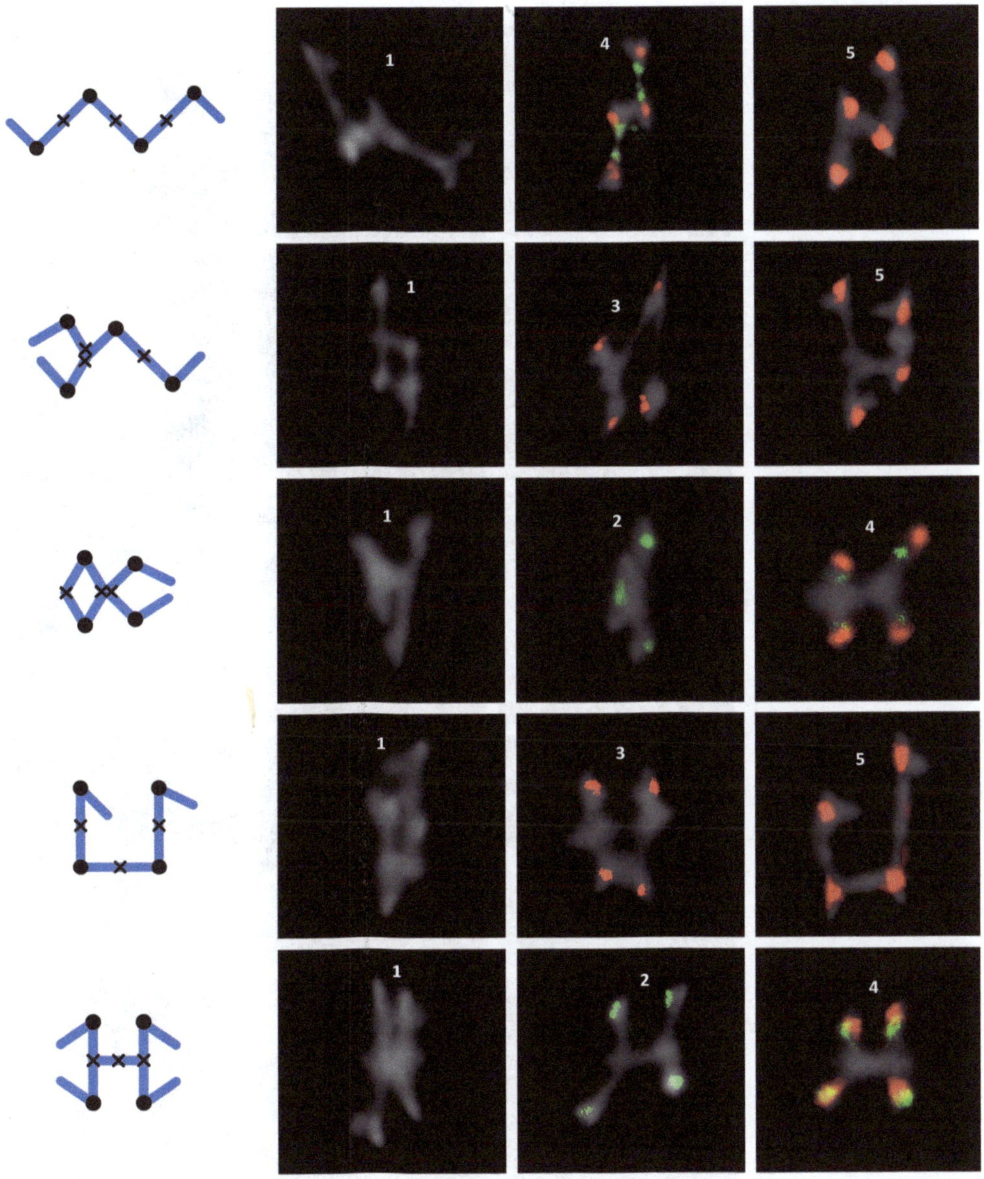

Fig. 2 Examples of quadrivalents with three chiasmata. Images of the same row show the same quadrivalent configuration formed by different chromosomes. On the left, there is a scheme of the quadrivalent indicating the number of chiasmata necessary for its formation (black crosses). For simplicity, chiasmata are considered distally located

10. The less time flower buds are in fixative solution, the better spreads are obtained. However, less than 5 days of fixation might lead to lose cells during the FISH protocol. Additionally, it is important to exchange the fixative solution after the first 24 h until it remains clear. This is essential in polyploids to obtain proper spreading.

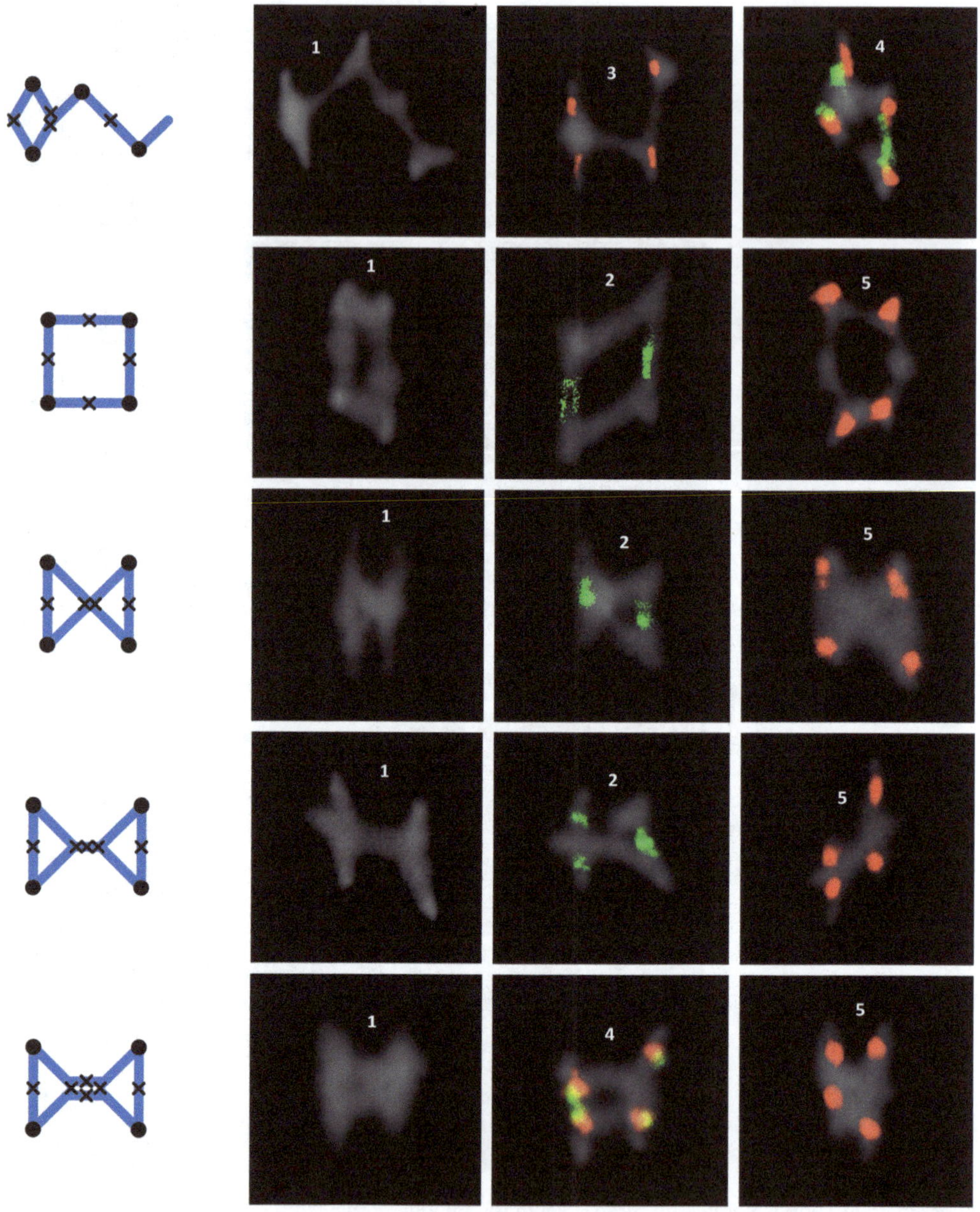

Fig. 3 Examples of quadrivalents with four or more chiasmata. Images of the same row show the same quadrivalent configuration formed by different chromosomes. On the left, there is a scheme of the quadrivalent indicating the number of chiasmata necessary for its formation (black crosses). For simplicity, chiasmata are considered distally located

11. There is a correlation between meiotic stages and bud size. Flower buds with yellow anthers contain pollen grains.
12. In the protocol designed for diploid plants, the digestion time is 2 h. For polyploids, it is necessary to increase the digestion time to get a better spread of the chromosomes.

13. Digested flower buds can be used immediately to continue the protocol or stored up to 2 days at 4 °C in a moist chamber.
14. Do not add too much buffer to be able to squeeze the flower bud easily, but at the same time be careful to ensure that the suspension does not dry out.
15. It is useful to make a hook bending a needle to squeeze better the flower buds.
16. Mark the region of the slide that contains the cell suspension using a diamond pen.
17. This is an optional step because the slides with metaphase I meiocytes can also be selected using a phase contrast microscope prior to the staining step.
18. Write down the microscope coordinates of metaphase I cells to find them easily after the FISH protocol.
19. Increase the time if chromosomes appear to be covered by cytoplasm.
20. Place the slides in a moist chamber to prevent evaporation.
21. Tilt slides to let the coverslip fall. Place the slides immediately into the 50% formamide–2× SSC to avoid getting the material dry.
22. It is important to remove air bubbles as much as possible.
23. To avoid the presence of background (nonspecific binding of the antibody), it is advisable to extend this time, as well as to shake the Coplin jars during the washes.

Acknowledgments

The authors acknowledge the support of the European Union by the advanced grant Marie Curie Initial Training Network (ITN) "COMREC" (Grant agreement number: 606956). This work has also been partially supported by a grant from the Ministerio de Economía y Competitividad of Spain (AGL2015-67349-P).

References

1. Masterson J (1994) Stomatal size in fossil plants: evidence for polyploidy in majority of angiosperms. Science 264:421–424. https://doi.org/10.1126/science.264.5157.421
2. Pozo JC, Ramirez-Parra E (2014) Deciphering the molecular bases for drought tolerance in *Arabidopsis* autotetraploids. Plant Cell Environ 37:2722–2737. https://doi.org/10.1111/pce.12344
3. Chao DY, Dilkes B, Luo H et al (2013) Polyploids exhibit higher potassium uptake and salinity tolerance in *Arabidopsis*. Science 341:658–659. https://doi.org/10.1126/science.1240561
4. Comai L (2005) The advantages and disadvantages of being polyploid. Nat Rev Genet 6:836–846. https://doi.org/10.1038/nrg1711

5. Sybenga J (1975) Meiotic configurations: a source of information for estimating genetic parameters. Springer, Berlin
6. Santos JL, Alfaro D, Sanchez-Moran E et al (2003) Partial Diploidization of meiosis in Autotetraploid *Arabidopsis thaliana*. Genetics 165:1533–1540
7. Laibach F (1907) Zur Frage nach der Individualität der Chromosomen im Pflanzenreich. Beih Bot Centralbl 22:191–210
8. Steinitz-Sears LM (1963) Chromosome studies in *Arabidopsis thaliana*. Genetics 48:483–490
9. Fransz P, Armstrong S, Alonso-Blanco C et al (1998) Cytogenetics for the model system *Arabidopsis thaliana*. Plant J 13:867–876. https://doi.org/10.1046/j.1365-313X.1998.00086.x
10. Sanchez Moran E, Armstrong SJ, Santos JL et al (2001) Chiasma formation in *Arabidopsis thaliana* accession Wassileskija and in two meiotic mutants. Chromosom Res 9:121–128. https://doi.org/10.1023/A:1009278902994
11. Sanchez-Moran E, Armstrong SJ, Santos JL et al (2002) Variation in chiasma frequency among eight accessions of *Arabidopsis thaliana*. Genetics 162:1415–1422
12. López E, Pradillo M, Oliver C et al (2012) Looking for natural variation in chiasma frequency in *Arabidopsis thaliana*. J Exp Bot 63:887–894. https://doi.org/10.1093/jxb/err319
13. Koorneef M, Fransz P, de Jong H (2003) Cytogenetic tools for *Arabidopsis thaliana*. Chromosom Res 11:183–194
14. Bouharmont J, Van De Hende J (1968) Inheritance of lethal chlorophyll mutants in tetraploid *Arabidopsis thaliana*. Arabidopsis Inform Serv 5:25–26
15. Bouharmont J (1969) Evolution of chromosome numbers in *Arabidopsis* polyploids. Chromosomes Today 2:197–201
16. Morris PC, Altmann T (1994) Tissue culture and transformation. In: Meyerowitz EM, Somerville CR (eds) Cold Spring Harbor laboratory press. Cold Spring Harbor, New York
17. Heslop-Harrison JS, Maluszynska J (1994) The molecular cytogenetics of Arabidopsis. In: Meyerowitz EM, Sommerville CR (eds) Cold Spring Harbor laboratory press. Cold Spring Harbor, New York
18. Weiss H, Maluszynska J (2000) Chromosomal rearrangement in autotetraploid plants of *Arabidopsis thaliana*. Hereditas 133:255–261. https://doi.org/10.1111/j.1601-5223.2000.00255.x
19. Gerlach WL, Bedbrook JR (1979) Cloning and characterization of ribosomal RNA genes from wheat and barley. Nucleic Acids Res 7:1869–1885. https://doi.org/10.1093/nar/7.7.1869
20. Campell BR, Song Y, Posch TE et al (1992) Sequence and organization of 5S ribosomal RNA-encoding genes of *Arabidopsis thaliana*. Gene 112:225–228. https://doi.org/10.1016/0378-1119(92)90380-8
21. Yu Z, Haage K, Streit VE et al (2009) A large number of tetraploid *Arabidopsis thaliana* lines, generated by a rapid strategy, reveal high stability of neo-tetraploids during consecutive generations. Theor Appl Genet 118:1107–1119. https://doi.org/10.1007/s00122-009-0966-9

Chapter 4

Cytological Characterization of *Arabidopsis arenosa* Polyploids by SIM

Chris Morgan and Eva Wegel

Abstract

Arabidopsis arenosa has recently become established as a model organism for investigating how meiosis has evolved to overcome the meiotic challenges faced by newly formed autotetraploids. Here, we describe a protocol for the preparation of spread, immunolabeled prophase I chromosomes from established *A. arenosa* autotetraploids for imaging with three-dimensional structured illumination microscopy (3D-SIM). This technique allows us to dissect the unique synaptic behavior in *A. arenosa* and identify synaptic partner switch sites that are unresolvable with conventional widefield microscopy.

Key words *Arabidopsis arenosa*, Meiosis, Synaptonemal complex, Synaptic partner switch, Structured illumination microscopy

1 Introduction

Autopolyploids arise from within-species whole genome duplication events and possess multiple (>2) homologous copies of each chromosome which are all equally capable of pairing, synapsing and recombining with each other during meiotic prophase I. Consequently, metaphase I multivalents are common in newly formed autopolyploids (neopolyploids) and are associated with an increased risk of homologue missegregation at anaphase I. Despite this, fertile autopolyploid species are found at high frequency within the plant kingdom, indicating that molecular mechanisms can evolve to overcome the meiotic challenges faced by neotetraploids [1–3].

Arabidopsis arenosa is a close relative of *A. thaliana* that possesses naturally established diploid and autotetraploid populations and, through work spearheaded by Kirsten Bomblies' lab, *A. arenosa* has emerged as a valuable model organism for uncovering the molecular mechanisms responsible for the stabilization of autotetraploid meiosis [3–5]. To investigate how meiotic stabilization is manifested in established autotetraploid *A. arenosa*,

Mónica Pradillo and Stefan Heckmann (eds.), *Plant Meiosis: Methods and Protocols*, Methods in Molecular Biology, vol. 2061, https://doi.org/10.1007/978-1-4939-9818-0_4,

cytological techniques have previously been adapted from protocols designed for use in *A. thaliana* to visualize how meiosis unfolds in both the diploid and established autotetraploid *A. arenosa* [6, 7]. However, these methods were designed to be used in combination with widefield epifluorescence microscopy. The resolution limit of widefield microscopy has meant that previous studies investigating detailed synaptic behavior in polyploids have relied upon transmission electron microscopy to detect structures such as synaptic partner switch (SPS) sites, where an individual chromosome desynapses with one homologous partner before resynapsing with a separate homologue [8].

In the last decade the development of a variety of different super-resolution technologies has allowed smaller cellular structures, which were previously considered un-resolvable using light-microscopy, to be visualized in much greater detail. In this chapter we present a modified sample preparation method for optimally imaging spread immunolabeled autotetraploid *A. arenosa* prophase I meiocytes using super-resolution three-dimensional structured illumination microscopy (3D-SIM). 3D-SIM is built upon a widefield fluorescence microscope but increases its resolution by a factor of two in all three dimensions. It utilises the moiré effect by which the overlay of an unknown small structure with a regular pattern moves high frequency components of the unknown structure into the observable space through interference of the two. In SIM, the regular pattern is usually composed of a sinusoidally striped laser illumination pattern that is rotated through three angles and moved through five phases so that for each final image 15 images are acquired. In Fourier space, the high-frequency components of each image are moved to the positions where they belong which results in a resolution increase in the final composite image [9]. Using 3D-SIM, both lateral elements of the synaptonemal complex (SC) can be discreetly visualized, meaning that the synaptic behavior of individual homologues can be more accurately defined and the localization of multiple SC-associated proteins can be simultaneously observed.

2 Materials

2.1 Plant Material

This protocol is specifically optimized using the autotetraploid Triberg (TBG) railway population of *A. arenosa*. Seeds are sown on a mix of 75% F2 compost + grit and 25% sand in 9 cm^2 pots and grown in controlled environment rooms at 70% humidity, 16 h day (20 °C, 125 μM × m^{-2} × s^{-1} light), 8 h night (16 °C). Plants usually flower 6–8 weeks after germination (*see* **Note 1**).

2.2 Equipment

1. Sterile 9 cm plastic petri dishes.
2. 9 cm diameter filter paper discs.

3. Dissecting stereomicroscope (preferably with a calibrated eyepiece graticule).
4. DIC light microscope.
5. Two pairs of fine (No. 5) forceps.
6. Thin brass rod.
7. Mounted needle.
8. High-precision number 1.5 coverslips (indispensable for SIM imaging).
9. Glass slides.
10. Moist chamber (we use a microfuge tube freezer storage box with a layer of damp paper towels in the bottom).
11. Hot plate set to 37 °C.
12. Parafilm.
13. Nail varnish.
14. 3D-SIM microscope (we use a Zeiss Elyra PS1).
15. Immersion oil (*see* **Note 2**).

2.3 Reagents

1. Digestion medium: 0.4% cytohelicase (Sigma), 1.5% sucrose and 1% polyvinylpyrrolidone (Sigma) in sterile deionized water. Store at −20 °C in 1 ml aliquots.
2. Spreading medium: 2% Lipsol detergent (SciLabware) in sterile deionized water.
3. Fixative: 4% paraformaldehyde in sterile deionized water, pH ~8. Dispense 16% methanol-free paraformaldehyde (Polysciences) into 50 μl aliquots and store at −20 °C. On the day of use, add 150 μl of sterile deionized water to the aliquot to make a 4% paraformaldehyde solution and add 5 μl of a 0.1 N NaOH solution to adjust the pH to ~8. All paraformaldehyde work should be performed in a fume hood.
4. Blocking buffer (10× stock solution): 3% bovine serum albumin in 1× washing solution (*see* **step 6**). Dissolve the stock 1:10 in sterile deionized water to make a 1× working solution.
5. Primary antibody: Dissolve the primary antibody to the desired final concentration (for most antibodies we use 1:500) in 1× blocking buffer.
6. Washing solution (10× stock solution): Dissolve 9.93 g of phosphate buffered saline (PBS) in 1 L of sterile deionized water to make a 10× PBS stock solution. Dissolve the stock solution 1:10 in sterile deionized water and add 0.1% Triton X-100 to make 1× washing solution.
7. Secondary antibody: Dissolve the secondary antibody to the desired final concentration (we tend to use 1:200) in 1× blocking buffer (*see* **Note 3**).

8. DAPI staining solution (100× stock solution): 1 mg/ml 4′,6-diaminido-2-phenylindole (DAPI) in sterile deionized water. Dissolve DAPI powder as a stock solution of 1 mg/ml in sterile deionized water and store in 1 ml aliquots at −20 °C. On the day of use dissolve the stock solution 1:100 (final conc. 10 μg/ml) in 1× washing solution to give 1× DAPI staining solution (*see* **Note 4**).
9. Antifade mounting medium: we use Vectashield (Vector Labs) (*see* **Note 5**).

3 Methods

3.1 Staging Anthers

1. Remove an inflorescence from your desired plant using a pair of forceps and place on a piece of moist filter paper in a petri dish ready for dissection. Take care not to let the filter paper dry out at any point during dissection.
2. Under the dissecting microscope, using fine forceps, remove each bud from the inflorescence by slicing through the pedicel below each bud. Retain the buds that have a diameter between 0.5 and 1 mm for staging and discard the others (*see* **Note 6**).
3. To stage the buds, carefully remove a single anther from each bud using fine forceps and place the anther in a small drop of water (~5 μl) on a glass slide. Add a coverslip to the drop and firmly press down on the coverslip using a piece of tissue. This will extrude the meiocytes from the anther so that their meiotic stage can be determined (Fig. 1).
4. Inspect the squashed anthers using DIC microscopy. Early prophase I (leptotene → early pachytene) meiocytes will extrude in long columns of connected cells. In late prophase I meiocytes (late pachytene → diplotene) the connected cells begin to break apart to form discreet cells and it is at this stage that antibodies targeting class I CO proteins (e.g., MLH1, HEI10) localize as bright, SC associated foci [10]. Anthers containing tetrads and pollen can also be identified and buds containing these are discarded (*see* **Note 7**).

3.2 Making the Slides

1. Using fine forceps remove the five remaining anthers from a single bud at the desired stage.
2. Place the anthers in a 5 μl drop of digestion medium on a high precision number 1.5 coverslip. To make the coverslips easier to handle, we first adhere them to a glass slide using a drop of water.
3. Gently macerate the anthers by softly tapping them with a brass rod for 1 min.

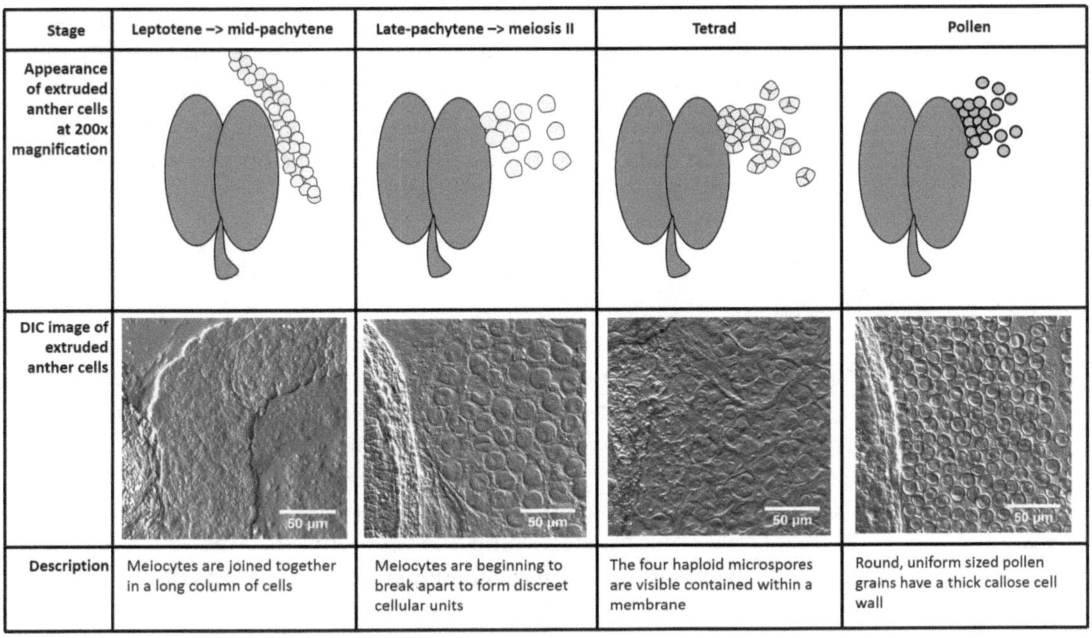

Fig. 1 A guide to determining the meiotic stage of squashed anthers from *A. arenosa*

4. Add another 5 μl of digestion mix to the coverslip and incubate in a moist chamber at 37 °C for 4 min.
5. Remove the coverslip from the moist chamber and add 10 μl of spreading medium. Using a mounted needle spread the mixture into a square shape that reaches 2–3 mm from the edge of the coverslip.
6. In the fume hood add 20 μl of fixative to the coverslip and stir the mixture briefly with the tip of the pipette. Leave the coverslip to dry in the fume hood for at least 3 h.
7. Add 50 μl of 1× blocking buffer to the coverslip, place a strip of Parafilm on top to seal in the liquid and incubate at room temperature for 15 min.
8. Add 50 μl of primary antibody to the coverslip, place a strip of Parafilm on top to seal in the liquid and incubate in a moist chamber, in the dark at 4 °C overnight.
9. Wash the coverslip 3× 5 min in 1× washing solution. We use petri dishes to wash our coverslips in, taking care to keep the side of the coverslips with the fixed cells on facing upward and we use a pair of fine forceps and a mounted needle to move the coverslips between petri dishes.
10. Add 50 μl of secondary antibody to the coverslip, place a strip of Parafilm on top to seal in the liquid and incubate in a moist chamber, in the dark at 37 °C for 90 min.

11. Add 50 μl of DAPI staining solution to the coverslip, place a strip of Parafilm on top to seal in the liquid and incubate in the dark at room temperature for 5 min.
12. Wash the coverslips 3× 5 min in 1× washing solution. Finally, wash the coverslips briefly in sterile deionized water and then blot the edges of the coverslip on a piece of tissue to remove any excess liquid.
13. Mount the coverslip in 7 μl of Vectashield in the center of a clean glass slide.
14. Seal the coverslips at the edges with nail varnish and store the slides in the dark at 4 °C. Slides can be stored in this way for several weeks prior to imaging but best results are achieved if the slides are imaged within a couple of days.

3.3 Imaging the Slides

1. Slides are imaged in 3D SIM mode with three stripe angles and five phases according to the microscope manufacturer's instructions. Switch channels at every *z*-position to achieve the best *z*-alignment between colors. The size of the *z*-stack needs to be big enough for the sample to be out of focus at both ends. Make sure you can clearly see the stripes in the sample provided there is enough structure to show them and do not overexpose your sample. If the stripes are not clearly visible slightly changing the refractive index of the immersion oil might help.
2. Use the SIMcheck plugin in ImageJ/Fiji [11] to inspect the quality of your raw data and reconstructions. This will allow you to find out whether a fluorophore bleaches too much (>30% intensity loss across the stack) during acquisition or shows too little contrast both of which will lead to poor quality images. If there is too much bleaching reduce the laser power and/or increase camera gain.
3. Suboptimal imaging conditions, microscope misalignment and the wrong Noise/Wiener filter can lead to reconstruction artifacts. A good overview of causes and effects can be found in [12].
4. Channels need to be aligned after acquisition to compensate for chromatic shifts and differences in magnification and rotation caused by diffraction, lensing effects and differences in the light path for each fluorophore. For this you need a medium density TetraSpeck microsphere slide (*see* **Note 2**). Ideally, you want to see 20 beads in a 512 × 512 pixel field of view. A bead stack needs to be imaged with every experimental session. An algorithm for channel alignment is included in the Zeiss Elyra software and presumably in other manufacturer's software as well. An aligned image of a pachytene cell from autotetraploid *A. arenosa* imaged with 3D-SIM is shown in Fig. 2.

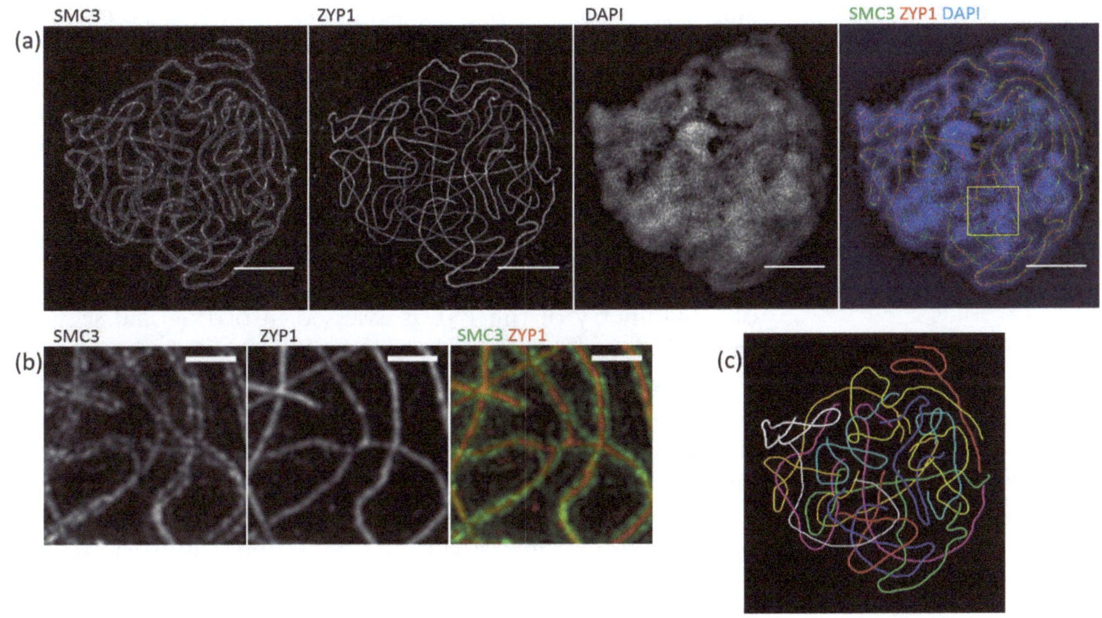

Fig. 2 A pachytene cell from autotetraploid *A. arenosa* prepared using this method, stained for SMC3 [15], ZYP1 [16], and DAPI and imaged using 3D-SIM (**a**). A yellow box indicates the position of a synaptic partner switch in the cell and a close-up image of this region is also shown (**b**). A 3D rendering of this cell generated using the Simple Neurite Tracer [13] plugin to imageJ/FIJI, in which each bivalent chromosome is labelled with a different color (**c**). Scale bars = 5 μm (**a**) and 1 μm (**b**)

3.4 Measuring the Length of Bivalent Pachytene Chromosomes

1. Lengths of individual synapsed bivalent pachytene chromosomes imaged using 3D-SIM can be measured semiautomatically in 3D through image stacks using the Simple Neurite Tracer plugin in ImageJ/FIJI [13] (*see* **Note 8**). A 3D rendering of a pachytene cell in which each bivalent chromosome has been labelled with a different color is shown in Fig. 2c.

4 Notes

1. Some *A. arenosa* ecotypes (e.g., SN, KZ, KA, KOWA) require 6 weeks vernalization at 4 °C before they will flower. It may also be necessary to modify this protocol for optimal results with other *A. arenosa* ecotypes. We suggest increasing/decreasing the digestion time or increasing/decreasing the Lipsol concentration of the spreading medium as the best places to start when attempting to modify this protocol.
2. For quality SIM reconstructions it is important to image the sample in such a way that each detected fluorophore has a symmetrical point spread function (PSF). To achieve this, we mix a series of our own immersion oils from laser liquids (Cargille) with two different refractive indexes (n = 1.510

and $n = 1.522$). A good starting selection is $n = 1.511$ to $n = 1.518$ in 0.001 intervals. Mixes should be shaken on a rotary shaker over night to achieve a good emulsion. To determine the best refractive index, 200 nm TetraSpeck microspheres (Invitrogen) are dried on a coverslip at low concentration and mounted in the mounting medium used for the experiment. The microspheres are imaged in 6 μm *z*-stacks with ca. 0.1 μm intervals with the same channel settings as for the experiment using the laser liquid mixes as immersion oils. Symmetry of the PSF is assessed in orthogonal sections through the stacks. The shape of the PSF depends on emission wavelength, temperature, and distance of the sample from the coverslip. Lower emission wavelengths, lower temperatures and samples closer to the coverslip will require lower refractive index oils to achieve a symmetrical PSF. Theoretically each fluorophore used in the experiment would require a different refractive index immersion oil. As this is impossible in practice a refractive index is chosen that works ok for the two most important fluorophores in the experiment, which have to be spectrally close (*see* **Note 3**).

3. We use exclusively secondary antibodies conjugated to Alexa Fluor dyes for staining our samples. Alexa Fluor dyes are amongst the most photostable commercially available dyes and the reduced photobleaching of these dyes enables superior image reconstruction. In practice, this means Alexa Fluor 488 and Alexa Fluor 568. Unless they are used as a fourth dye, far red fluorophores are not recommended for SIM because the gain in resolution is small compared with lower wavelengths.
4. We do not advise adding DAPI directly to the mounting medium. This can cause a fluorescent background haze in the DAPI channel as not all the fluorophores intercalate with nucleic acids.
5. Commercial hard-setting antifade mounting media with optimal refractive indexes are also available such as Prolong Diamond or Prolong Glass. We have previously tried mounting coverslips in Prolong Diamond but found the hard-setting mounting medium had an adverse effect on sample quality when compared with the soft-setting Vectashield. We have also successfully used a mounting medium consisting of 20% Vectashield in TRIS-glycerol (TRIS 1 M, pH 8, to a final conc. of 5% (v/v) in glycerol) [14] in combination with Alexa Fluor 555 and 647 to image cells with dSTORM that were otherwise prepared using this same spreading protocol.
6. Although bud size can be used as a rough estimate of meiotic stage [6], we find it is always better to stage an anther from

each bud individually to increase the chances that the meiocytes are at the desired meiotic stage. This not only saves time, as fewer slides need to be made and then screened to find cells of the desired stage, but it also prevents valuable primary antibody from being used up on making suboptimum slides.

7. Instead of using DIC microscopy to visualize the extruded meiocytes, cells can also be squashed in a drop of dye (e.g., aceto-orcein) and observed using brightfield.
8. It is advised that chromosome measurements are always taken in 3D using this method, rather than by tracing chromosomes in 2D along maximum intensity projections, as measuring chromosome lengths in 2D will lead to the exclusion of valuable *z*-axis information. Chromosomes from cells imaged using widefield epifluorescence microscopy can also be measured using the Simple Neurite Tracer plugin [13]; however, low resolution in the *z*-axis means it can be difficult to identify the correct path of individual bivalent chromosomes at points where two chromosomes overlap in the *xy* orientation.

Acknowledgments

We thank Dr. Kirsten Bomblies (John Innes Centre) for supporting the development of this protocol and Prof. Chris Franklin (University of Birmingham) for kindly supplying antibodies. This work was supported by a European Research Council Consolidator grant to K.B. (CoG EVO-MEIO 681946) and a UK Biological and Biotechnology Research Council (BBSRC) studentship to C.M. (DTP BB/MO1116 x/1 M1BTP). This work has also been supported by the BBSRC via grant BB/P013511/1 to the John Innes Centre.

References

1. Lloyd A, Bomblies K (2016) Meiosis in autopolyploid and allopolyploid Arabidopsis. Curr Opin Plant Biol 30:116–122
2. Bomblies K, Higgins JD, Yant L (2015) Meiosis evolves: adaptation to external and internal environments. New Phytol 208:306–323
3. Bomblies K, Jones G, Franklin C, Zickler D, Kleckner N (2016) The challenge of evolving stable polyploidy: could an increase in "crossover interference distance" play a central role? Chromosoma 125:287–300
4. Yant L, Hollister JD, Wright KM, Arnold BJ, Higgins JD, Franklin FCH, Bomblies K (2013) Meiotic adaptation to genome duplication in Arabidopsis arenosa. Curr Biol 23:2151–2156
5. Hollister JD, Arnold BJ, Svedin E, Xue KS, Dilkes BP, Bomblies K (2012) Genetic adaptation associated with genome-doubling in Autotetraploid Arabidopsis arenosa. PLoS Genet 8:e1003093
6. Higgins JD, Wright KM, Bomblies K, Franklin FCH (2014) Cytological techniques to analyze meiosis in Arabidopsis arenosa for investigating adaptation to polyploidy. Front Plant Sci 4:546
7. Pawlowski WP, Grelon M, Armstrong S (2013) Plant meiosis: methods and protocols. In: Methods in Molecular Biology, vol 990. Humana Press, New York
8. Grandont L, Cunado N, Coriton O, Huteau V, Eber F, Chevre AM, Grelon M, Chelysheva L, Jenczewski E (2014) Homoeologous chromosome sorting and progression of meiotic recombination in Brassica napus: Ploidy does matter. Plant Cell 26:1448–1463

9. Gustafsson MGL, Shao L, Carlton PM, Wang CJR, Golubovskaya IN, Cande WZ, Agard DA, Sedat JW (2008) Three-dimensional resolution doubling in wide-field fluorescence microscopy by structured illumination. Biophys J 94:4957–4970
10. Lloyd A, Morgan C, Franklin FCH, Bomblies K (2018) Plasticity of meiotic recombination rates in response to temperature in Arabidopsis. Genetics 208:1409–1420
11. Ball G, Demmerle J, Kaufmann R, Davis I, Dobbie IM, Schermelleh L (2015) SIMcheck: a toolbox for successful super-resolution structured illumination microscopy. Sci Rep 5:15915
12. Demmerle J, Innocent C, North AJ, Ball G, Muller M, Miron E, Matsuda A, Dobbie IM, Markaki Y, Schermelleh L (2017) Strategic and practical guidelines for successful structured illumination microscopy. Nat Protoc 12:988–1010
13. Longair MH, Baker DA, Armstrong JD (2011) Simple Neurite tracer: open source software for reconstruction, visualization and analysis of neuronal processes. Bioinformatics 27:2453–2454
14. Olivier N, Keller D, Rajan VS, Gonczy P, Manley S (2013) Simple buffers for 3D STORM microscopy. Biomed Opt Express 4:885–899
15. Lam WS, Yang XH, Makaroff CA (2005) Characterization of Arabidopsis thaliana SMC1 and SMC3: evidence that AtSMC3 may function beyond chromosome cohesion. J Cell Sci 118:3037–3048
16. Higgins JD, Sanchez-Moran E, Armstrong SJ, Jones GH, Franklin FCH (2005) The Arabidopsis synaptonemal complex protein ZYP1 is required for chromosome synapsis and normal fidelity of crossing over. Genes Dev 19:2488–2500

Chapter 5

Targeted Analysis of Chromatin Events (TACE)

Jason Sims, Changbin Chen, Peter Schlögelhofer, and Marie-Therese Kurzbauer

Abstract

Visualization of meiotic chromatin from pollen mother cells has become an essential technique to study meiosis in the model plant *Arabidopsis thaliana*. Here we present an advanced cytogenetic method that combines improved immunocytology with chromosome painting, thereby generating a tool to quantitatively analyze localization of proteins to any given genomic region. Proteins involved in different processes such as DNA double-strand break formation and recombinational repair can be visualized on meiotic chromatin with the additional feature of assessing their abundance at specific chromosomal locations.

Key words *Arabidopsis thaliana*, Chromosome painting, FISH, Immunocytology, Meiosis

1 Introduction

Arabidopsis thaliana is an excellent model organism to study meiosis, particularly suited for cytological analysis due to its compact genome size and organization (135 Mb, diploid, five chromosomes) and good accessibility of meiotic tissue. Furthermore, it is closely related to several important crop plants like *Brassica napus* and therefore the organism of choice for developing novel techniques that may be adopted by plant breeders. With the recent advances in manipulating meiotic double-strand break (DSB) formation and recombination (reviewed in [1, 2]) it was necessary to develop a method to visualize and quantify meiotic proteins at specific genomic regions.

Several techniques are currently available to monitor meiotic recombination in defined regions by analysis of fluorescent marker segregation in pollen grains or seeds [3, 4] or evaluating the segregation patterns of DNA markers such as single-nucleotide polymorphisms (SNPs, e.g., [5]). While these are powerful tools to assess recombination rates, they fall short of revealing which protein factors localize to and act on the chromosomal regions of interest. The presented protocol enables researchers to identify

Mónica Pradillo and Stefan Heckmann (eds.), *Plant Meiosis: Methods and Protocols*, Methods in Molecular Biology, vol. 2061, https://doi.org/10.1007/978-1-4939-9818-0_5, © Springer Science+Business Media, LLC, part of Springer Nature 2020

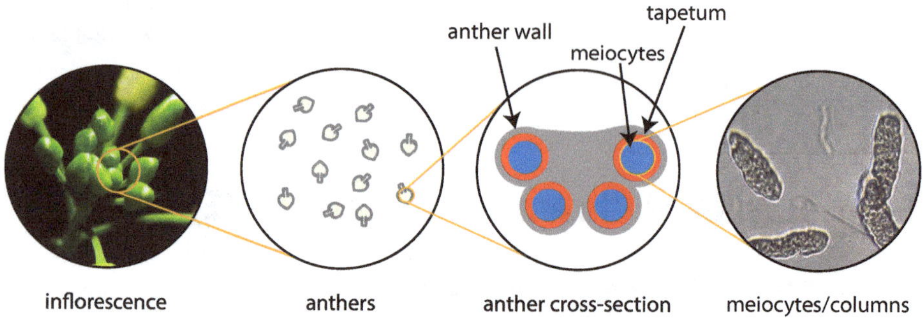

Fig. 1 Locating meiocytes within inflorescences. Meiotic nuclei develop within four elongated syncytia ("columns") per anther that can be separated from the surrounding tissue

and quantify protein factors in specific chromosomal regions and thereby attribute local recombination frequencies to the presence or absence of certain DSB or crossover (CO) promoting factors.

Chromatin spreading followed by immunohistochemical staining is currently widely used for a more detailed analysis of genome-wide events during early–mid meiotic prophase. It enables quantitative investigation of the localization of proteins to chromatin when antibodies directed against meiotic proteins are available. A technique using the detergent Lipsol® for spreading of plant pollen mother cells (PMCs) was developed years ago [6, 7] and has been optimized over time [8, 9]. The original basic procedure involves preparation of anthers containing meiotic cells, digestion with the enzyme cytohelicase (digestive enzyme from *Helix pomatia* containing several enzymatic activities required to digest the callose surrounding meiotic cells), chromatin spreading, and formaldehyde fixation on glass slides. It yields differing amounts of meiotic nuclei and the non-meiotic tissue of anthers present on the slides often causes considerable background staining. To overcome this, we took advantage of the capillary collection method (CCM) developed to specifically isolate meiotic tissue from floral buds [10, 11]. Meiotic nuclei develop within four elongated syncytia per anther (six anthers per flower bud) that can be separated from the surrounding tissue (Fig. 1). Mechanical extraction and subsequent collection of those "columns" with a glass capillary allows for enrichment of meiotic nuclei in very small volumes and removes all non-meiotic cells, thereby improving digestion and spreading efficiency, reducing background staining and increasing the number of spread nuclei per slide.

This technique allows detecting a wide range of meiotic proteins and analyzing their numbers within a nucleus. It is, however, not possible to distinguish between specific genomic loci. In order to address protein abundance on selected chromosomal regions, we combine the immunocytological procedure with fluorescent in situ hybridization (FISH).

Originally developed more than 35 years ago, FISH is still widely applied to mark specific nucleic acid sequences in chromosome preparations [12]. In short, DNA or RNA (or locked nucleic acids, LNA [13], and peptide nucleic acids, PNA [14]) sequences are amplified, labeled, and then hybridized to DNA. Microscopical detection of the labeled probes now bound to their complementary genomic counterparts allows identifying individual chromosomes or even specific regions on chromosomes (reviewed in [15]). Chromosome painting involves the use of composite pools of FISH probes, enabling staining of entire chromosome arms or even complete chromosomes [16, 17].

The presented protocol combines optimized sample collection for Lipsol® spreading with a streamlined FISH technique to study the localization and abundance of meiotic proteins in relation to specific DNA sequences/chromosomal positions. This chapter describes the newly established technique with special emphasis on sample collection, slide preparation and probe labeling for FISH.

2 Materials

2.1 Preparation and Labeling of FISH Probes

1. Identify bacterial artificial chromosome(s) (BAC) harboring the sequence(s) of interest (BAC clones can be found using TAIR's DNA search feature at https://www.arabidopsis.org/servlets/Search?action=new_search&type=dna) and order respective bacterial culture(s).
2. Solution 1: 25 mM Tris–HCl pH 8 and 10 mM EDTA pH 8. Use ice-cold.
3. Solution 2: 0.2 M NaOH and 1% SDS. Prepare fresh before use.
4. Solution 3: 3 M KOAc in 11.5% acetic acid. Use ice-cold.
5. Probe labeling: Use the Nick Translation mix from Roche following the manufacturer's instructions. Label each BAC individually and concentrate them in a single Eppendorf tube. We recommend to use either fluorescently labeled nucleotides such as ChromaTide Alexa Fluor 488-5-dUTP (Thermo Fisher), ChromaTide Texas Red-5-dUTP (Thermo Fisher) or digoxigenin-labeled nucleotides digoxigenin-11-dUTP (Roche). If using digoxigenin-labeled probes, antibodies for detection are additionally required (e.g., anti-digoxigenin-FITC; Roche). We experienced best results with the mentioned products but others might work as well.

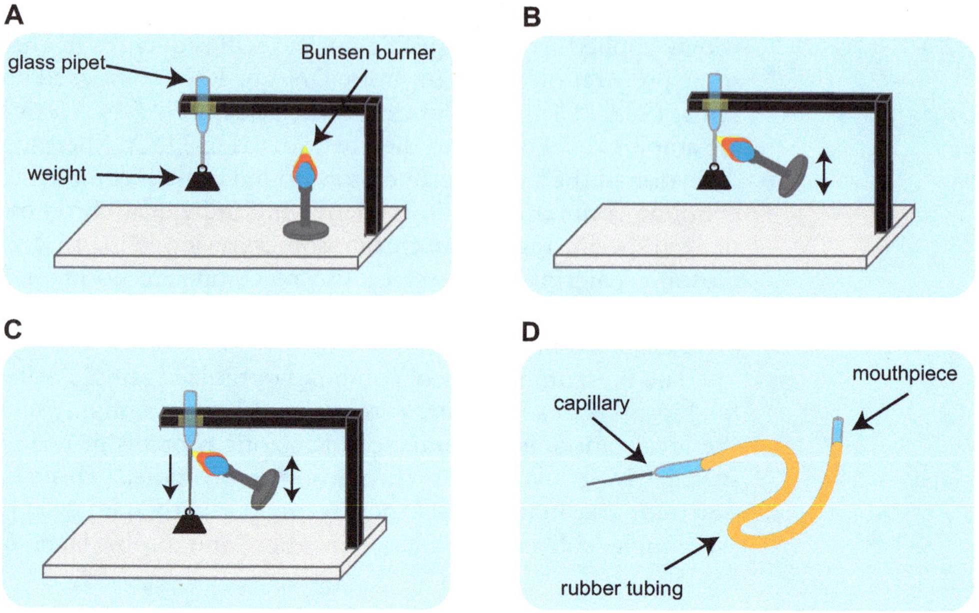

Fig. 2 Preparation and assembly of glass capillary and mouthpiece. In order to obtain a glass capillary of appropriate size and length, mount a glass Pasteur pipette to a stand. Make sure that the holders are of nonflammable material and attach a small weight (e.g., a liquid-filled 15 ml conical tube mounted to a bulldog-clip) to the bottom of the pipette (**a**). Move the flame of a Bunsen burner up and down along the thin part of the glass pipette to heat it up slowly (**b**). The pipette will melt and be pulled downward, thereby forming a thin capillary. Immediately remove the flame from the glass once the weight starts to move downward (**c**). Let the glass cool down, clip it at the thinnest part and connect it to the mouthpiece with a soft rubber tube (**d**)

2.2 Sample Collection and Preparation of Slides

1. Glass capillary connected to tubing with mouthpiece: mount a glass Pasteur pipette (preplugged, 150 mm, open jet) to a stand facing downward and use a bulldog clip or similar to attach a small weight (e.g., a liquid-filled 15 ml conical tube) to the pipette tip. Carefully apply heat to the middle section of the thin pipette part using a Bunsen burner until the weight pulls the glass to form a thin capillary. Once the glass has cooled down, clip it at the thinnest part with forceps and connect the capillary with the rubber tubing to a cut and rounded-off Pasteur pipette serving as mouthpiece (Fig. 2). The capillary can be rinsed and reused several times.
2. Collect inflorescences with at least two open flowers from the primary shoot of the plant and place them in a Petri dish with wet 2 mm filter paper. They will remain hydrated for a maximum of 2 h.
3. Digestion mix: 1% cytohelicase from *Helix pomatia*, 1.5% sucrose, 1% polyvinylpyrrolidone (MW 40,000–50,000) in deionized water. Aliquot 15 μl of mix into PCR tubes and keep at −20 °C with light protection.

4. Spreading solution: 2% Lipsol® detergent (Bibby Sterilin) in deionized water buffered to pH 9 with 500 mM borax (sodium tetraborate) solution.
5. Artificial pond water (APW; [18]): 0.5 mM NaCl, 0.2 mM $NaHCO_3$, 0.05 mM KCl, 0.4 mM $CaCl_2$ in deionized water. Autoclave solution and store at room temperature (RT).
6. 10× PBS: 1.37 M NaCl, 27 mM KCl, 100 mM Na_2HPO_4, 18 mM KH_2PO_4 in deionized water; pH 7.4.
7. Fixing solution: 4% formaldehyde in 1× PBS.
8. Charged microscope slides.
9. Coverslips 24 × 32 mm.
10. Two 0.5 × 16 mm needles and 1 ml syringes as needle holders.
11. Ultrafine forceps.
12. Coplin jars.
13. Dissecting microscope.

2.3 Immunocytology

1. Washing buffer: 2× SSC, diluted from a stock of 20× SSC: 3 M NaCl, 0.3 M Na_3-citrate in deionized water; pH 7–8.
2. Blocking solution: 1% BSA, 0.01% NaN_3 (w/v) in 1× PBS.
3. Primary antibodies: prepare working dilution in blocking solution.
4. Secondary antibodies: prepare working dilution in blocking solution.

2.4 FISH/Chromosome Painting

1. Fixing solution: 4% formaldehyde in 1× PBS.
2. Alcohol series: prepare three Coplin jars containing 70%, 90%, and absolute ethanol for slide dehydration.
3. Hybridization mix: 10% dextran sulfate (use high MW 50,000), 50% formamide, 2× SSC. Aliquot into 1 ml microcentrifuge tubes and store at −20 °C.
4. Formamide SSC: 50% formamide in 2× SSC.
5. Counterstaining/mounting solution: 1–10 μg/ml 4,6-diamidino-2-phenlyinidole (DAPI) in VECTASHIELD® (Vector Laboratories) or a similar antifade mounting medium.
6. Rubber cement.
7. Hot plate.

3 Methods

3.1 Preparation of FISH Probes

1. Grow *E. coli* cultures carrying the appropriate BACs in 3 ml of standard medium supplemented with the appropriate antibiotics over night at 37 °C.

2. Pellet the bacteria and resuspend them in 100 μl of ice-cold solution 1.
3. Add solution 2, gently invert the tube until the solution appears clear and incubate on ice for 5 min.
4. Add 150 μl of ice-cold solution 3, vortex for 10 s and incubate on ice for 5 min.
5. Add 450 μl of phenol–chloroform–isoamyl alcohol (25:24:1), vortex and centrifuge for 10 min at maximum speed at 4 °C.
6. Carefully retrieve the clear supernatant and transfer to a fresh tube containing two times the supernatant's volume of absolute ethanol and 1/10 of the volume of 3 M NaOAc (pH 5.2). Gently mix the solution by inversion and incubate at −80 °C for at least 1 h.
7. Centrifuge at maximum speed for 30 min at 4 °C.
8. Discard the supernatant and wash the pellet with 70% ethanol. Centrifuge at maximum speed for 5 min at RT. Remove the supernatant and let the pellet dry.
9. Dissolve the pellet in 25 μl of deionized water and incubate with 100 μg/ml of RNase I for 10 min at 37 °C.
10. Use the Nick Translation kit from Roche following the product specifications for labeling of the probes. It is ideal to label each BAC individually with directly labeled dNTPs (*see* **Note 1**).
11. In case of combining several BACs to visualize longer chromosomal regions, pool the equally labeled BACs and precipitate by adding two times the volume of absolute ethanol and 1/10 of the volume of 3 M NaOAc (pH 5.2). Gently mix the solution by inversion and incubate at −80 °C for at least 1 h.
12. Centrifuge at maximum speed for 30 min at 4 °C.
13. Discard the supernatant and wash the pellet with 70% ethanol. Centrifuge at maximum speed for 5 min at RT. Remove the supernatant and let the pellet dry.
14. Dissolve the pellet in 30 μl of deionized water and store at −20 °C in the dark.

3.2 Sample Collection and Preparation of Slides

1. Collect five to ten young inflorescences (from main shoots) on moist filter paper in a Petri dish (*see* **Note 2**).
2. Pipet 10 μl of APW onto the center of a charged glass slide.
3. Under a dissecting microscope, dissect buds with needles and transfer all anthers with transparent lobes to the droplet. It may be useful to tap open one anther per bud and check for the release of meiotic syncytia ("columns") before transferring all of them (Fig. 3). If free microspores (meiotic products) are released, discard the anther and all the other ones from the same bud and proceed with smaller buds.

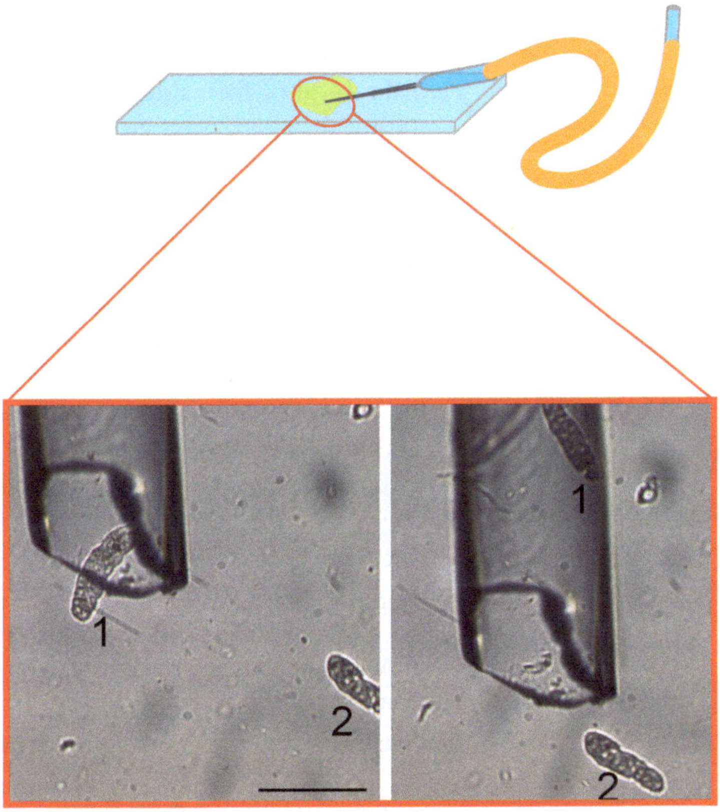

Fig. 3 The capillary collection method. Transfer slide to a light microscope, lower the tip of the capillary into the liquid droplet and apply gentle suction via the mouthpiece to collect columns inside the capillary. Size bar: 50 μM

4. After collection of anthers of all selected inflorescences into the APW droplet, squash them between the tips of the dissection forceps to release all meiocyte columns. When no intact anthers are left, rinse the forceps with some microliters of APW to collect all columns that might stick to the inner surfaces of the forceps.
5. Transfer the slide to a light microscope equipped with a 5× objective. Dip the tip of the capillary into the liquid and apply gentle suction to collect columns within the capillary that is connected to the mouthpiece. Release columns into a clean Eppendorf tube. Keep the tube on ice until all meiocytes are collected (*see* **Notes 3–5**).
6. Add 15 μl of digestion solution to the meiocyte solution and mix carefully. Incubate at RT for 5 min in the dark (*see* **Note 6**).
7. Stop digestion by putting the tube on ice. Transfer 5–8 μl of digested meiocytes to a clean and charged microscope slide. Check for the presence of digested columns (material should appear decayed).
8. Add 20 μl of spreading solution and mix carefully by tilting the slide. Incubate for 4 min at RT. Check for bursting of the

columns and presence of spread chromatin in phase contrast. Spread nuclei appear as grey discs.

9. Add 24 μl of formaldehyde fixing solution and mix gently by tilting the slide. After air-drying, the slides are ready for staining and should be processed as soon as possible (*see* **Note** 7).

3.3 Immunocytology

All washings are performed in Coplin jars at RT except otherwise noted.

1. Before rehydration of the slides, the area with the spread cells may be marked with a diamond needle to facilitate orientation during microscopy.
2. Wash the slides in 2× SSC for 5 min, add the desired primary antibody mix (20 μl) to the center of the defined region and cover the slides with a piece of laboratory film or autoclave bag in the size of a coverslip (*see* **Note 8**).
3. Incubate the slides in a humidity chamber at 4 °C overnight.
4. Remove the film/autoclave bag cover, wash the slides in 2× SSC for 5 min and add the appropriate secondary antibody mix (20 μl). Cover the slide again and incubate in a humidity chamber at 37 °C for 1 h.
5. Wash the slides in 2× SSC for 5 min.
6. At this stage, slides can be mounted with DAPI-VECTASHIELD® and analyzed by conventional or super-resolution fluorescence microscopy without FISH labeling. If FISH is omitted, all previous washings may be performed in 1× PBS instead of 2× SSC.

3.4 FISH/Chromosome Painting

All washings are performed in Coplin jars at RT except otherwise noted.

1. Add 15 μl of 4% formaldehyde fixing solution to the slide, cover it with a strip of laboratory film or autoclave bag and incubate for 10 min in the dark.
2. Dip the slides quickly into deionized water and dehydrate them by passing through an alcohol series 70%, 90% and absolute ethanol, 2 min each.
3. Let the slides air-dry in the dark.
4. Prepare the probe mix by adding 14 μl of hybridization mix to 2–3 μl of BAC probes (2–3 μg) and bring to 20 μl with deionized water.
5. Denature the probe mix at 95 °C for 10 min and then cool on ice for 5 min.
6. Add 20 μl of the denatured probe mix to the slide and cover with a glass coverslip. Seal the borders of the coverslip with

rubber cement and place them on a hot plate at 70–80 °C (*see* **Note 9**) for 4 min in the dark.

7. Incubate the slides in a humidity chamber at 37 °C overnight.
8. Carefully remove the rubber cement and the coverslip and wash the slides in formamide SSC at 42 °C for 5 min in the dark.
9. Wash the slides twice in 2× SSC for 5 min.
10. If the probe was labeled with digoxigenin, proceed to detection with the appropriate antibodies.
11. Add 15 μl of DAPI-VECTASHIELD® to the marked area on the slide and cover with a coverslip.
12. Place the slide in a drying block and gently press to absorb all excess liquid.
13. Seal the borders of the coverslip with nail polish.
14. Slides can be stored at 4 °C for a few weeks. For best results, proceed with the analysis as soon as possible.
15. Examine slides under a conventional or superresolution fluorescence microscope using appropriate light sources and filters (Fig. 4).

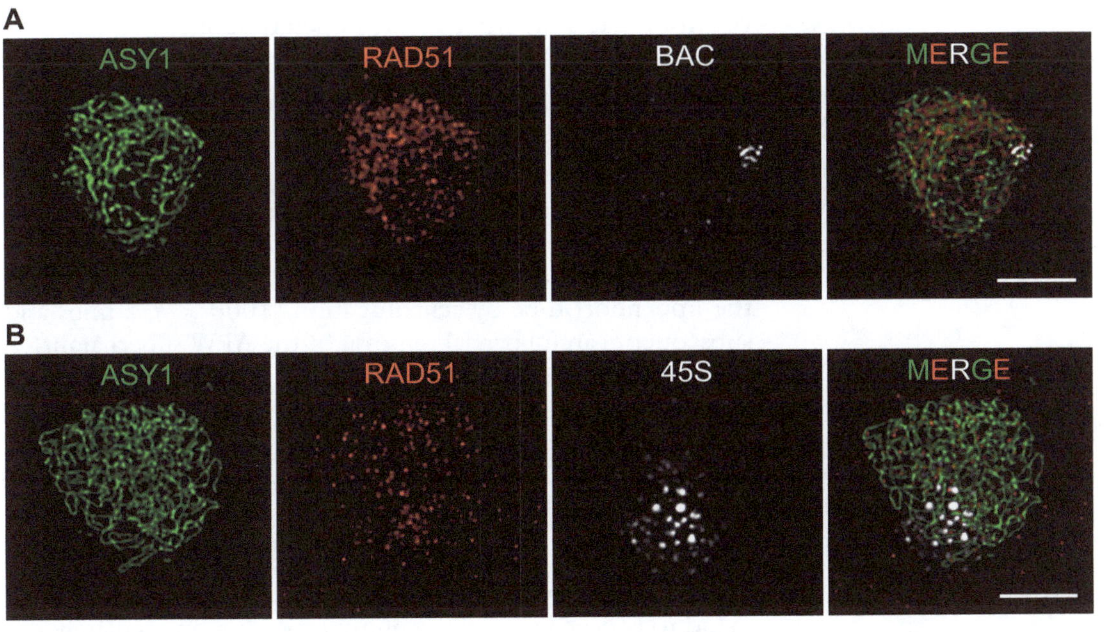

Fig. 4 TACE preparations. TACE preparations of PMCs in zygotene stage stained with antibodies directed against the axial element protein ASY1 (green; [8]) and the recombinase RAD51 (red; [9]) and hybridized with a FISH probe corresponding to a 500 kb region of chromosome five (**a**; white) or marking the repetitive 45S rDNA (**b**; white; [20]). Slides were imaged with a conventional fluorescence microscope. *Z*-stacks with 100 nm intervals were acquired and deconvolved and are presented as maximum intensity projections. Size bars: 5 μm

4 Notes

1. In order to reduce the workload and save resources, one may pool two to three purified BAC DNA samples and proceed with labeling of the mixture when attempting to paint larger chromosomal regions.
2. Only young inflorescences still harboring many small buds should be selected. The number of inflorescences for a single experiment depends on the skills of the researcher. Five are usually sufficient for two to three slides and processing of more than ten at a time might be too time-consuming. Only small buds (usually 0.2–0.4 mm long or at flower stage 9 [19]) harbor anthers with transparent lobes containing PMCs in prophase I. Note that bud size is subject to substantial variations and depends on growth conditions and genotype. It takes some experience to select the right anthers and there is only one safe distinguishing mark—the anther lobes. Only anthers having transparent and clear lobes should be selected.
3. The stages of most microscopes are operated by the right hand causing problems for right-handed persons who may find it hard to control the capillary with their left hand. We therefore recommend to convert the microscope to left-handed stage control to facilitate right-handed operation of the capillary.
4. Avoid aspiration of air bubbles with the capillary, they might clog it. Be careful to keep all meiotic cells in one droplet at the bottom of the collection tube and try to avoid air bubbles in the droplet.
5. Try to collect all columns in the least possible amount of liquid (not more than 5 μl in total). It is possible to reduce the amount of liquid by collecting the columns at the bottom of the Eppendorf tube by centrifugation (1000 × *g*, 2 min) and subsequent careful partial removal of the APW supernatant.
6. Successful spreading will only work on well-digested material and experimental problems are mostly caused by insufficient digestion. It is instrumental to use high-quality enzyme to receive optimal results. We achieve best results with cytohelicase from *Helix pomatia* supplied by Sigma.
7. Fixed and dried slides can be kept at RT for a minimum of 2 weeks, but quality might decrease over time and we recommend to proceed with the staining procedure immediately.
8. Specific and reproducible staining results largely depend on the antibodies and FISH probes used. Make sure that they are well-tested and carefully check for undesirable background staining that might distort results. Incomplete digestion may lead to the

presence of varying amounts of nucleoplasm/cytoplasm around the chromatin and cause staining artifacts.

9. Hybridization should be performed as short as possible and at the lowest possible temperature in order to avoid damage to chromatin. In case of probing for repetitive elements, 70–75 °C are usually sufficient, hybridization to more complex regions may require higher temperatures. Different probes may require different combinations of time and temperature to achieve ideal results.

Acknowledgments

We thank the European Union (FP7-ITN 560 606956), the University of Vienna (I031-B), and the Austrian Science Fund (SFB F3408; P18036) for funding (J.S., P.S., and M.T.K.). C.C. is supported by the National Science Foundation (IOS:1025881 and IOS:1546792) and the Grant-in-Aid program at the University of Minnesota. We furthermore thank Chris Franklin for providing the ASY1 antibody and Sue Armstrong, James Higgins, and Katja Schneider for very helpful technical demonstrations and discussions.

References

1. Baltes NJ, Gil-Humanes J, Voytas DF (2017) Genome engineering and agriculture: opportunities and challenges. Prog Mol Biol Transl Sci 149:1–26. https://doi.org/10.1016/bs.pmbts.2017.03.011
2. Lambing C, Heckmann S (2018) Tackling plant meiosis: from model research to crop improvement. Front Plant Sci 9(829). https://doi.org/10.3389/fpls.2018.00829
3. Berchowitz LE, Copenhaver GP (2008) Fluorescent Arabidopsis tetrads: a visual assay for quickly developing large crossover and crossover interference data sets. Nat Protoc 3 (1):41–50. https://doi.org/10.1038/nprot.2007.491
4. Melamed-Bessudo C, Yehuda E, Stuitje AR, Levy AA (2005) A new seed-based assay for meiotic recombination in *Arabidopsis thaliana*. Plant J 43(3):458–466. https://doi.org/10.1111/j.1365-313X.2005.02466.x
5. Rowan BA, Patel V, Weigel D, Schneeberger K (2015) Rapid and inexpensive whole-genome genotyping-by-sequencing for crossover localization and fine-scale genetic mapping. G3 (Bethesda) 5(3):385–398. https://doi.org/10.1534/g3.114.016501
6. Albini SM, Jones GH, Wallace BM (1984) A method for preparing two-dimensional surface-spreads of synaptonemal complexes from plant meiocytes for light and electron microscopy. Exp Cell Res 152(1):280–285. doi:0014-4827(84)90255-6 [pii]
7. Albini SM, Jones GH (1984) Synaptonemal complex-associated centromeres and recombination nodules in plant meiocytes prepared by an improved surface-spreading technique. Exp Cell Res 155(2):588–592
8. Armstrong SJ, Caryl AP, Jones GH, Franklin FC (2002) Asy1, a protein required for meiotic chromosome synapsis, localizes to axis-associated chromatin in Arabidopsis and Brassica. J Cell Sci 115(Pt 18):3645–3655
9. Kurzbauer MT, Uanschou C, Chen D, Schlogelhofer P (2012) The recombinases DMC1 and RAD51 are functionally and spatially separated during meiosis in Arabidopsis. Plant Cell 24(5):2058–2070. https://doi.org/10.1105/tpc.112.098459
10. Chen C, Farmer AD, Langley RJ, Mudge J, Crow JA, May GD, Huntley J, Smith AG, Retzel EF (2010) Meiosis-specific gene discovery in plants: RNA-Seq applied to isolated

Arabidopsis male meiocytes. BMC Plant Biol 10:280. https://doi.org/10.1186/1471-2229-10-280

11. Chen C, Retzel EF (2013) Analyzing the meiotic Transcriptome using isolated Meiocytes of *Arabidopsis thaliana*. Methods Mol Biol 990:203–213. https://doi.org/10.1007/978-1-62703-333-6_20
12. Bauman JG, Wiegant J, Borst P, van Duijn P (1980) A new method for fluorescence microscopical localization of specific DNA sequences by in situ hybridization of fluorochrome labelled RNA. Exp Cell Res 128(2):485–490
13. Silahtaroglu A, Pfundheller H, Koshkin A, Tommerup N, Kauppinen S (2004) LNA-modified oligonucleotides are highly efficient as FISH probes. Cytogenet Genome Res 107(1–2):32–37. https://doi.org/10.1159/000079569
14. Paulasova P, Pellestor F (2004) The peptide nucleic acids (PNAs): a new generation of probes for genetic and cytogenetic analyses. Ann Genet 47(4):349–358. https://doi.org/10.1016/j.anngen.2004.07.001
15. Levsky JM, Singer RH (2003) Fluorescence in situ hybridization: past, present and future. J Cell Sci 116(Pt 14):2833–2838. https://doi.org/10.1242/jcs.00633
16. Lysak MA, Fransz PF, Ali HB, Schubert I (2001) Chromosome painting in *Arabidopsis thaliana*. Plant J 28(6):689–697
17. Ried T, Schrock E, Ning Y, Wienberg J (1998) Chromosome painting: a useful art. Hum Mol Genet 7(10):1619–1626
18. Miller AL, Gow NA (1989) Correlation between root-generated ionic currents, pH, Fusicoccin, Indoleacetic acid, and growth of the primary root of *Zea mays*. Plant Physiol 89(4):1198–1206
19. Smyth DR, Bowman JL, Meyerowitz EM (1990) Early flower development in Arabidopsis. Plant Cell 2(8):755–767. https://doi.org/10.1105/tpc.2.8.755
20. Wanzenbock EM, Schofer C, Schweizer D, Bachmair A (1997) Ribosomal transcription units integrated via T-DNA transformation associate with the nucleolus and do not require upstream repeat sequences for activity in *Arabidopsis thaliana*. Plant J 11(5):1007–1016

Chapter 6

Whole-Mount Immuno-FISH on *Arabidopsis* Meiocytes (WhoMI-FISH)

Jason Sims, Jihed Chouaref, and Peter Schlögelhofer

Abstract

Imaging cells, nuclei, and DNA in their natural spatial contexts and configurations is challenging yet required to understand the biology of genome organization, maintenance, and transmission. Live-cell imaging allows capturing dynamic changes of chromosomes in their nuclear and cellular context but lacks resolution. In contrast, imaging of fixed, spread chromosome samples provides unmatched resolution but potentially distorts configurations and spatial relations. Fixed whole-mount samples preserve chromosome configurations and cellular contexts and allow high-resolution imaging. Importantly the latter method allows simultaneous visualization of specific genomic regions (via fluorescent in situ hybridization—FISH) and proteins (via immune-localization using antibodies or tags). Here we present an advanced "whole-mount immuno-FISH" (WhoMI-FISH) method based on the published protocol by Bey Till et al. (Methods Mol Biol 1675:467–480, 2018) specifically optimized for pollen mother cells (PMCs) of *Arabidopsis thaliana*. It focuses on (1) specimen preparation that maintains meiocyte nuclei positions and genome organization in anthers and also on (2) simultaneous detection of specific genomic regions and meiotic proteins.

Key words *Arabidopsis thaliana*, Immunocytology, Meiosis, Whole-mount fluorescence in situ hybridization

1 Introduction

The spatiotemporal organization of the genetic material within the nucleus changes dynamically during the cell cycle, depending on several processes including gene expression, DNA repair and chromosome condensation and segregation during mitosis and meiosis.

The term "cytology" encompasses major techniques to study chromosome organization and dynamics. However, recent developments of the 3C-based techniques (Chromosome Conformation Capture) in particular Hi-C, its high-throughput sequencing

---

Electronic supplementary material: The online version of this chapter (https://doi.org/10.1007/978-1-4939-9818-0_6) contains supplementary material, which is available to authorized users.

Mónica Pradillo and Stefan Heckmann (eds.), *Plant Meiosis: Methods and Protocols*, Methods in Molecular Biology, vol. 2061, https://doi.org/10.1007/978-1-4939-9818-0_6,

derivative, took 3D nuclear organization studies to the next level, depicting genome-wide DNA interactions with kilobase resolution [2, 3]. Applying Hi-C techniques to meiotic cells, especially considering higher eukaryotes, is challenging, but single-cell Hi-C techniques have successfully been developed [4–7]. A further development is the combination of Hi-C with ChIP (HiChIP) [8], this protein-centric chromatin conformation assay monitors the association of chromatin with certain protein complexes. These exciting developments also demand verification and complementary approaches.

Fluorescent in situ hybridization (FISH) allows direct visualization of the position of a gene sequence, a gene region or an entire chromosome in a single nucleus. It can be combined with immunolabeling to gain insights into the localization of proteins (on the labeled chromatin or elsewhere) and the architecture of the nucleus and its subnuclear structures. FISH on meiotic chromosomes is generally applied on spread chromatin but the spreading procedure results in loss of the spatial organization of the nucleus, and hence crucial information about the relative position of genomic loci or protein complexes. The visualization of genomic targets by whole-mount FISH allows analyzing them in their original spatial relation in 3D within intact nuclei. Several protocols for whole-mount FISH are available for different tissues: maize meiotic tissue [9], *Arabidopsis* roots [10, 11], ovule primordia [12], and leaf fragment [13]. Recently, Bey et al. [1] developed a protocol for whole-mount FISH for *Arabidopsis* seedlings combined with immunolabeling of proteins. Due to differences in cell wall composition and thickness, the whole-mount FISH protocol has to be optimized for each cell type. Here we describe an efficient protocol for whole-mount immuno-FISH to be applied to *Arabidopsis* meiocytes. It combines the detection of DNA targets in intact anthers with immunohistochemical detection of meiotic proteins (Fig. 1, Videos 1 and 2). It represents an important and complementary approach to study specific spatial DNA-DNA and DNA-protein relations during plant meiosis.

2 Materials

2.1 Preparation of BAC Probes

1. Solution 1: 25 mM Tris–HCl pH 8 and 10 mM EDTA pH 8 (ice-cold).
2. Solution 2: 0.2 M NaOH and 1% SDS. The solution has to be prepared fresh for each extraction.
3. Solution 3: 3 M KOAc and 11.5% glacial acetic acid (ice-cold).

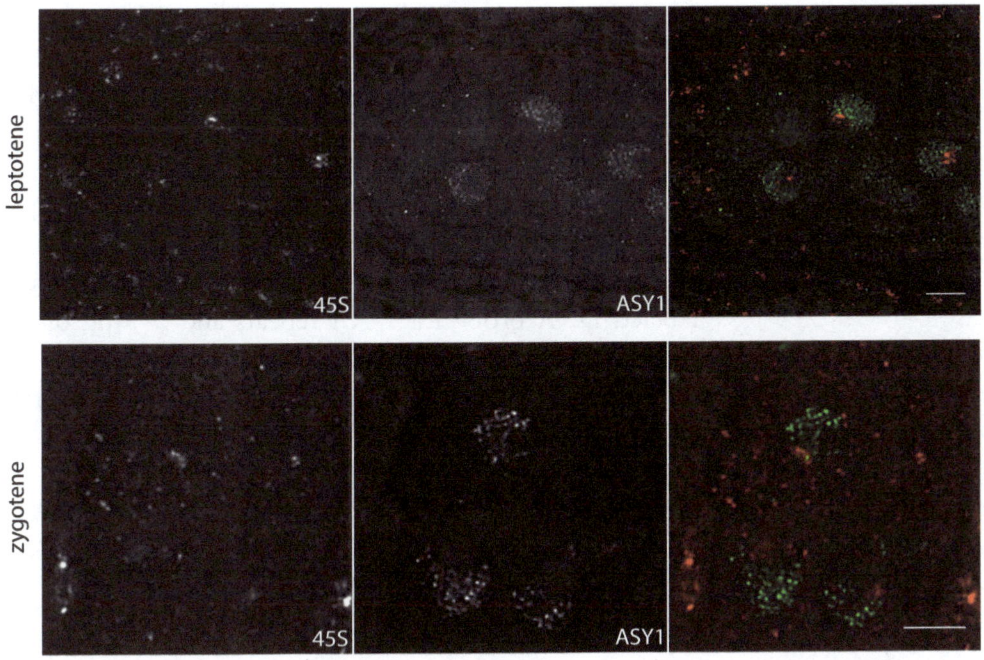

Fig. 1 Single slice of a WhoMI-FISH preparation. Depiction of a single slice of a whole-mount immuno-FISH preparation of pollen mother cells at leptotene (**a**) and zygotene (**b**) stages. Samples were hybridized with a 45S rDNA probe (red) and stained with antibodies directed against the axial element protein ASY1 (green). The images were acquired with an LSM 710 confocal microscope supplied with an Airy Scan unit for 160 nm resolution in *x*, *y* and *z*. The image corresponds to a single slice acquired every 160 nm in *z*. Size bar = 5 μm

2.2 Preparation of Whole-Mount Samples

1. Methanol.
2. 96% ethanol
3. Xylene.
4. Citrate buffer: 4.45 mM $C_6H_8O_7$ (citric acid) 5.55 mM $Na_3C_6H_5O_7$ (trisodium citrate), pH 6.
5. 1× PBS: 10 mM Na_2HPO_4, 2 mM KH_2PO_4, 2.7 mM KCl, and 137 mM NaCl, pH 7.4
6. Fixative solution: 1% formaldehyde, 10% dimethylsulfoxide, 1× PBS, 60 mM EGTA.
7. Enzyme solution: 0.6% cytohelicase, 0.6% pectolyase, and 0.6% cellulase in citrate buffer pH 4.5.

2.3 Whole-Mount FISH

1. Cut edges frosted end microscopy slides.
2. Coverslips 24 × 32 mm, #0.17.
3. Colorless nail polish.
4. 1× PBS
5. DNase-free RNase A.
6. Labeled FISH probes.

7. 1× PBT:: 10 mM Na_2HPO_4, 2 mM KH_2PO_4, 2.7 mM KCl, and 137 mM NaCl, pH 7.4, 0.01% Tween 20.
8. 2× SSC: 30 mM $Na_3C_6H_5O_7$ (sodium citrate), 300 mM NaCl, pH 7.
9. Formamide.
10. HB50 buffer: 50% formamide, 2× SSC, 50 mM NaH_2PO_4 (sodium phosphate).
11. Hybridization solution: Labeled DNA probe (0.5–1.5 μg of labeled DNA probe for DNA repeats and 2–5 μg of labeled DNA for unique sequences), 50% formamide in 2× SSC.
12. 5 μg/ml DAPI (4′,6-diamidino-2-phenylindole) in water.
13. Heating block.
14. Binocular (magnification ×6.4–16).

2.4 Immunolabeling

1. Blocking solution: 1% BSA, 1% Triton X-100 in 1× PBS.
2. Primary antibody diluted in blocking solution.
3. Secondary antibody diluted in blocking solution.

3 Methods

3.1 Preparation of BAC Probes

1. Grow *E. coli* carrying the desired BAC clone in 3 ml of standard medium supplemented with the appropriate antibiotics over night at 37 °C.
2. Pellet the bacteria of the entire culture and resuspend them in 100 μl of ice-cold solution 1.
3. Add 100 μl of solution 2, gently invert the tube until the solution appears clear and incubate on ice for 5 min.
4. Add 150 μl of ice cold solution 3, vortex for 10 s and incubate on ice for 5 min.
5. Add 450 μl of phenol–chloroform–isoamyl alcohol (25:24:1), vortex and then centrifuge for 10 min at maximum speed at 4 °C.
6. Transfer the supernatant carefully to a new 1.5 ml tube.
7. Add 2 volumes of EtOH 96% and one-tenth of the volume of 3 M NaOAc pH 5.2. Gently mix the solution by inversion and incubate at −80 °C for at least 1 h.
8. Centrifuge at maximum speed for 30 min at 4 °C.
9. Discard the supernatant and wash the pellet with 70% EtOH. Centrifuge at maximum speed for 5 min at room temperature (RT). Remove the supernatant and let the pellet dry.
10. Dissolve the pellet in 25 μl of sterile water and incubate with 100 μg/ml of RNase I for 10 min at 37 °C.

3.1.1 Labeling of BAC Probes

Use the Nick Translation kit from Roche following the product specifications for labeling of the probe (*see* **Note 1**).

1. Ideally, each BAC clone is individually processed. We recommend using directly labeled dNTPs (*see* **Note 2**). For painting larger genomic regions, pool the desired BAC clones (from **step 10** above) in one tube and concentrate them by adding 2 volumes of EtOH 96% and 1/10 of the volume of 3 M NaOAc pH 5.2. Gently mix the solution by inversion and incubate at −80 °C for at least 1 h.
2. Centrifuge at maximum speed for 30 min at 4 °C.
3. Discard the supernatant and wash the pellet with 70% EtOH. Centrifuge at maximum speed for 5 min RT. Remove the supernatant and let the pellet dry.
4. Dissolve the pellet in 30 μl of sterile water and store at −20 °C in the dark.

All washes and incubations are performed at room temperature (RT) unless otherwise specified.

3.2 Preparation of Whole-Mount Samples for *Arabidopsis* meiocytes

1. Collect inflorescences from the primary shoots of an *Arabidopsis* plant and remove all open flowers and pollen-containing buds with forceps and a needle under a dissection microscope.
2. Open up all remaining buds with the help of two needles to increase accessibility of anthers for the enzyme solution and probe.
3. Place the prepared inflorescence in 500 μl fixation solution and take care that all buds are submerged (max. 3 inflorescences per 1.5 ml plastic tube).
4. Place the solution and the buds under vacuum for 10 min at RT (until all cellular interstitial space is evacuated and filled with fixation solution) and incubate for further 30 min at RT.
5. Remove the fixation solution and incubate the tissue 2 × 10 min in 500 μl methanol.
6. Remove methanol and incubate for 2 × 5 min in 500 μl ethanol.
7. Remove ethanol and incubate in 500 μl xylene–ethanol (1:1) for 15 min.
8. Incubate in xylene at 50 °C for 30 min.
9. Wash 2× in 500 μl ethanol.
10. Wash 2× in 500 μl methanol.
11. Wash 3× in 500 μl PBT (fixed buds can be stored in PBT at 4 °C overnight).

3.3 Sample Digestion and Postfixation

1. Incubate the buds in the enzyme solution at 37 °C for 1 h.
2. Gently wash the sample twice in 500 μl 2× SSC.
3. Incubate the sample in 500 μl of 0.1 mg/ml of RNase A in 2× SSC at 37 °C for 1 h.
4. Wash twice in 500 μl 1× PBT.
5. Fix the samples in 1% formaldehyde (in 1× PBT) for 30 min.
6. Wash twice in 500 μl 1× PBT.

3.4 Whole-Mount FISH

1. Wash sample in 500 μl 2× SSC.
2. Incubate sample in 500 μl of a 1:1 mix of HB50 and 2× SSC for 30 min.
3. Incubate sample in 500 μl HB50 for 30 min.
4. Incubate samples in 30 μl of hybridization solution in the dark (*see* **Note 3**) for 1 h.
5. From now on, samples should be protected from light.
6. Place the sample on a heating block for 4 min at 85 °C and afterward on ice for 3 min.
7. Incubate the sample at 37 °C overnight in the dark to let the probe anneal.
8. Wash samples in 500 μl HB50.
9. Incubate in fresh 500 μl HB50 at 42 °C for 1 h.
10. Wash in 500 μl PBT for two times 10 min each.

3.5 Immuno-detection of Meiotic Proteins

All steps must be performed in the dark if samples have been previously hybridized with a fluorescent probe.

1. Place the samples in an Eppendorf tube with 30 μl of primary antibody solution and incubate at 37 °C for 3 h with 30 min under vacuum (or O/N at 16 °C) (*see* **Note 4**).
2. Wash 3× with 200 μl in PBT for 10 min.
3. Place samples in 30 μl of secondary antibody solution and incubate at 37 °C for 3 h (or O/N at 16 °C) with 30 min under vacuum.
4. Wash three times with 200 μl PBT for 10 min.
5. Incubate samples in 200 μl PBT supplemented with 5 μg/ml DAPI for 30 min.
6. Collect the samples with forceps and place them on a microscope slide.
7. Add 15 μl of DAPI-Vectashield and cover with a coverslip.
8. Seal with nail polish.
9. Image slides with a confocal microscope (Videos 1 and 2).

4 Notes

1. Use pure and highly concentrated DNA for nick translation (above 1 µg/µl).
2. We use directly labeled dUTP with Alexa Fluor dyes (e.g., Alexa Fluor 488, Alexa Fluor 594, Thermo Fisher Scientific) and cyanine dyes (e.g., Cy3, Cy5, GE Healthcare).
3. Optimal results are achieved when a directly labelled DNA probe is used. Repetitive targets can already be detected with a final probe concentration of 0.5 µg/µl.
4. As a general rule, the primary antibodies need to be ten times more concentrated compared to the concentration used for regular cytological spread preparations.

Acknowledgments

We thank the European Union (FP7-ITN 606956) and the Austrian Science Fund (SFB F34) for funding. We also thank Dr. Paul Fransz for his valuable suggestions and support during developing this protocol.

References

1. Bey TD, Koini M, Fransz P (2018) Fluorescence in situ hybridization (FISH) and immunolabeling on 3D preserved nuclei. Methods Mol Biol 1675:467–480. https://doi.org/10.1007/978-1-4939-7318-7
2. Li R, Liu Y, Hou Y et al (2018) 3D genome and its disorganization in diseases. Cell Biol Toxicol 34:351–365. https://doi.org/10.1007/s10565-018-9430-4
3. Dong P, Tu X, Chu PY et al (2017) 3D chromatin architecture of large plant genomes determined by local A/B compartments. Mol Plant 10:1497–1509. https://doi.org/10.1016/j.molp.2017.11.005
4. Flyamer IM, Gassler J, Imakaev M et al (2017) Single-nucleus Hi-C reveals unique chromatin reorganization at oocyte-to-zygote transition. Nature 544:110–114. https://doi.org/10.1038/nature21711
5. Nagano T, Lubling Y, Stevens TJ et al (2013) Single-cell Hi-C reveals cell-to-cell variability in chromosome structure. Nature 502:59–64. https://doi.org/10.1038/nature12593
6. Nagano T, Lubling Y, Várnai C et al (2017) Cell-cycle dynamics of chromosomal organization at single-cell resolution. Nature 547:61–67. https://doi.org/10.1038/nature23001
7. Ramani V, Deng X, Qiu R et al (2017) Massively multiplex single-cell Hi-C. Nat Methods 14:263–266. https://doi.org/10.1038/nmeth.4155
8. Mumbach MR, Rubin AJ, Flynn RA et al (2016) HiChIP: efficient and sensitive analysis of protein-directed genome architecture. Nat Methods 13:919–922. https://doi.org/10.1038/nmeth.3999
9. Bass HW, Marshall WF, Sedat JW et al (1997) Telomeres cluster de novo before the initation of synapsis: a three-dimensional spatial analysis of telomere positions before and during meiotic prophase. J Cell Biol 137:5–18. https://doi.org/10.1083/jcb.137.1.5
10. Costa S, Shaw P (2006) Chromatin organization and cell fate switch respond to positional information in Arabidopsis. Nature 439:493–496. https://doi.org/10.1038/nature04269
11. Berr A, Schubert I (2007) Interphase chromosome arrangement in Arabidopsis thaliana is similar in differentiated and meristematic tissues and shows a transient mirror symmetry

after nuclear division. Genetics 176:853–863. https://doi.org/10.1534/genetics.107.073270

12. She W, Grimanelli D, Rutowicz K et al (2013) Chromatin reprogramming during the somatic-to-reproductive cell fate transition in plants. Development 140:4008–4019. https://doi.org/10.1242/dev.095034
13. Tirichine L, Andrey P, Biot E et al (2009) 3D fluorescent in situ hybridization using arabidopsis leaf cryosections and isolated nuclei. Plant Methods 5:1–7. https://doi.org/10.1186/1746-4811-5-11

Part II

Cytological and Imaging Approaches to Study Meiosis in Other Species

Chapter 7

Using Genome In Situ Hybridization (GISH) to Distinguish the Constituent Genomes of *Brassica nigra* and *B. rapa* in the Hybrid *B. juncea*

Zeeshan Shamim and Susan J. Armstrong

Abstract

The genome in situ hybridization (GISH) technique has become important for deciphering the organization of the constituent genomes in the allopolyploid plants that comprise many of the crop species. This technique comprises using the nuclear DNA from the constituent genomes as probes that have been labeled separately with different nucleotides that can be identified by using secondary antibodies. The Brassica family includes a range of mustard species with diverse phytochemical and morphological profile, hence making it an important plant family in agriculture. Meiosis is a specialized cellular division which brings the homologous chromosomes together and creates recombinants via pairing and synapsis during its early phase. Transfer of the genetic material within homoelog pairs creates novelty in subsequent generations which hold promise for improving the agriculture sector. This chapter is concerned with developing a GISH technique that discriminates between the constituent genomes in the allopolyploid *B. juncea,* in order to study meiosis.

Key words DNA probe, In situ hybridization, Chromosomes, Allopolyploid, *Brassica*, Meiosis

1 Introduction

Genome in situ hybridization (GISH) is a special form of the DNA hybridization technique where the whole nuclear genome is used as a probe to identify the constituent genomes and to study the evolutionary basis in plants. The methodology has been extended from certain DNA fragments to a whole chromosome to discriminate between natural and artificial hybrids and is referred to as GISH [1]. The method has been widely used in polyploid organisms and in plants particularly where most of the edible cereals and crops have more than one progenitor. The nuclear DNA must be broken in short fragments by using enzymes, sonication or autoclaving. Sonication gives the precise size of fragments, but it is expensive therefore a cheaper option, autoclaving, is widely used for truncation of the fragments. These fragments are then labeled

Mónica Pradillo and Stefan Heckmann (eds.), *Plant Meiosis: Methods and Protocols*, Methods in Molecular Biology, vol. 2061, https://doi.org/10.1007/978-1-4939-9818-0_7, © Springer Science+Business Media, LLC, part of Springer Nature 2020

with nucleotide conjugates by nick translation that in turn are secondarily labeled with fluorescent antibodies so that they may be viewed under a fluorescent microscope. Infertility in genomes can be studied by analysing chromosomes at meiotic metaphase-I. In the Birmingham meiosis laboratory we have studied this technique specifically to investigate the nature of meiosis in an artificial hybrid between *Allium cepa* and *A. fistulosum*. Using GISH we demonstrated that the gametes contain unbalanced products [2], because the diploid nature of the hybrid contained only the constituent genome, which although they paired up in meiosis, the outcome of this process was shown to consist of unbalanced products because the *A. cepa* component had 30% more DNA than the corresponding *A. fistulosum* component, which in turn led to unbalanced products due to the nature of the crossovers between the two.

More recently we have paid attention to the Brassica genus and we have used this technique to show the cytogenetic relationship of the economically important species of *B. napus* [3, 4] where blocking DNA is used distinguish subgenomes otherwise in its absence the probe can have higher degree of homology on subgenomes and the hybridization can occur in all subgenome chromosomes. The technique is robust and reproducible so that the constituent genomes, which even share homology up to 95%, can be discriminated using the blocking agent. We have demonstrated that it is possible to distinguish the component genomes despite them separating at around 10,000 years ago [3, 4]. Archaeological evidence has revealed that *B. rapa* (turnip rape) and *B. nigra* (black mustard) were the first domesticated diploid species in this genus besides the amphidiploid *B. juncea* (Indian mustard) that originated from the cross between them. Since *B. nigra* and *B. rapa* belong to different lineage, blocking DNA is not used because of the phylogenetic distance between them. Cytogenetic and molecular studies reveal the diphyletic origin which succeeded after triplication of the ancestral genome that had been branched into two distinct lineages, first the *B. nigra* lineage and the second *B. oleracea/rapa* lineage [5–7]. Together these two genomes are found in the allopolyploid *B juncea* and with GISH we have detected the discrimination between them to analyze their behavior in meiosis. Part of the thesis of Zeeshan Shamim is concerned with investigating if it is possible to demonstrate homeology pairing in the artificial hybrids *B. rapa* and *B. nigra*. The overall aim of investigating if this can be achieved in the hybrid *B. juncea is* so that we can be in a position to introgress genes into the species.

This chapter is based on the preliminary work that Zeeshan has carried out to identify the component species in the hybrid *B. juncea*.

The method in this chapter is based on [3, 4]. It differs from the method described in these papers in that it does not use blocking DNA and both constituent genomes are labeled with nucleotide conjugates so that both of them are labeled with different secondary antibodies for visualization under the microscope.

2 Materials

2.1 Materials for Mitotic and Meiotic Slide Preparation

2.1.1 For Mitotic Fixation

1. The seeds of *B. nigra*, *B. rapa*, and *B. juncea* are sterilized by placing them into 4% of a commercial bleaching agent for 15 min, after which they are washed in sterile water four times for 5 min each time. They are spread out on an agar plate in a growth room with a minimum of 15 °C. The roots should be at a suitable stage after 4–5 days of planting.

2.1.2 For Meiotic Fixations

1. The seeds are sown in sterilized compost in 5 cm flower pots and transferred to a dedicated brassica growth room maintained at 18 °C with a 16 h light cycle. The plants are ready at the primary inflorescence stage by 7–8 weeks.

2.2 Extraction of DNA for Probes

1. Sow seeds of *B. nigra* and *B. rapa* following the protocol for sterilizing seeds for root tips. The plants are grown under the conditions already listed for meiotic preparations to provide young leaf material only and may be harvested after 3–4 weeks (*see* **Note 1**).

   Day 1: Fixation of material, extraction of DNA and making it suitable for nick translation.

   Day 2: Preparation of slides. Production of GISH probes for hybridization to slides. Make sure all of the solutions are available for day 3.

   Day 3: Hybridization of slides with nicked DNA of the constituent genomes. Make sure that all of the solutions are made up ready for posthybridization washes on day 4.

   Day 4: Posthybridization washes and analysis of slides.

2.3 Materials Required for Mitotic and Meiotic Preparations

1. Two pairs of fine watchmaker's forceps.
2. Lacto-propionic stain: take 5 gm synthetic orcein dissolve it into equal amounts 50 ml lactic and propionic acids overnight with stirring. Following on from this filter the solution with No. 1 Whatman filter paper. For a working solution dilute it 1:1 with distilled water. It may be kept on the bench.
3. Freshly prepared fixative (3 parts absolute ethanol and 1 part glacial acetic acid) and is made daily as required and stored on ice during the day (*see* **Note 2**).
4. Citrate buffer: 0.01 M, pH 4.5 (using 0.1 N HCl) is made up freshly before slide preparation can be carried out. Stock solutions of 0.1 M citric acid and 0.1 M sodium citrate are

autoclaved and stored in the refrigerator. Make up the buffer to the correct pH as follows: 555 μl of citric acid is combined with 445 μl of sodium citrate and made up to 10 ml with sterile deionized water. Any unused citrate buffer is discarded at the end of the day.

5. Stock digestion medium: dissolve 1% cellulose and 1% pectolyase (Sigma) in 0.01 M citrate buffer pH 4.5. Dispense in aliquots 100 μl and 300 μl and store at −20 °C.
6. Working solution for digestion of root tips: take 100 μl aliquot and dilute it with 900 μl of citrate buffer to make up to 0.1%.of the enzyme mix in 1 ml.
7. Working solution for digestion of meiotic material: take 300 μl aliquot and dilute it with 700 μl of citrate buffer to make up to 0.1% of the enzyme mix in 1 ml.
8. Hot block set at 45 °C.
9. DAPI: prepare a stock solution of 4-6-diamidino-2-phenylindole (DAPI) in sterile deionized water at 1 mg/ml, dispense in aliquots and store at −20 °C.
10. Working DAPI solution. For use, add 10 μl of this stock to 1 ml of an antifade mounting medium, Vectashield (Vector Laboratories, UK) and store at 4 °C.
11. Fluorescence microscope with appropriate filters and an image capture system.

2.4 Probes for GISH

Total genomic DNA of both diploid genomes (*B. nigra* and *B. rapa*) is used to identify constituent genomes in the allotetraploid *B. juncea*.

1. Genomic DNA from both progenitors (*B. nigra* and *B. rapa*) is used as probes in GISH.
2. RNases.
3. Nick translation kits for labeling DNA fragments of both sub genomes.

2.5 Slide Pretreatment and GISH Procedure

1. Pretreatment washing solution: 2× SSC buffer (0.3 M NaCl, 0.03 M sodium citrate, pH 7.0 prepare a stock solution 20× SSC Using 175 g sodium chloride And 88.3 g sodium citrate. Dissolve in 1 L of deionized water.) in 100 ml 0.01 M HCl. Prepare freshly and preheat for an hour at 37 °C before use. Autoclave and store at room temperature. Make up to 2× SSC as required.
2. Digestion medium: 0.01 g pepsin (ICN 102598).
3. Paraformaldehyde fixative: Weigh out 4 g paraformaldehyde (EM grade) in the fume hood. Dissolve in 100 ml of sterile deionized water and four drops of 1 M NaOH, prewarmed to

60 °C in the microwave. Stir the mixture on a magnetic stirrer for 1 h, or until it dissolves, before filtering through Whatman paper. Adjust the pH to 8.0. The fixative can be stored for up to 1 week at 4 °C.

4. Alcohol series with absolute ethanol and sterile water: 70%, 90%, and 100% ethanol for slide dehydration.
5. Hybridization mix: add 5 ml of formamide (molecular biology grade) and 1 ml of 20× SSC to 1 g dextran sulfate (high MW500,000) and make up to 7 ml with sterile deionized water. Dissolve at 65 °C, cool and adjust pH to 7.0. Aliquot this into 1 ml Eppendorf tubes and store at −20 °C.
6. Prepare 20 μl of probe mixture per slide. Add 14 μl of hybridization mix and around 90 ng of labeled *B. nigra* and *B. rapa* in 3 μl respectively. The final hybridization mix consists of 50% deionized formamide, 2 SSC and 10% dextran sulphate pH 7.0, and up to 100 ng of each of the two probes.
7. Vulcanizing rubber solution (found in bicycle wheel repair kits).
8. Hot lock set at 75 ° C.

The next day set up as follows.

2.6 Post-hybridization

1. Posthybridization washes: Three staining jars with 50% deionized formamide, 2× SSC, pH 7.0. We use 200 ml formamide (Analar grade) with 10 g mixed bed resin (Amberlite IRN-150L) stirred for 1 h in a fume hood and filtered. Mix 150 ml of this deionized formamide with 30 ml of 20× SSC and 120 ml sterile deionized water. Warm to 45 °C in a water bath and adjust to pH 7.0 with 1 N HCl being careful not to go below this pH. Allow to cool and pour into three staining jars. Also make up one staining jar of 2× SSC and one staining jar of 4× SSC, 0.05% Tween 20 (4T). Store the five jars at 4 °C between uses and replace solutions monthly.
2. Make up a solution of 4× SSC, 0.05% Tween 20 (e.g., 1 L with 200 ml 20× SSC, 800 ml sterile deionized water and 0.5 ml Tween 20) for all subsequent washes on the day.
3. For detection of digoxigenin probes make up antibodies as anti-digoxigenin-FITC (Roche UK) at (5 ng/μl) made up in digoxigenin blocking solution shortly before use. The blocking solution is made from 100 ml 4× SSC, 0.05% Tween 20, and 0.5% Roche digoxigenin blocking reagent, centrifuge at 19,000 × *g* for 5 min and store the supernatant in 1 ml aliquots at −20 °C.
4. For biotin labeled probes use Cy3-streptavidin (Sigma Aldrich) made up in biotin blocking solution. This is made from 100 ml 4× 4SSC, 0.05% Tween 20, and 5 g dried skimmed milk.

Centrifuge at 19,000 × *g* for 5 min and store the supernatant in 1 ml aliquots at −20 °C.

5. Counterstaining (DAPI) solution.

3 Methods

3.1 Fixation of Root Tips and Meiotic Material and Slide Preparation

1. The primary roots are washed with water before keeping them at 4 °C for 24 h to get good metaphase preparations of chromosomes before fixation. Following 24 h in a refrigerator to collect mitotic metaphases select creamy root tips that should have a white area immediately behind the tip. This area will contain the meristematic tissues that are dividing.
2. Place root tips about 75 mm in length into ice cold freshly made up fixative. Change fix after about 15 is then leave on bench overnight and change fix for the last time. The fixed roots although used preferably after 24 h in fixation immediately for slide preparation, they can also be used for several months later provided they are stored in the cold room.
3. Fixation of anthers: place young terminal inflorescences on damp filter paper in a petri dish. Using a stereomicroscope and two pairs of fine watchmaker's forceps, dissect out individual buds and place them in size order on the filter paper, keeping each inflorescence separate. Alternatively, take buds directly from the plant onto damp filter paper, leaving smaller buds to develop. The pollen mother cells in anthers of buds approximately 0.5–2 mm in length will be the ones undergoing meiosis. Check that the buds at the upper end do not contain pollen, as these will have passed meiosis, and can be discarded.
4. Tapetal cells will be in mitosis in buds at the lower end of this size range. Remove one anther from a bud, place it on a microscope slide with lacto-propionic orcein stain, tap with a brass rod, add a coverslip, and examine with a phase-contrast microscope to determine the meiotic stage. Place the remaining in fresh, ice-cold fixative in a microfuge tube. Store anthers from several anthers 1 bud at the same stage in the same tube. Leave it at room temperature and replace the fixative after 2–3 h. Leave overnight and change the fixative. Use after 24 h or store at 4 °C for up to 6 months.

3.2 Slide Preparation

1. Place fixed root tips and fixative into a black watch glass. Using Pasteur pipettes, replace the fixative with 0.01 M citrate buffer. Wash with the same buffer 3 × 2 min to remove all traces of fixative which may interfere with the digestion medium. Replace the citrate buffer with the thawed digestion medium made up to 0.1% before use. Incubate in a moist box at 37 °C for 60–75 min.

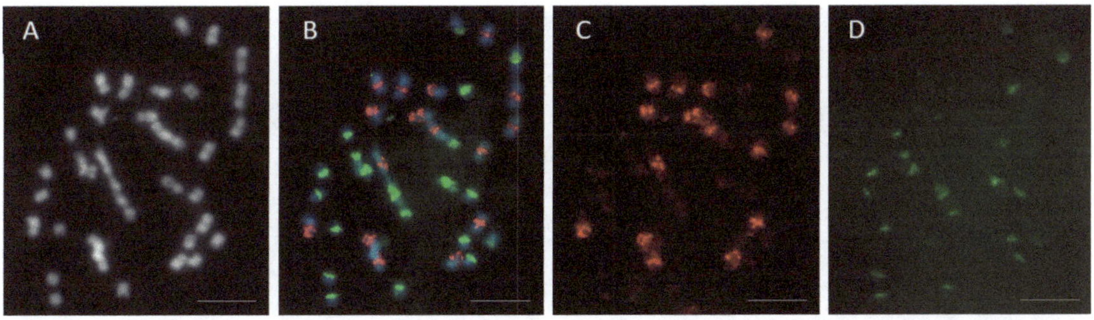

Fig. 1 GISH of mitotic metaphase of *B. juncea* (AABB) species. (**a**) DAPI spread of *B. juncea* chromosomes from root tip. It contains 36 chromosomes altogether and are counted under a microscope (**b**) GISH with total genomic DNA from the *B nigra* (red fluorescence) and *B. rapa* (green fluorescence). *B. rapa* chromosomes can be distinguished from 16 pairs of *B. nigra* chromosomes based on different colors detected with a microscope in mitotic cells (**c**) *B. nigra*. (**d**) *B. rapa*. Bar = 10 μm

2. After this time place a single root tip onto a slide with a minimum of liquid; we generally use slides with frosted ends so that we can write onto the slide with a pencil. Macerate the root tip with a mounted needle quickly without letting the material dry. Add 10 μl of 60% acetic acid. With a diamond pen, mark a region around the sample, no greater than the size of a circle with a diameter of around 18–20 mm around 22 × 22 mm coverslip.
3. Place the slide on a hot block at 45 °C for about 10s and during this time add another 10 μl of 60% acetic acid and mix gently once with the mounted needle. Remove the slide and add 100 μl of ice-cold fixative. Tip the slide to drain it and dry with a hair dryer. Check slides with a phase-contrast microscope to select those with good quality spreads of the required mitotic stage (Fig. 1).
4. Alternatively drain each slide out of the 2× SSC (but do not dry), add 7 μl of DAPI and a coverslip. Keep in the dark at 4 °C. Check the slides using the fluorescence microscope for well spread mitotic metaphase cells (*see* **Note 3**).
5. Unstained slides are usually kept overnight at room temperature before in situ hybridization.
6. A similar procedure is carried out for the meiotic staged anthers (Fig. 2), except that the digestion mix used is at 0.3% and they are left in this solution for between 75 and 120 min.

3.3 Preparation of Probes for GISH

1. The DNA of *B. nigra* and *B. rapa* is extracted from the young leaves of *B. nigra* and *B. rapa* using a DNA extraction kit, such as the Illustra Nucleon PhytoPure, according to the manufacturer's instruction.

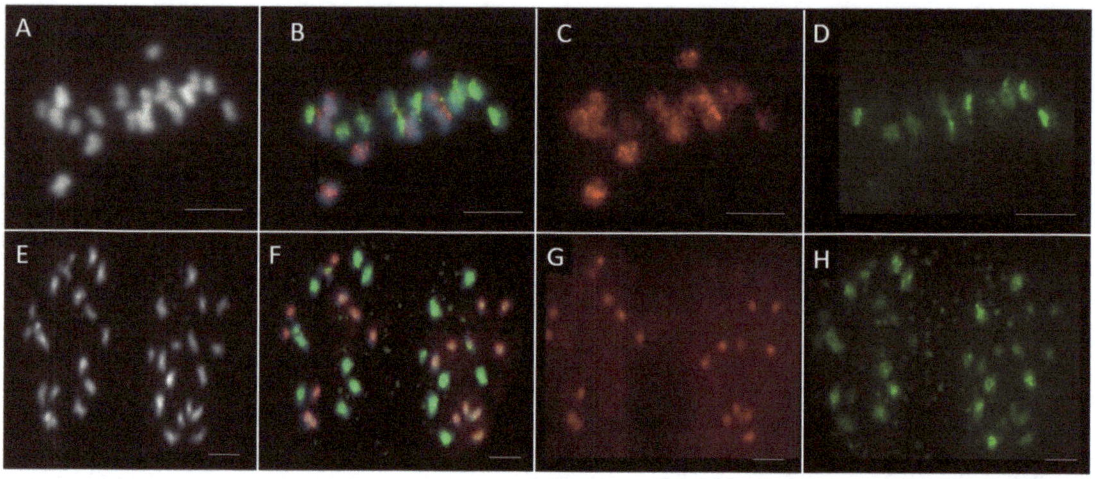

Fig. 2 GISH of meiotic metaphases of *B. juncea* (AABB) species. (**a**, **e**) DAPI spread of meiotic cell at M-I and telophase-I stage. (**b**, **f**) GISH with total genomic DNA from *B. nigra* (red fluorescence) and *B. rapa* (green fluorescence). *B. rapa* chromosomes can be distinguished from 16 pairs of *B. nigra* chromosomes based on different colors detected with a microscope during the metaphase-I stage of Meiosis I and telophase I (**c**, **g**) *B. nigra* detected red. (**d**, **h**) *B. rapa* FITC labeled detected green under the fluorescent microscope. Bar = 10 μm

2. Treatment with RNases and quantification on NanoDrop method using a spectrophotometer.
3. Before carrying out the nick translation steps we need to obtain fragments of 100–500 bp. In order to do this we generally place 150 μl of the DNA (which may have been diluted or concentrated according to our earlier results) into a screwable topped tube and this is placed in a bench top autoclave at 25 psi. In our hands, autoclaving twice is sufficient. Although the fragment length can be checked on a mini gel with appropriate markers.
4. GISH probes: label DNA of *B. nigra* with the biotin-16-dUTP (Roche) and DNA of *B. rapa* with digoxygenin-11-dUTP (Roche) following the manufacturer's instructions.

3.4 The GISH Hybridization Step

1. The slides are placed in Coplin jar containing 2× SSC for 5 min to remove any traces of fixative or ethanol that may prevent pepsin digestion.
2. The cells on the slide are digested with pepsin solution for 90 s in a coplin jar, now kept at room temperature. This will digest cytoplasm and help in accessing the target sequences in the genome for the probe.
3. Two times washing is carried out in SSC solution for 5 min each to remove any debris left on the slide.

4. The slides are fixed 4% paraformaldehyde at room temperature for 10 min under fume hood which helps in improving cross-linking ability in the cells on the slides.
5. The slides are washed in sterile distilled water three times.
6. The slides are dehydrated in ethanol series (i.e., 70%, 85%, and 100%) each for 2 min before left for drying for at least half an hour.
7. The probe mixture is prepared as follows: *B. nigra* DNA Biotin labeled 3 μl (~ 90 ng). *B. rapa* DNA Digoxigenin labeled 3 μl (~90 ng). Master mix 14 μl. Total volume 20 μl.

   20 μl of a probe containing probe mix is used for a single slide and thus up to a 100 ng of the genomic DNA of each constituent genome is used per slide for the GISH probe to hybridize with the DNA in the cells and covered with a coverslip.
8. Vulcanizing rubber solution is used to seal the coverslip and it is placed on a hotplate at 75 °C for 3.5 min. This allows for the denaturation of both the DNA in the probe and the cells.
9. The slides are kept in a moist box at 37 °C for overnight to allow reannealing of the DNAs before they are washed next day.

3.5 Post-hybridization Washes

1. The water bath is set at 45 °C with the wash jars containing formamide solutions and as well as the jars containing 2× SSC and 4T next morning. Once the jars are at the correct temperature the rubber solution and coverslips are removed from the slides using a pair of fine forceps.
2. Washing of the slides is carried out in the three formamide solution jars for 5 min in each jar. The slides were washed after that in 2× SSC for 5 min and finally for 5 min in the 4T solution.
3. The slides are transferred to 4T solution at room temperature before being subjected to application of secondary antibodies on them.
4. The secondary labels are applied to the slides separately on the slide with a square of Parafilm and left in the dark at 37 ° C for 30 min in a moist box.
5. Wash 4T three times in the dark for 5 min each at room temperature between application of the second antibody in the same way as in four above and also wash at the end of this application.
6. The slides are stained with 7 μl DAPI Vectashield, and a coverslip is placed over this.
7. The Images are captured and analyzed using the Smart Capture software on the Olympus fluorescent microscope (GISH images in Figs. 1 and 2).

4 Notes

1. The whole procedure from fixation to analysis of the GISH slides takes a minimum of 4 days. It is preferable to devote a whole week to familiarize yourself with this complex protocol.
2. It is essential that fixative is made up freshly for a day, and kept on ice. This fixative must be discarded at the end of the day, as we find that it undergoes esterification if kept any longer, with water as a by-product of this reaction thus destroying it as a fixative.
3. Less experienced cytogeneticists may prefer to use DAPI for prescreening. These slides need to be destained before the GISH protocol as follows: 10 min 2× SSC to remove coverslips, wash slides through an alcohol series 2 min each 70%, 85% and 100%. The slides now are ready for GISH protocol.

Acknowledgments

We would like to thank Eugenio Sanchez Moran and Elaine Howell for the help and discussion in the laboratory. For technical assistance we would thank Steve Price and Karen Staples.

References

1. Schwarzacher T, Leitch AR, Bennett MD, Heslop-Harrison JS (1989) In situ localization of parental genomes in a wide hybrid. Ann Bot 64(3):315–324
2. Stevenson M, Armstrong SJ, Ford-Lloyd BV, Jones GH (1998) Comparative analysis of crossover exchanges and chiasmata in *Allium cepa x fistulosum* after genomic *in situ* hybridization (GISH). Chromosome Research 6: 567–574.
3. Howell EC, Kearsey MJ, Jones GH, King GJ, Armstrong SJ (2008) A and C genome and chromosome identification in *Brassica napus* by sequential FISH and GISH. Genetics 180:1849–1857
4. Howell EC and Armstrong SJ (2013) Using sequential fluorescence and genomic in situ hybridization (FISH and GISH) to distinguish the A and C genomes in Brassica napus Wojciech. In: Pawlowski P et al. (eds) Plant meiosis: methods and protocols, methods in molecular biology, vol. 990. Springer Science+Business Media, New York, pp 38–47. https://doi.org/10.1007/978-1-62703-333-6_3
5. Song KM, Osborn TC, Williams PH (1990) Brassica taxonomy based on nuclear restriction fragment length polymorphism (RFLPs) 3. Genome relationship in Brassica and related genera and the origin of B. oleracea and *Brassica rapa* (syn. Campestris). Theor Appl Genet 79:497–506
6. Pradhan AK, Prakash S, Mukhopadhyay A, Pental D (1992) Phytogeny of Brassica and allied genera based on variation in chloroplast and mitochondrial DNA patterns: molecular and taxonomic classifications are incongruous. Theor Appl Genet 85(2–3):331–340
7. Warwick SI, Black LD (1991) Molecular systematics of Brassica and allied genera (subtribe Brassicinae Brassicae) chloroplast genome and cytodeme congruence. Theor Appl Genet 82:81–92

Chapter 8

Preparing Maize Synaptonemal Complex Spreads and Sequential Immunofluorescence and Fluorescence In Situ Hybridization

Stephen M. Stack, Lindsay A. Shearer, Leslie D. Lohmiller, and Lorinda K. Anderson

Abstract

Immunofluorescence and fluorescence in situ hybridization (FISH) can be used to locate specific proteins and DNA sequences, respectively, in chromosomes by light microscopy. Here we describe sequential use of these techniques on spreads of maize synaptonemal complexes (SCs) to determine whether crossing over can occur in knob heterochromatin. We used antibodies to AFD1, an SC protein, and MLH1, a class I (interference-sensitive) crossover protein found in most recombination nodules (RNs) to identify crossovers (COs) along SCs. Next, we used FISH to localize a 180 bp knob-specific tandem repeat. Combining immunofluorescence and FISH images of the same SC spreads showed that heterochromatic knobs do not prohibit class I COs. This technique is broadly applicable to investigations of plant prophase I chromosomes where meiotic recombination takes place.

Key words AFD1, Crossing over, Fluorescence in situ hybridization, Heterochromatin, Immunofluorescence, Knobs, Maize, MLH1, Recombination nodules, Synaptonemal complex

1 Introduction

Whether crossing over occurs in heterochromatin has long been in question both because there are few genes in heterochromatin suitable for linkage studies and because cytological analysis of crossing over using chiasmata lacks sufficient resolution to distinguish between a crossover in euchromatin or in heterochromatin, especially near their borders [1]. Subsequently, electron microscopy of late recombination nodules (RNs) on synaptonemal complexes (SCs = pachytene chromosomes) greatly improved cytological resolution of crossover sites [2]. RNs are ellipsoidal protein complexes that are about 100 nm in their longest dimension, which mark sites of crossing over along SCs [3]. Most RNs contain MLH1, a protein involved in class I (interference-sensitive)

Mónica Pradillo and Stefan Heckmann (eds.), *Plant Meiosis: Methods and Protocols*, Methods in Molecular Biology, vol. 2061, https://doi.org/10.1007/978-1-4939-9818-0_8, © Springer Science+Business Media, LLC, part of Springer Nature 2020

crossovers in plants and animals, and antibodies to MLH1 are often used as a fluorescent marker of crossover sites on SCs [4]. SCs are 200 nm-wide ladder-like protein complexes that link pairs of homologous chromosomes (pachytene bivalents) during prophase I of meiosis [5]. The proteins making up SCs are not well conserved evolutionarily, but several SC-specific proteins have been identified, including AFD1 in maize [6]. Simultaneous immunofluorescence of MLH1 and an SC protein (such as AFD1) on spreads of SCs prepared by hypotonic bursting of pachytene cells provides an efficient method to reveal the location of interference-sensitive crossovers sites along SCs [4, 7]. We were interested in determining whether heterochromatin per se (and not the centromere) interferes with crossing over. Addressing this question is difficult because most heterochromatin is located around centromeres that are known to strongly suppress crossing over [8]. However, maize (*Zea mays*) has several dense masses of constitutive heterochromatin called knobs that are located in distal euchromatin far from centromeres [9]. This provides an opportunity to investigate the occurrence of crossing over in heterochromatin beyond interference from centromeres. While hypotonic bursting disperses chromatin along SCs so that knob heterochromatin is not readily distinguishable from euchromatin, knob heterochromatin can be visualized in SC spreads using fluorescence in situ hybridization (FISH) of a 180 bp knob-specific sequence [10]. Using this combination of immunofluorescence and FISH, it was possible to demonstrate that crossing over occurs within maize knobs, and the frequency of interference-sensitive crossing over appears to be indistinguishable along SCs in knob heterochromatin compared to SC in euchromatin adjacent to knobs [1]. This general approach should be useful for other investigations of protein and chromatin relationships in meiotic chromosomes.

2 Materials

2.1 Growing Maize

1. 5–100 or more KYS (or other variety) maize seeds (caryopses).
2. Petri dishes.
3. Fungicide for seeds. We use a 0.5% aqueous solution of Captan. Depending on the number of seeds to be treated, make 5 ml or 10 ml total volume.
4. Filter paper to cover the bottom of a petri dish.
5. 15–20 one gallon (~4 L) or larger planting pots.
6. Trays to put under pots of maize.
7. Potting soil.
8. Pot tags.

9. Osmocote or some other complete, slow-release fertilizer.
10. Access to a greenhouse, preferably with supplemental lighting for year-round use.

2.2 Maize DNA Isolation

1. Young maize plants 8–20 cm in height.
2. Vinyl (or similar) gloves to handle plants.
3. Single-edge razor blade.
4. Forceps to transfer leaf segments.
5. Lid or bottom of a clear plastic petri dish to use as a cutting board.
6. 10 ml aqueous DNA extraction buffer: 200 mM Tris–HCl, pH 7.5 (from 1M Tris–HCl, pH 7.5 stock solution), 25 mM EDTA, pH 8.0 (from 500 mM EDTA, pH 8.0 stock solution), 250 mM NaCl (from 5M NaCl stock solution), 0.5% sodium dodecyl sulfate (SDS; from 10% SDS stock solution), 5 mM dithiothreitol (DTT; from 0.5 M DTT stock solution that was prepared in water and stored at −20 °C). All other stock solutions can be stored indefinitely at room temperature.
7. 1.5 ml plastic microfuge tubes.
8. Disposable plastic pestle for leaf maceration in microfuge tubes.
9. −80 °C freezer.
10. 65 °C water bath.
11. Microcentrifuge.
12. Isopropanol.
13. 70% ethanol.
14. Sterile deionized water.
15. 50 mg/ml aqueous stock of RNase A.
16. NanoDrop or other spectrophotometer for quantifying DNA.

2.3 FISH Probe Preparation

1. 20 bp forward (F) and reverse (R) primers for the 180 bp knob repeat (prepared by a commercial or on-campus facility). Add sterile deionized water to each primer to make 100 μM stock solutions and store at −20 °C.
   (a) F: caacgcccatttttatcgaa.
   (b) R: cgaccagaggatcgtacacc.
2. Standard PCR kit.
3. PCR purification kit.
4. Digoxygenin labeling kit (*see* **Note 1**).
5. Spectrophotometer (*e.g.*, NanoDrop).

2.4 Preparing SC Spreads

1. Healthy maize plants 8–10 weeks old with anthers about 2 mm long.

2. Single edge razor blade, sharp pointed steel dissecting needles, chisel-shaped steel dissecting needles, and microscalpels (*see* **Note 2**).
3. New, prewashed glass microscope slides frosted on one end.
4. 18 × 18 mm or 22 × 22 mm cover glasses for preparing quick squashes to assess meiotic stage.
5. Clear plastic dissecting platform (*see* **Note 3**).
6. Kimwipes (Kimberly-Clark) or some other brand of lint-free paper wipe.
7. Dissecting microscope with an ocular micrometer calibrated in tenths of millimeters and equipped with incident and substage illumination.
8. Compound microscope with bright field and phase contrast illumination.
9. 2% aceto-orcein (*see* **Note 4**).
10. Small level and modeling clay (to make a level surface for slides).
11. A plastic box for 25 microscope slides (such as supplied by Carolina Biological Supply) for use in glow discharging slides. The bottom of the box needs to be cut out with a fine toothed saw or a hot needle while leaving a rim under the separation ridges on either side to keep slides from falling through.
12. Glow-discharge apparatus with a chamber large enough to hold the plastic box for 25 slides (*see* **Note 5**).
13. Glass depression slide (3 × 25 × 75 mm).
14. 15 × 100 mm Polystyrene plastic petri dish with a "V"-shaped glass rod that is about 5 mm in diameter in the bottom or petri dishes with dividers that make three or four compartments (*e.g.*, Falcon type 1004) in which slides can be incubated on the dividers over water to maintain a humid environment during digestion. Hereafter these will be referred to as incubation petri dishes.
15. Small butane torch for pulling micropipettes.
16. Siliconized glass micropipettes with aspirator tube assembly (*see* **Note 6**).
17. Digestion medium: 0.56 mM KH_2PO_4, 0.1 mM acid PIPES, 0.5 mM $CaCl_2$, 0.2% potassium dextran sulfate, 0.7 M mannitol, 2% polyvinylpyrrolidone (PVP, average MW 10,000), pH 4.1 (*see* **Note 7** for preparation of stock and working solutions). Store digestion medium on ice before and during the experiment. Both the potassium dextran sulfate solution and the digestion medium need to be made fresh for each SC spreading experiment.

18. 3 mg of desalted, lyophilized cytohelicase (Sigma) and 3 mg of desalted, lyophilized pectinase (Sigma) for digesting the special callose walls from primary microsporocytes at pachytene (*see* **Note 8**).
19. 4% aqueous formaldehyde solution, pH 8.2–8.5 (*see* **Note 9).**
20. Bursting solution: 0.046% IGEPAL® CA-630, 0.00092% potassium dextran sulfate, 0.3% aqueous formaldehyde (pH 8.2–8.5) (*see* **Note 10** for preparation of stock and working solutions). Keep the solution on ice before and during use. Bursting solution needs to be made fresh for each experiment.
21. Nebulizer (Enviro Safety Products, 3M Nebulizer FT-13) with air pump (*e.g.*, MEDNEB).
22. 0.2% Photoflo 200 (Kodak) prepared with deionized (or distilled) water.

2.5 Silver Staining of SCs

1. Aqueous 4% paraformaldehyde pH 8.2–8.5 (*see* **Note 9**).
2. Plastic incubation petri dishes (*see* Subheading 2.4, **item 14** above).
3. Nylon screen cut to ~25 × 50 mm to cover the clear part of slides with frosted ends (*see* **Note 11**).
4. Silver nitrate dissolved in deionized or distilled water to make a 33% w/v solution (1 gm $AgNO_3$ in 2 ml water, *see* **Note 12**).

2.6 Immunofluorescent Labeling

1. Plastic incubation petri dishes (*see* Subheading 2.4, **item 14** above).
2. Coplin jars (vertical orientation of slides) and staining dishes (horizontal orientation of slides).
3. ~35 × 55 mm coverslips cut from plastic autoclave bags to cover the nonfrosted part of a 25 × 75 mm microscope slide with about 5 mm overhanging the upper edge. These are referred to as plastic coverslips.
4. Plastic container with a sealable lid that is large enough to hold a staining dish. We use a 16 cup (3.8 l) plastic food container.
5. Tris-buffered saline with Triton X-100 (TBST): 10 mM Tris base, 150 mM NaCl, 0.05% Triton (from 20% stock solution of Triton stored at 4 °C), pH adjusted to 8.0 with 1 N HCl and brought to volume with water.
6. Antibody dilution buffer (ADB): 10% goat serum, 3% bovine serum albumen (BSA, fraction V, Sigma), 0.05% Triton X-100 (from 20% stock stored at 4 °C), 0.05% sodium azide (made from 10% stock solution) in TBST divided into 1 ml aliquots and stored at −20 °C. ADB is used at full concentration with the antibody incubation steps and at 10% concentration (in TBST) for blocking and intermediate wash steps.

7. 0.1 M ammonium chloride (NH_4Cl) in TBST (prepared fresh before use).
8. DNase I digestion buffer: 10 mM Tris, 2.5 mM $MgCl_2$, 0.5 mM $CaCl_2$, pH 7.5 (adjusted using HCl). We make this solution at 10× concentration and store it at room temperature.
9. DNase I stock solution: 1 mg/ml in 50% glycerol and stored at −20 °C.
10. Primary antibodies to maize AFD1 (from rat serum) and tomato MLH1 (affinity-purified from rabbit serum). We often affinity-purify the antibodies from serum to reduce background problems [11] (*see* **Note 13**).
11. Secondary antibodies (from goat) conjugated to green or red fluorescent probes. We use Dylight 488 and Dylight 549 (Jackson ImmunoResearch Laboratories, Inc.) or similar fluorescent markers. We usually buy goat secondary antibodies because they are often cheaper than those from other hosts (as is goat serum for preparing blocking solutions).
12. 4′,6-diamidino-2-phenylindole (DAPI) DNA stain: stock solution of 1 mg/ml in water, store indefinitely at −20 °C. Dilute 1:100 in water to use.
13. Vectashield (Vector Laboratories) or similar antifade, nonhardening mounting medium.
14. Aqueous 0.2% Photo-Flo 200.
15. 24 mm × 50 mm cover glasses.

2.7 Imaging (for MLH1 and AFD1 Immunofluorescence and for FISH)

1. Microscope equipped for epifluorescence microscopy with digital imaging. It is essential that there is no pixel shift between the different fluorescent signals. We use a Leica 5000 epifluorescence microscope equipped with:
   (a) DAPI filter for blue fluorescence (excitation 359 nm with emission at 461 nm).
   (b) Narrow band pass filters for red fluorescence [cube ET-DSRed (TRITC/Cy3); excitation 545 nm with 30 nm band width and emission 620 nm with 60 nm band width].
   (c) Narrow band pass filters for green fluorescence [cube ET-EGFP(FITC-Cy2); excitation 470 nm with 40 nm band width and emission 525 with 50 nm band width].
   (d) A motorized stage (Prior).
   (e) A cooled Hamamatsu monochrome 1344 × 1044 pixel camera.
   (f) IP Lab software (ver. 4).

2.8 Fluorescence In Situ Hybridization (FISH)

1. Speed vac (or lyophilizer) to concentrate DNA samples.
2. Microscope slide heating block for denaturing DNA on slides.
3. Water bath (42 °C).
4. Rotator (40–50 rpm).
5. 37 °C Incubator/Oven.
6. 24 mm × 50 mm cover glasses.
7. Plastic incubation petri dishes (*see* Subheading 2.4, **item 14** above).
8. Coplin jars and staining dishes (*see* Subheading 2.6, **item 2** above).
9. 45% aqueous acetic acid. This solution may be reused about ten times before replacing.
10. 1:3 acetic ethanol (1 part glacial acetic acid to 3 parts 95%–100% ethanol). This solution must be made fresh and used within an hour because ethanol and acetic acid react slowly to form ethyl acetate (that is not a fixative).
11. 100% ethanol (not denatured).
12. 2× SSC. Made by diluting 20× SSC 1:10 with water. The stock solution of 20× SSC (saline sodium citrate) is aqueous 3 M NaCl and 300 mM trisodium citrate, pH 7.0 (adjusted with 1 N HCl). 20× SSC can be stored in a closed bottle at room temperature for months.
13. RNase A solution. A stock solution of 10 mg/ml RNase A in 1 M Tris–HCl, pH 7.5 is prepared and then boiled for 15 min to inactivate DNase contaminants. The stock solution is stored in 20 ml aliquots at −20 °C. For use on the slides, dilute the stock solution with 2× SSC, pH 7.0 to a final concentration of 100 μg/ml.
14. Digoxygenin-labeled probe DNA (*see* Subheading 2.3, **item 4** above). Use 25 ng/slide.
15. 10 mg/ml herring sperm DNA (Promega) dissolved in water and stored at −20 °C. Use 5 μg/slide.
16. 0.01 N HCl.
17. Pepsin solution is prepared by diluting a 500 μg/ml pepsin stock solution 1:100 with 0.01 N HCl to a final concentration of 5 μg/ml. The pepsin stock solution is prepared in 0.01 N HCl and stored in 1 ml aliquots in 1.5 ml microtubes at −20 °C.
18. 4T is prepared by making 4× SSC (diluted with water 1:5 from 20× SSC stock—*see* Subheading 2.8, **item 12** above) and

adding 20% Tween 20 to a final concentration of 0.05% Tween 20.

19. TN is prepared by making a tenfold dilution of 10× TN stock that is 1 M Tris–HCl and 1.5 M NaCl, pH 7.5.
20. TNT is TN (diluted from stock 10× TN) that is made 0.05% Tween 20 using an aqueous 20% Tween 20 stock.
21. TNB is TN that is 5% Roche blocking reagent (Cat. No 1096176). Prepare 30 ml by placing 3 ml 10× TN stock, 1.5 ml Roche blocking reagent, and 25.5 ml water in a 50 ml test tube with a cap and incubating the tube in a water bath at ~65–70 °C for an hour with occasional mild agitation to get the reagent into solution. Place 1 ml aliquots of the TNB into microfuge tubes and store at −20 °C. Add enough donkey serum to TNB to make 5% donkey serum for diluting donkey secondary antibodies.
22. 1% formalin fixative: Make 100 ml just before use by combining 10 ml of 10× PBS (1.27 M NaCl, 70 mM dibasic sodium phosphate, 30 mM monobasic sodium phosphate. Adjust pH to 7.0 with 1 N HCl), 10 ml of 10× $MgCl_2$ (= 500 mM $MgCl_2$), 3 ml of 37% formalin (a reagent grade formaldehyde solution stabilized with methanol), and 77 ml deionized water. Dilute the 10× PBS and 10× $MgCl_2$ solutions with some of the water before mixing them together to avoid forming a precipitate. All stock solutions are stored at room temperature.
23. 70%, 90% and 100% ethanol series for dehydration in vertical Coplin jars. Use pure 100% ethanol for these solutions, not denatured ethanol.
24. 50% formamide in 2× SSC (stringent wash solution) prepared by mixing five parts 100% formamide, one part 20× SSC stock, and four parts water.
25. 50% sodium dextran sulfate stock solution (in water).
26. 10% sodium dodecyl sulfate (SDS) in water. This can be made by diluting a 20% SDS stock with an equal part of water.
27. Hybridization solution: 50% formamide, 10% dextran sulfate (diluted from 50% stock), 0.25% SDS (diluted from 10% or 20% stock), 2× SSC (diluted from 20× SSC stock), sterile water as needed to make a final volume of 25 μl per slide. (This solution will be added to the knob probe DNA and herring sperm DNA that will have been previously measured and dried in a microtube.)
28. Plastic container with a sealable lid that is large enough to hold a horizontal staining dish. We use a 16 cup (3.8 l) plastic food container. This will be used for overnight incubation for FISH and for later antibody incubations.

29. Primary antibody: sheep anti-digoxigenin diluted 1:125 in TNB.
30. Secondary antibody: donkey anti-sheep conjugated to tetramethyl rhodamine isothiocyanate (TRITC), diluted 1:100 in TNB with 5% donkey serum.
31. Vectashield (Vector Labs) containing 5 μg/ml DAPI (1:200 dilution of a 1 mg/ml DAPI stock prepared in water and stored at −20 °C). The Vectashield/DAPI can be stored at 4 °C for at least 6 months.

2.9 Fluorescent Image Processing

1. Computer program for overlaying, aligning, and annotating fluorescent light microscope images taken after AFD1 and MLH1 immunolabeling and after FISH. We use Adobe Photoshop CS6.

3 Methods

3.1 Growing Maize

1. To avoid damping off, treat maize seeds with a general purpose fungicide (such as Captan). Put 50–100 maize seeds in an open petri dish with 5–10 ml (just enough to cover the seeds) 0.5% Captan solution. Be sure the seeds are completely coated with the Captan, and then leave the dish open to air dry. In more humid climates, the drying process can be shortened using a fan. Captan-coated seeds can be stored in a refrigerator at ~4 °C for months.
2. To begin planting, place two dry, Captan-coated seeds on wet filter paper in a petri dish, close the dish, and incubate at room temperature. After several days, germination should occur with the primary root emerging first, followed a day or two later by the coleoptile. Keep the filter paper damp and do not let the seeds and seedlings dry out during germination. Start two seeds every 2 days for a month (or more).
3. After the coleoptile in the first dish is a centimeter or more in length, transfer the two seedlings to a 1 gallon (~4 l) pot filled with damp potting soil. Plant the two seedlings about 10 cm apart with the coleoptiles just showing above the surface of the soil. Insert a tag in the pot to indicate the date of planting. During the long days of summer, maize can be grown to flowering in a greenhouse without supplemental lighting, but in winter supplemental lighting may be needed.
4. Keep the soil damp, watering pots once or twice a day as needed, and add fertilizer according to instructions (*e.g.*, Osmocote) throughout growth. It is helpful if the pots are in trays that collect runoff water that can be taken up to keep the soil damp.

3.2 Maize DNA Isolation

1. Wearing vinyl gloves, use a single edge razor blade to vertically slit open the shoot of a young maize plant (8–20 cm in height). Gently open the shoot, and use forceps to remove the young whitish leaves from the center of the shoot. The cells are still relatively small, and the cell walls are thin in these leaves so the effective concentration of nuclei is higher per weight/volume than in more mature, expanded leaves.
2. Using a clean plastic petri dish as a cutting board, dice the leaves into <0.5 cm segments with a clean razor blade.
3. Use forceps to transfer ~100 mg of leaf segments into a 1.5 ml plastic microfuge tube containing 500 μl of DNA extraction buffer, and freeze the tissue overnight at −80 °C.
4. The next day heat the tissue suspension to 65 °C for 10 min in a water bath. Then gently macerate the suspension in the microfuge tube for 20–30 seconds using a disposable pestle.
5. In a microcentrifuge at room temperature spin the maceration at maximum speed (14,000–16,000 × *g*) for 5 min.
6. Transfer the supernatant to a new microtube, and discard the pellet.
7. To precipitate the DNA, add an equal volume of isopropanol to the supernatant and mix by hand inverting the microtube several times. At room temperature centrifuge the mixture for 10 min at maximum speed.
8. Carefully decant the supernatant, leaving the DNA pellet in the bottom and along one side of the tube.
9. Carefully add 1 ml of 70% ethanol. Over a 5 min period gently invert the microtube a few times. The goal is to wash some of the contaminants from the pelleted DNA, while leaving the DNA on one side and bottom of the tube.
10. Centrifuge the sample for 5 min at maximum speed. Keep the side of the tube with the pelleted DNA facing the outside of the centrifuge to avoid disturbing the pellet.
11. Decant the 70% ethanol, invert the open tube, and allow the pellet to dry ~15 min.
12. Add 100 μl of sterile, deionized water and flick the tube to dissolve the DNA pellet.
13. Add 1 μl of 10 mg/ml RNase A stock solution, mix, and incubate at 37 °C for 1 h.
14. To precipitate the DNA, add 300 μl 70% ethanol, invert several times to mix, and incubate for 10 min. at −20 °C.
15. Centrifuge at maximum speed for 10 min.
16. Carefully decant the supernatant, leaving the DNA pellet in the tube.

17. Invert the tube on a paper towel to drain the ethanol completely, and allow the pellet to dry for about 10 min (*see* **Note 14**).
18. Resuspend the pellet in 200 μl of sterile, deionized water.
19. Use a NanoDrop or other suitable spectrophotometer to quantify the DNA, and then adjust the concentration to 100 ng/μl. We found that a 260/280 ratio of 1.5 indicated sufficient purity for FISH probe preparation.

3.3 FISH Probe Preparation

1. Amplify the 180 bp knob repeat using maize genomic DNA, knob repeat primers, and a PCR kit for standard PCR according the manufacturer's instructions. For a 25 μl reaction, we used 2.5 μl of a 10 μM solution of forward primer and 2.5 μl of a 10 μM solution of reverse primer. This step yields an amplified 164 bp knob repeat.
2. Clean the PCR product using a PCR cleanup kit according to the manufacturer's instructions.
3. Using the amplified 164 bp knob repeat, repeat primers, and a DIG PCR kit, amplify and label the 164 bp knob DNA with digoxygenin according to the manufacturer's instructions.
4. Purify the labeled probe DNA with a PCR cleanup kit.
5. Quantify the labeled probe with a suitable spectrophotometer (*e.g.*, a NanoDrop). We usually obtained about a microgram of DIG-labeled DNA per reaction.
6. Store the labeled DNA in sterile, deionized water at −20 °C.

3.4 Preparing SC Spreads

1. Select plants that are 8–10 weeks old (after planting the seedlings). Use a single edge razor blade to slit the stalk vertically from the apex (where the leaves are separating) down one third to one half the length of the shoot toward the base. Cut no deeper than half the width of the stalk to avoid damaging the tassel in the center of the stalk. Then as gently as possible, pry open the stalk with your fingers to reveal the tassel (male inflorescence) in the middle of the stalk. Be careful not to crush or bruise the tassel (anthers) that consists of a tightly packed central yellowish-white rachis (main spike) with less mature side spikelets coming off the base of the rachis. More mature florets are on the upper half of each spike grading to less mature florets at the base. A male floret consists of translucent bracts surrounding six anthers, three larger more mature anthers and three smaller less mature anthers. We find that anthers about 2–2.1 mm in length contain primary microsporocytes (PMCs) at or near pachytene. Florets near each other on a spike are almost of the same length and in the same stage of development. If florets from the main spike are too long, florets from side spikes, which sometimes will have anthers at the right

stage, are suitable. If the plant is too young, the cut in the stalk can be closed and sealed using Parafilm for a day or more before retesting for pachytene-stage florets again.

2. Look for anthers containing PMCs at pachytene.
   (a) Remove a floret from a spike by pinching the floret off at its base with sharp forceps and place the floret on a clear plastic dissecting platform on a dissecting microscope stage.
   (b) At ~10× magnification, push aside the bracts with dissecting needles, and cut an anther off at its base. Be gentle with the anthers.
   (c) Place the anther in a drop of 2% aceto-orcein on a glass slide that was wiped with a Kimwipe to remove dust and lint that could interfere with a squash. Dipping a dissecting needle in water to make it wet helps anthers adhere to the needle for transfer.
   (d) Under a dissecting microscope, use the chisel-like dissecting needle to cut the anther in half (transverse to the anther's long axis). Squeeze out the contents of the anther (PMCs and tapetal cells) into the aceto-orcein and remove the anther walls.
   (e) Cover the drop of aceto-orcein and anther contents with a cover glass and gently warm the slide over an alcohol lamp until the water condensate (fog) that forms on the underside of the slide toward the flame just disappears. Do not boil the aceto-orcein solution because boiling damages the structure of the chromosomes.
   (f) Fold a Kimwipe or some other paper wipe to form a pad, and place the pad over the cover glass. Then gently press out any excess aceto-orcein. Be sure not to let the cover glass slip, because this will roll the cells and ruin the squash.
   (g) Examine the squashed PMCs (the big cells compared to the smaller tapetal cells) with a compound microscope using a dry (not oil immersion) objective at a total magnification of 200–400× (*i.e.*, a 20 or 40× objective with 10× eyepieces). Verify that the PMCs are in pachytene [12]. Phase contrast illumination will be helpful to visualize prophase I chromosomes and determine meiotic stage. Alternatively, bright field contrast can be increased by adjusting the condenser diaphragm and/or by adding a green filter over the light source. If pachytene is found in this first anther, we do not check the meiotic stage of other anthers because we have found in several maize cultivars that anthers between 2.0 and 2.1 mm in length generally contain microsporocytes at pachytene.

3. Glass microscope slide preparation for SC spreads. (This should be done prior to removing the tassel from the plant.)
   (a) Wipe each slide immediately before glow discharging, even if slides are labeled precleaned. For this, place a glass slide on a table-top with the frosted side up. While securely holding the frosted end on a table top with the fingers of one hand, firmly wipe the clear portion of the frosted side ~5–10 times with a pad of dry Kimwipes or some other clean, lint-free paper towel. At first there is some resistance, but as wiping continues, there is less resistance until the surface seems comparatively slick. After wiping, do not touch the clear glass surface. Place the slides in a plastic 25-slide box with the bottom cut out to facilitate exposure of the slides to glow discharge. We usually fill the box with 25 wiped slides.
   (b) Glow-discharge slides to make the slide surfaces hydrophilic. This causes the bursting medium and bursting protoplasts to disperse actively over the surface of the slide to help spread the SCs.
      - Place the open slide box between the two aluminum plates (electrodes) in the vacuum chamber of the glow-discharge apparatus, and close the lid.
      - Close the air inlet valve to the vacuum chamber, and turn on the vacuum pump.
      - When the vacuum reaches about 0.2 mm of mercury (0.2 Torr), adjust the needle valve to let just enough air into the chamber so the vacuum pump just holds this vacuum. In practice once the needle valve has been adjusted properly, it can be left at that position without further adjustment for subsequent use. Dim the room lights so you can see the glow in the vacuum chamber. Turn on the power source, and increase the voltage. At around 350 V and 50 mA of current, glow discharge between the aluminum plates will start. No further increase in voltage is needed. The glow between the plates will be a pinkish violet with room air leaking through the needle valve into the chamber. The blue component is due to water vapor, and generally, the higher the humidity, the more purple the glow and the better the slides are glow discharged, that is, negatively charged and hydrophilic. Glow discharge the slides for 10 min.
      - Turn off the current on the power unit to avoid being shocked when touching the aluminum plates (electrodes). Close the vacuum chamber to the pump, release the vacuum in the vacuum chamber, remove the lid, and remove the slide box.

- Open the vacuum pump to the vacuum chamber, and quickly turn off the vacuum pump. This will prevent the vacuum pump from shutting down while still under vacuum, which can make it difficult to restart the vacuum pump.
- Place the slide box on a nonconducting surface such as a cardboard box to slow the loss of charge on the slides. The slides should remain hydrophilic and usable for chromosome spreading for several hours. To determine whether glow discharge has been successful, place a drop of water on a slide. The drop should spread like a spider web over the surface. This is in contrast to the behavior of a drop on a non–glow-discharged, hydrophobic glass surface where a water drop beads up and does not spread much. If unused slides are left over from an experiment, they can be reused later, but they need to be wiped and glow discharged again.

4. Preparing spreads of maize SCs.
   (a) Place a depression slide in an incubation petri dish that contains about 10 ml of water. Add 0.2 ml of digestion medium to the depression, and close the dish to reduce the rate of evaporation of the digestion medium (*see* **Note 15**).
   (b) To collect anthers, transfer a segment of a spike with buds containing anthers ~2 mm long to the plastic dissecting platform on a dissecting microscope stage (*see* **Note 16**). Cut a floret from the spike with a chisel-shaped dissecting needle. Gently separate the bracts to reveal the anthers. We usually use only the larger trio of anthers in a floret, but the smaller trio also can be used if they are the right length (in which case the longer trio will be far too long and contain pollen grains). Cut the short filament at the base of each anther and gently transfer the anthers to the digestion medium in a depression slide (*see* **Note 17**).
   (c) Collect 50–100 anthers in the digestion medium, being careful to minimize evaporation of the medium, and add water as needed to keep up the volume.
   (d) Put 3 mg of desalted and lyophilized cytohelicase and 3 mg of desalted and lyophilized pectinase in the digestion medium with the anthers.
   (e) Transfer the depression slide to the stage of a dissecting microscope with substage illumination. Use a sharp dissecting needle in one hand to orient anthers and a microscalpel in the other hand to slice all 50–100 anthers in half transversely to their long axes. Draw the microscalpel

across the anthers to slice them with minimal pressure. Cutting anthers with an ax-like motion can bruise anthers and cause cytomixis and death of the microsporocytes. It is important for the microscalpel to be sharp (*see* **Note 2**). During cutting, the digestion medium will be agitated enough to mix the enzymes. At the end of cutting, a 30 μl drop of distilled water can be added to make up for evaporation.

(f) Return the depression slide to the petri dish, and cover the dish with a cardboard box or some other light proof box so digestion takes place in the dark for 45 min at room temperature (no higher than 20 °C). The enzymes are somewhat light sensitive so subdued light helps maintain enzyme activity.

(g) After 45 min put the depression slide on a dissecting microscope stage. At 10–25× magnification, use a sharp dissecting needle in one hand to position anther halves and a chisel-shaped dissecting needle in the other hand to press the contents out of several anther halves. Remove anther halves that have been emptied. By this time the special cell walls around living primary microsporocytes will have been digested (removed) by the enzymes. Living protoplasts come out as translucent spheres, while dead protoplasts look shriveled, dark, and dense. Initially the pachytene protoplasts are held together by cytoplasmic bridges in grape-like clusters, but these bridges break soon after protoplasts are squeezed out of the anther halves, and the protoplasts fall apart as individual spheres lying on the bottom of the depression. To observe the protoplasts clearly as translucent spheres, substage illumination needs to be adjusted properly, and higher magnification (25–40×) will help as well.

(h) When 20–30 spherical protoplasts are visible, stop squeezing out protoplasts, and use a siliconized micropipette to suck up as many protoplasts as possible in no more than 0.5 μl of digestion medium. Larger volumes of digestion medium interfere with SC spreading. Sucking up a half microliter or less with 20–30 protoplasts will take some practice.

(i) Return the depression slide to the petri dish, and close the dish to reduce evaporation.

(j) Place a glow discharged slide with the frosted side up on a flat surface (*see* **Note 18**). Use a pencil to number the slides on their frosted ends in order of preparation.

(k) Draw 10 μl of bursting medium into a mechanical micropipette (*e.g.*, a P10 or P20 Gilson Pipetman) and carefully expel the droplet so that it remains attached to the pipette tip. In our experience, this works best with yellow tips that are more hydrophilic than uncolored tips (*see* **Note 19**).

(l) Gently expel the protoplast suspension into the 10 μl of bursting medium at the end of the plastic pipette tip. Take care to leave the droplet attached to the end of the plastic tip of the pipette, and do not blow air into the suspended droplet.

(m) Wait 5 s with gentle shaking of the mechanical pipette to help mix the cell suspension with the bursting medium. Take care not to shake the droplet off the pipette tip. During this time the protoplasts will swell in the hypotonic bursting medium, and some may burst.

(n) Touch the mixture off at the center of the nonfrosted part of a glow discharged microscope slide that is lying flat on the table top. If the slide has been properly glow discharged, the droplet will spread in a spider web-like pattern over the surface of the slide.

(o) Immediately touch off an additional 10 μl of bursting medium at the center of the slide where the first drop was touched off. To see if the protoplasts are swelling and bursting, without a cover glass the slide can be examined for 10–30 s at a magnification of 100–200× using a compound microscope with phase contrast illumination. Protoplasts should be visible as clear spheres with granules inside, which are swelling and bursting. This delay before the next step of fixation is often useful to give the SCs time to spread better.

(p) Carefully carry the slide to a hood, keeping it as level as possible, so the liquid does not run to one side of the slide. Lay it flat, and spray the slide with 30 sweeps of nebulized aqueous 4% paraformaldehyde. The opening in the nebulizer from which the mist sprays out should only be about 2–3 cm from the surface of the slide, and the mist should noticeably dampen the frosted end when the sweep passes over that end of the slide. The total volume added to the slide will be ~50–100 μl.

(q) Allow the slides to dry 1–2 h in the hood (*see* **Note 20**). It is best not to move the slides much until they are dry, as moving them tends to increase the number of free, single SCs. A slide should have a thin film of dried mannitol over much of its surface. If a slide has a dense accumulation of mannitol in one small location, there will be few, if any, good SC spreads.

(r) Repeat **steps k–q** above until enough slides have been made, or all the anthers on the depression slide have been used, or until the all the protoplasts on the depression slide appear to be dead (*see* **Note 21**).

(s) Set aside any slides to be used for silver staining. Seal the remaining air-dried slides in a 25-slide plastic box using black plastic electrician's tape. On the box write the number of slides, and store the box at −80 °C. Slides should remain usable for at least a year.

3.5 Silver Staining of SCs

1. To see how well an SC spreading experiment worked before using the slides for immunofluorescence and/or FISH, silver stain a few slides (*e.g.*, one each from the beginning, middle, and end of slide making) (*see* **Note 22**).
2. Prepare three 140 ml beakers each containing fresh deionized (DI) or distilled water, and briefly rinse (dip for about 5 s without agitation) the slides sequentially in each beaker so that each slide gets three washes to remove mainly mannitol. Air dry the slides.
3. Lay the slides flat on a rack in a hood, and fix the slides by flooding the frosted sides of the slides with aqueous 4% paraformaldehyde. After 2 min, pour off the fixative into a waste container, and immediately flood each slide again with 4% paraformaldehyde.
4. After 8 min, pour off the fixative, and transfer each slide sequentially through a series of three 140 ml beakers filled with distilled or deionized water. Each wash step is 10–15 s without agitation. Air dry the slides.
5. Place each slide (frosted side up) in an incubation petri dish containing several milliliters of water. (The water should not contact the bottom of the slide.)
6. Cover the clear glass of each slide (where the SCs spreads should be) with a roughly 25 × 50 mm piece of Tetko nylon screen. Add two to three drops of 33% $AgNO_3$ to wet the nylon on each slide.
7. Close the petri dishes, and float them in a closed water bath at 40 °C for 25 min. At the end of this time the nylon should have turned light tan in color. This results in lightly stained SCs that are best viewed by phase contrast illumination. If incubation time in the water bath is extended, the nylon grows darker, and SC staining is progressively darker and coarser so bright field illumination can be used.
8. Rinse nylon screens off the slides with a gentle stream of distilled or DI water, and air dry the slides. Then examine the slides by phase contrast or bright field microscopy using a

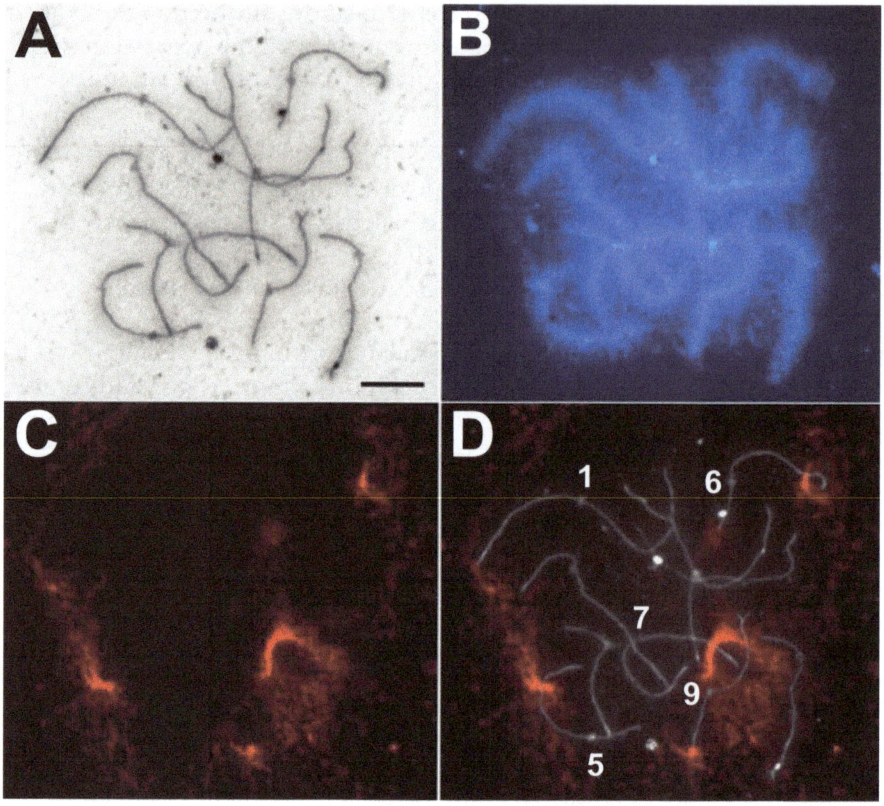

Fig. 1 The same maize KYS SC spread stained first with (**A**) silver, next with (**B**) DAPI, and then (**C**) labeled by FISH with the probe for the180 kb knob repeat (red). In (**D**), the silver stained image has been inverted and the FISH image superimposed on it. The five SCs (SC1, SC5, SC6, SC7, and SC9) that carry knobs are indicated with the appropriate number near their kinetochores (roundish, white structures near the middle of most SCs). The kinetochore of SC7 is not easily visible, but it is partially fused to an adjacent kinetochore of another SC. The knobs of SC1 and SC9 are located at the tips of the short arms, and the knob of SC9 is larger than that of SC1. The knobs for SC5 and SC7 are large, and both occur at interstitial locations in the long arms. SC5 and SC7 both have the same relative length and arm ratio, and the knobs are similar in size and location, so the identification of these two SCs in this image is arbitrary. SC6 has two knobs, a terminal one located in the short arm and an interstitial knob located in the long arm. Bar equals 10 μm

20–40× objective. Addition of a dry cover glass will improve resolution. SCs look like long dark threads along which kinetochores appear as dark fuzzy balls about 1 μm in diameter (Fig. 1A). Silver-stained slides can be used for FISH, but we did not use silver-stained slides for immunofluorescence.

9. Trouble-shooting (*see* **Note 23**).

3.6 Immuno-fluorescent Labeling

1. Remove a box of slides with SC spreads from the −80 °C freezer, and allow the box to warm to room temperature without opening it (to avoid condensation of water on the slides). Remove the lid and take the slide(s) to be used from the slide box. Remove any pencil marks that may be present on the

frosted end by wiping it with a damp Kimwipe (*see* **Note 24**). Mark each slide (*e.g.*, 1–8) with a permanent felt-tip marker on the frosted part. We typically do no more than eight slides in one immunolabeling experiment. If there are unused slides in the box, reseal the box with tape, write the number of slides remaining on the box, close and tape the box, and replace the box in the −80 °C freezer for later use.

2. Prepare a Y-plate petri dish for each slide by adding ~1 ml deionized (or distilled) water in the bottom of the dish. This will maintain a humid atmosphere during incubation steps while keeping the slide out of the water (*see* **Note 25**).
3. Rinse and hydrate each slide by gently squirting 1–2 ml TBST from a plastic transfer pipet onto the frosted end of the slide and letting it drain down the entire surface of the slide as it is slightly tilted. This method of washing the slide is used throughout the procedure.
4. Place a slide in a petri dish, and add about 500 μl TBST on top of the slide. Unless otherwise noted, each slide should be kept level in the damp plastic petri dish during incubation steps, and the back (bottom) of the slide (that is often damp or wet) should be wiped with a Kimwipe each time after removing one solution and prior to adding the next solution. We have noted an increase in background if the backs of the slides are not wiped consistently between steps throughout the procedure. The petri dishes used for incubation also need to be kept clean to reduce background.
5. Pour TBST off the slide, and immediately add 500 μl of 0.1 M ammonium chloride solution. Cover the petri dish, and incubate the slide for 10 min (*see* **Note 26**).
6. Pour the solution off the slides, and rinse with TBST.
7. Add 500 μl of 10% ADB, place the slide in a petri dish, cover the dish, and incubate for 15 min. Repeat twice more for a total of three 10% ADB treatments over 45 min (*see* **Note 27**).
8. While the slides are incubating in 10% ADB, dilute the primary antibodies (in this case, rat anti-ZmAFD1 (from serum) and affinity-purified rabbit anti-SlMLH1) in ADB (*see* **Note 28**). Prepare 100 μl of antibody solution for each slide plus an additional 10–20 μl in a 1.5 ml microtube. Centrifuge at maximum speed (16,000 × *g*) for 30 min at 4 °C (*see* **Note 29**).
9. Pour off the last 10% ADB wash, and place 100 μl of primary antibody solution on each slide. Use the primary antibody solution from the top of the microtube, being careful not to disturb the lower 10–20 μl layer where any protein aggregates may remain. Cover each slide with a plastic coverslip; avoiding bubbles under the slip (*see* **Note 30**). The plastic slip should be

aligned along the back edge of the slide so that the plastic coverslip extends a few millimeters beyond the front edge of the slide.

10. Insert the slides upside down in a horizontal staining dish that is resting on its side. The extending lip of plastic coverslip should be oriented outward. Place the staining dish on its side into a sealable chamber with wet paper towels in the bottom. The dish should be placed on a slight wedge made out of paper towels to keep the slides angled to the back (bottom) of the dish that is resting on its side, so the slides do not fall out. Close the chamber and incubate the slides overnight in a refrigerator at 4 °C (*see* **Note 31**).
11. The next morning, take the closed container with the slides out of the refrigerator, open the container, and let the slides warm up for 30–60 min at room temperature.
12. Wipe water condensate from the back (bottom) of the slide.
13. Wash the plastic coverslip off the front of the slide with TBST (*see* **Note 32**). Immediately rinse the front of the slide by gently squirting 1–2 ml TBST from a plastic transfer pipet onto the frosted end of the slide and letting the TBST drain over the entire surface of the slide.
14. Immediately place the slide in an incubation petri dish, add 500 μl 10% ADB to the slide, cover the petri dish, and incubate for 15 min. at room temperature. Repeat two more times for a total of three 10% ADB washes for 15 min each.
15. During these wash steps, prepare the fluorescently labeled secondary antibodies, in this case, goat anti-rat 549 (red) and goat anti-rabbit 488 (green) each diluted 1:500 in ADB. Prepare 100 μl of antibody solution for each slide plus an additional 10–20 μl in a 1.5 ml microtube. Centrifuge at maximum speed for 30 min at 4 °C (*see* **Note 33**).
16. Pour off the last 10% ADB wash, and place 100 μl of fluorescent secondary antibody solution on each slide. Use the antibody solution from the top of the microtube, being careful not to disturb the lower 10–20 μl layer where any protein aggregates may remain. Cover each slide with a plastic coverslip, avoiding bubbles. Place the slides upside down in a horizontal staining dish on its side. Put the staining dish on its side with coverslips down in a damp chamber on a paper towel wedge. Close the chamber, and incubate the slides for 1 h at room temperature in a dark environment.
17. Open the chamber, remove the plastic coverslip from the slide (*see* **Note 32**), rinse the slide with a gentle stream of TBST, wipe the bottom of the slide, and place it in a Y petri dish with water in the bottom.

18. Add 400 μl of DAPI staining solution without a cover glass to each slide, and incubate for 2–3 min.
19. Pour the DAPI solution off, rinse with a stream of TBST, and place each slide in a vertical Coplin jar containing TBST. Incubate for 5 min, pour off TBST and add fresh TBST to the Coplin jar. Repeat for a total of three 5 min TBST washes without agitation.
20. Remove a slide from the Coplin jar, wipe the back of the slide with a Kimwipe, and drain the fluid from the slide until the slide is barely damp, but do not allow the slide to dry.
21. Place 1–2 drops of Vectashield antifade mounting medium on the slide, and cover with a clean 24 mm × 50 mm cover glass, being careful to avoid air bubbles under the glass (*see* **Note 34**).
22. Remove extra mounting medium by covering the slide with a Kimwipe and gently pressing, being careful not to move the cover glass. Any large bubbles in the area of the slide containing most of the spreads can be removed by using the blunt end of a pair of forceps and gently pressing on the cover glass to "chase" the bubble out from under the nearest edge of the cover glass.
23. Store the slides in darkness in a plastic slide box in a refrigerator at 4 °C. A folded Kimwipe or other paper towel in the bottom of the slide box will absorb any excess mountant.

3.7 Immunofluorescence Imaging

1. Transfer one immunolabeled slide from the refrigerator to a separate light-tight slide box, close the box, and bring the slide to room temperature. Alternatively, the slide can be brought to room temperature in a dark environment (*see* **Note 35**).
2. Using a microscope with UV illumination, a motor-driven stage (or a stage with a Vernier scale), a dry 40× objective, and filters for DAPI, scan the slide to locate good SC spreads (and even single SCs, a good proportion of which have an interstitial knob) using the blue fluorescent DAPI signal (that stains DNA around SCs). Record the stage coordinates of each SC spread (or individual SC). It is important that the slide be firmly seated with the stage clips so stage coordinates accurately guide return to the same SC spreads later.
3. After scanning the slide for SCs, use the recorded position of each SC spread and a Plan-APO 100× oil objective to image the blue, red, and green signals for DAPI, AFD1, and MLH1 fluorescence, respectively (*see* **Note 36**, Fig. 2B–D).
4. Once imaging is complete, hold the edges of the cover glass with the fingers of one hand and gently remove immersion oil from the surface of the cover glass with a dry Kimwipe. Be careful not to move the cover glass. If any smear of oil remains,

wipe the cover glass with a clean Kimwipe that has been dampened with 95% ethanol. Return the slide to the original slide box, and store in the refrigerator.

3.8 Fluorescence In Situ Hybridization (FISH)

1. Remove slides with immunolabeled SC spreads with cover glasses from the refrigerator, and place them in a horizontal staining dish filled with deionized water. Allow the slide(s) to incubate at room temperature long enough to loosen the cover glasses. Remove the cover glasses using a pair of forceps, being careful not to scratch the slides' surface. Rinse the slide with water until all of the mounting medium has been removed, then air dry the slides.
2. Place dry slides into a vertical Coplin jar containing enough aqueous 45% acetic acid (about 50 ml) to cover the slides, and incubate for 1 min at room temperature.
3. Pour off the acetic acid solution, pour on enough freshly prepared 1:3 acetic ethanol to cover the slides, then incubate for 1 min at room temperature. This fixation step helps to preserve chromosome structure through the following FISH steps.
4. Pour off the acetic ethanol, rinse the slides in 100% ethanol in the Coplin jar, remove slides one at a time, and allow them to dry in a rack.
5. Lay each slide frosted slide up in an incubation petri dish with water in the bottom. Add 100 μl of RNase solution to each slide, cover with 24 × 50 cover glass (although plastic coverslips may also be used throughout this procedure if they are prepared from autoclave bags), close the petri dish, and incubate for 1 h at 37 °C. During this incubation time, the digoxigenin-labeled probe should be mixed with the hybridization solution and denatured (*see* **step 15** below) so the probes will be ready when the slides are dry after the ethanol dehydration series (*see* **step 14** below).
6. Remove the cover glasses by placing the slides in a Coplin jar containing 2× SSC for about 3 min. When the cover glasses are loose, remove them with forceps, being careful not to scratch the surface of the slide, and place the slides into another Coplin jar filled with fresh 2× SSC. We save and reuse the cover glasses used throughout the procedure after rinsing them with deionized water and drying them on paper towels.
7. Wash the slides three times for 3 min each with 2× SSC at room temperature on a rotator with gentle agitation (*see* **Note 37**).
8. After pouring off the last 2× SSC wash, cover the slides with aqueous 0.01 N HCl, and incubate at room temperature for 2 min. Pour off the HCl solution. Remove one slide at a time from the Coplin jar, place 200 μl of pepsin solution on each

slide, cover with a 24 × 50 mm cover glass, and incubate for 8 min at 37 °C in a closed, incubation petri dish with water in the bottom.

9. After incubation, place the slides in a staining dish containing deionized water, and remove the cover glasses using a pair of forceps, being careful not to scratch the surface of the slide. It is important not to over-digest with pepsin, therefore the cover glasses should be removed immediately.
10. Place slides into fresh deionized water in a Coplin jar and incubate for 3 min. Pour off the water, and wash slides twice with 2× SSC for 3 min each.
11. Pour off the 2× SSC, and add ~50 ml of 1% formalin fixative to the Coplin jar, and fix for 10 min at room temperature.
12. Pour off the formalin fixative (into a container for later appropriate disposal), and wash the slides three times for 3 min each with 2× SSC using gentle rotation at room temperature.
13. Rinse slides briefly in deionized water.
14. Dehydrate with a series of 3 min ethanol washes starting with 70% ethanol, followed by 90% ethanol, and finally 100% ethanol. Air-dry slides in a drying rack (*see* **Note 38**).
15. Probe preparation.
    (a) Prepare the volume of hybridization mixture needed for the total number of slides to be used for FISH (plus the amount needed for one extra slide to allow for measuring errors).
    (b) Place a volume equal to 25 ng of labeled probe DNA plus 5 μg of herring sperm DNA per slide (plus one extra slide) into a 1.5 ml microtube (*see* **Note 39**). For example, for eight slides, this would be 225 ng of probe DNA plus 45 μg of herring sperm blocking DNA. Vortex to mix and then dry the mixture using a speed vac (or freeze the mixture and lyophilize it).
    (c) To the microtube containing dried probe and herring sperm DNA, add 25 μl of hybridization mixture for each slide plus one; for example, for eight slides, add 225 μl of hybridization mixture. Vortex vigorously to mix the DNA and other components together.
    (d) Incubate the DNA–hybridization mixture solution in a boiling water bath for 10 min to denature the DNA, and then place the microtube immediately into a water–ice slurry to cool the probe mixture quickly and prevent the DNA from reassociating. Keep the tube on ice for at least 5 min to be sure the mixture is thoroughly cooled (*see* **Note 40**).

16. After the slides are dry and the probes are ready, put 25 µl of the DNA/hybridization mixture in the middle of the clear area on the frosted side of each slide (where the SC spreads are located), and cover the droplet with a 24 × 50 mm cover glass.
17. Place each slide on an aluminum block at 80 °C for 2.5 min to denature the chromosomal DNA on the slide (*see* **Note 41**).
18. Place each slide cover glass side down into a staining dish that is placed on its side so that the slides are held horizontally. Once all of the slides are in the dish, place the dish on its side with cover glasses down in a moist chamber and incubate overnight (or at least 12 h) at 37 °C.
19. The next morning (before **step 20** below), fill three Coplin jars with 50% formamide in 2× SSC, and bring the solutions to 42 °C in a water bath in a hood. Use a thermometer to check the temperature of the solution in each Coplin jar before introducing the slides. Each jar represents one stringent wash step (**step 21** below, *see* **Note 42**).
20. Fill three Coplin jars with 2× SSC at room temperature. Place the slides in the first jar for about 3 min to remove cover glasses and DNA/hybridization solution. When the cover glasses have fallen off (or been removed carefully with a pair of forceps), put the slides into 2× SSC in the next Coplin jar, and wash for 3 min. Repeat a 3 min wash in 2× SSC in the final Coplin jar. If necessary, the slides can be held at this point while the stringent wash solutions are being brought up to the correct temperature.
21. Place slides into first stringent wash (*see* **step 19** above) at 42 °C, and incubate for 5 min. Transfer the slides to the second stringent wash and incubate for 5 min. Repeat for the third stringent wash.
22. Place the slides into a fresh Coplin jar containing 2× SSC at room temperature, and wash the slides for 3 min on a rotator.
23. Wash slides twice with 4T for 3 min each at room temperature on a rotator.
24. Remove slides one at a time from the 4T, wipe the back of each slide, add 100 µl of blocking solution to the front of the slide, and cover with 24 × 50 mm cover glasses. As described in the immunolabeling procedure, be sure to wipe the back of the slides after removing them from wash solutions in the following antibody incubation steps.
25. Place slides into a horizontal staining dish placed on its side so that the slides remain horizontal with the cover glasses facing down. Put the staining dish in a plastic container containing a damp paper towel, close it, and transfer the container to an incubator at 37 °C. Incubate the slides in the blocking solution

for at least 30 min. All subsequent antibody incubation steps will use this same slide arrangement and incubation chamber.

26. Put slides into a staining dish containing TNT, and remove the cover glass from each slide.
27. Primary antibody incubation: Add 100 μl of sheep anti-digoxigenin IgG (diluted 1:125 in TNB) to each slide, cover with a 24 × 50 mm cover glass, and incubate as described in **step 25** for 1 h at 37 °C.
28. Repeat the TNT wash to remove cover glasses. Then place slides into a Coplin jar with fresh TNT, and wash slides three times in TNT for 3 min each at room temperature on a rotator.
29. Secondary antibody incubation: Remove slides from TNT, add 100 μl of donkey anti-sheep conjugated to TRITC (fluoresces red; diluted 1:100 in TNB plus 5% donkey serum) to each slide, cover with a 24 × 50 mm cover glass, and incubate as described in **step 27** for 1 h at 37 °C. From this point forward, the procedure should be done in the dark or in subdued light to avoid possible bleaching of the fluorescent probe.
30. Repeat **step 28**.
31. Dehydrate slides in an ethanol series (70%, 90%, 100%) as described in **step 14** above, and air dry. Be sure to protect the slides from direct light.
32. Add 15 μl of Vectashield containing DAPI, and cover with a clean 24 × 50 mm cover glass. Remove any air-pockets under the cover glass. Store slides at 4 °C.

3.9 Fluorescent Imaging

1. Bring FISH-labeled slides to room temperature in a dark environment, and continue to protect the slides from strong light through the rest of the procedure. It is also a good idea to warm only as many slides as one can image in one session to avoid multiple warming cycles that can decrease the quality of the preparations.
2. Using the same fluorescent microscope and stage coordinates previously recorded from the immunolabeling step, find SC spreads and evaluate FISH labeling for each spread (*see* **Note 43**).
3. If an SC spread has not been damaged and is clearly labeled with the FISH probe, capture the corresponding DAPI and TRITC signals using a 100× objective and the appropriate filters (Figs. 1B–D and 2A, D–E). Save the images as TIF files.

3.10 Fluorescent Image Processing

1. Open all of the immunolabeled TIF images of the same SC spread using Photoshop or some similar computer program. Because all of the images have been taken with the same microscope using the same settings, assembling the images is

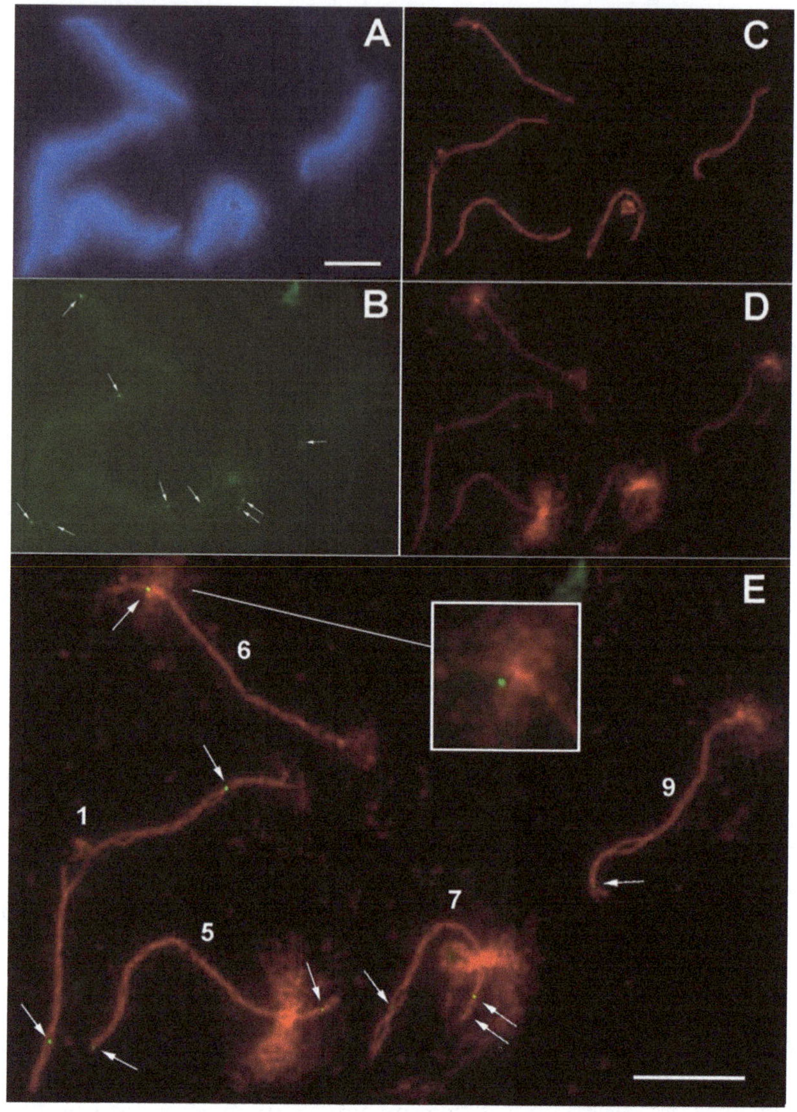

Fig. 2 A portion of an SC spread from KYS maize in which the five SCs with knobs were fortuitously located together and shown (**A**) stained with DAPI, (**B**) immunolabeled with anti-MLH1 (green foci marked with arrows) to show the location of interference-sensitive RNs, (**C**) immunolabeled with anti-AFD1 (red lines) to show SCs, and (**D**) labeled by FISH with the 180 kb knob probe (red). In our experience, some of the AFD1 immunofluorescence on SCs is retained after the FISH procedure (notice faint red lines in **D** in same positions as those shown in **C**), and this remaining SC label can be used to align the FISH and immunolabeled images. (**E**) Larger image of the same group of SCs showing MLH1 (from **B**) and FISH signals (from **D**) superimposed on SCs (from **C**). Now the relative positions of knobs and MLH1 foci on the same SCs are clear. Each SC has at least one MLH1 focus, but most MLH1 foci are not localized within knob FISH signals except for the MLH1 focus on SC6. In the 2× magnified inset image, the MLH1 signal on SC 6 has been moved 20 pixels down with respect to the FISH signal to show the edges of the FISH label more clearly. The MLH1 signal overlaps with the edge of the knob, showing that crossing over can occur within heterochromatic knobs of maize. The bar in (**A**) equals 10 μm for (**A–D**) and the bar in (**E**) equals 10 μm

relatively easy, but care is required to get the images properly aligned. Note that when we refer to immunolabeled images, we are specifically referring to those labeled with anti-MLH1 and anti-AFD1 and appropriate secondary antibodies. Although secondary antibodies are also used to visualize FISH probes, we refer to those as FISH images. We use Photoshop CS6 in our lab, so all of the following instructions are based on that program.

2. Select the entire TRITC (red) SC image (Ctrl + A), copy (Ctrl + C), and close the window (Ctrl + W). Select the DAPI (blue) image, and paste the red image into it (Ctrl + V). Repeat the process for the FITC (green) MLH1 labeled image, so that now the original DAPI file contains the following layers: background (DAPI image), layer 1 (red SC image) and layer 2 (green MLH1 image). If desired, each layer can be renamed to reflect the labeling type.
3. Save the file with three layers as a Photoshop (PSD) file. The original DAPI file can now be closed.
4. Add a new (empty) layer that will be used to help align the immunolabeled and FISH images.
5. Link all of the fluorescence layers and the alignment layer by highlighting each layer (hold Control while clicking on each layer in the "Layers" window) and then click the chain link icon at the bottom of the "Layers" window. Because all of these images were captured at the same time, they are all aligned properly.
6. Open the (red) FISH image. In practice, we have found that a faint red image of the previous AFD1 labeling is retained in addition to the new red FISH signal. If so, this can be very helpful in aligning the FISH image with the immunolabeled images. If not, it may be necessary to open the FISH DAPI image and the immunofluorescent DAPI image and try to align the two DAPI images. In either case, it can be helpful to use the pen tool on the alignment layer to make a dotted line that follows two or three SCs from the AFD1 (SC) layer that is visible (but not active). This alignment layer can help to align the immunolabeled and FISH images more easily.
7. Select, copy and paste the FISH image into the immunolabeled set of images (follow **step 2** above).
8. Select all layers except the DAPI background layer, and change the setting from "Normal" to "Screen" on the top left of the Layers menu. This allows each layer to be seen as a composite on top of the DAPI image. It will probably be clear that the SCs of the FISH image are not exactly aligned with the SCs from the original AFD1 label. While the orientation may not be exactly correct, the magnifications are identical.

9. Shift (and, if necessary, rotate) the FISH image until the SC images in the FISH and AFD1 layers line up properly. It is often useful to increase magnification of the image to ensure proper alignment, and the marks on the alignment layer can help align the two groups of images. It may be helpful to change contrast or brightness of the FISH and/or AFD1 signals to facilitate their alignment. It may also be helpful to turn off the other layers while aligning these two images. Since the AFD1 immunolabeled SC images are linked to both the DAPI and MLH1 layers, there is no danger of mistakenly moving the AFD1 image relative to the other two.
10. Crop the image to eliminate excess nonuseful space, but retain a border space around the SC spread for later annotation (if desired).
11. Change the resolution of the file from 72 DPI to 150 DPI by selecting "Image" from the top drop-down menu, then "Image Size," then "Resolution." Other resolutions can also be selected, but this is the maximum resolution that can be utilized by the MicroMeasure 3.0 program. Alternatively, the file can be left at its original resolution.
12. Save the completed file with multiple layers as a Photoshop (PSD) file (Figs. 1 and 2).

4 Notes

1. Kits with alternative DNA labels such as biotin or dinitrophenol (DNP) can also be used, but then appropriate antibodies will be needed.
2. These needles are modified stock dissecting needles with wooden or plastic handles that are available from most biological supply houses. Usually the steel in dissecting needles is too soft to hold a sharp edge, so they need to be hardened before putting the final point or edge on them. Sharp pointed dissecting needles can be prepared by rolling and rubbing the end of a needle at a narrow angle (~5–10°) in a drop of light machine oil (*e.g.*, 3-IN-ONE oil) on a coarse sharpening stone until the end is in the form of a gradually beveled point. Then harden the steel by heating the distal centimeter of the needle red hot in a Bunsen burner or an alcohol lamp and then immediately cooling the needle by plunging it in water. To make a sharp pointed tip, rub/roll the end of the needle at 15–20° in oil on a fine sharpening stone. The finished point should look sharp when viewed under a dissecting microscope at 10× magnification.

To make a chisel-shaped dissecting needle, cut the beveled end off of a dissecting needle with wire cutting pliers. Rub the cut end perpendicularly to the long axis of the needle in oil (*e.g.*, 3-IN-ONE oil) on a rough sharpening stone to leave a flat tip. Then rub one side and then the other side of the tip at 10–15° to form bevels that come to a straight edge at the end of the needle where the flat end used to be. The edge of the end of the needle is perpendicular to the long axis of the needle. Flame-harden the end of the needle as described above, and then use a fine sharpening stone and oil to smooth the bevels and finish the flat edge, which should look sharp when viewed on its side at 10× under a dissecting microscope.

To make a microscalpel, remove any oil from a new double-edge razor blade, (*e.g.*, Gillette) by wiping it with an organic solvent such as acetone. Break the blade in half lengthwise with your fingers. The blades are brittle and break easily, although be careful not to cut yourself. Then use flat-nose pliers (glass breaking pliers are best), to break off rhombus-shaped segments from half a blade with a sharp side 2–3 mm in length. Using a single edge razor blade split the end of a wooden applicator stick lengthwise for 4–5 mm, and insert a blade fragment with the short, sharp edge parallel to the long axis of the stick and the other short side buried in the stick. Glue the blade on the stick with epoxy or some other glue that binds to metal and wood. Microscalpels can be used repeatedly if sharpened on a very fine sharpening stone. For this put a drop of light machine oil (*e.g.*, 3-IN-ONE oil) on a very fine sharpening stone. Then angle the blade at ~20° from the surface of the stone and draw it along the surface of the stone (not into the stone) several times on both side of the blade. To remove oil and debris, wipe the blade carefully with a paper towel dampened with a little acetone or ethanol followed by water. If oil is not available sharpening can be performed in drop of water or even dry, but oil is the best. Alternatively, a scalpel with a #11 blade (Feather is our preferred brand) can be used.

3. A top or a bottom of a plastic petri dish makes a good platform for a dissecting microscope stage because the plastic is clear for substage illumination, hard enough to cut against, but soft enough not to dull dissecting needles and microscalpels. We break off the sides of the petri dish to make a flat platform that is easier to dissect on.

4. 2% aceto-orcein is prepared by mixing 2 g of powdered orcein with 98 ml of aqueous 45% acetic acid. Boil this mixture for an hour, preferably in a water-cooled reflux condenser. If a reflux condenser is not available, then boil the solution in a hood and continually add enough 45% acetic acid to maintain the volume. After the boiling time is completed, cool the dye solution

to room temperature, filter the solution through medium-grade filter paper, and then filter the solution again through the same filter paper. The dye solution can be kept indefinitely in a closed dropper bottle at room temperature. Over time precipitate inevitably forms in the solution, but this can be removed by refiltering the solution as described above.

5. Glow discharge apparatuses are commercially available (*Electron Microscopy Sciences*, Ladd Research, *Ted Pella, Inc.*, *Quorum Technologies*). Alternatively, a functional glow discharge apparatus can be assembled from parts [13].
6. We use Drummond disposable 1 × 100 mm glass micropipettes with aspirator tube assembly designed for collecting blood samples (see Fisher and VWR as distributors). To draw out the micropipette tip, hold either end of a 100 mm pipette, and heat the middle of the micropipette in a flame from a small butane torch until the glass softens, that is, begins to melt. Roll the pipette with your fingers to heat the center evenly. Remove from the flame, then immediately pull both ends of the pipette to form a thin tube of glass between the end segments. Bend and break the thin tube to form two potential micropipettes. Pull a dozen or more micropipettes. If a tip is smooth and has an outside diameter near 0.15 mm, keep it as a useable micropipette. If a pipette has a diameter larger than 0.15 mm, discard it. If a tip has a diameter smaller than 0.15 mm, break off a bit of the tip to obtain a larger diameter tip near 0.15 mm in diameter. If the tip is jagged, it can be smoothed by inserting the micropipette into the aspirator tube assembly, drawing water into the micropipette (to stabilize the tip during polishing and to help prevent grindings from stopping up the tip) and gently stroking (in one direction) the tip in a drop of water on a fine sharpening stone. A beveled rather than a flat tip obtained in this way is acceptable as long as the tip is smooth and 0.15 mm in diameter. To mark the point on the bore at which 0.5 μl of digestion medium and protoplasts have been drawn up, place 0.5 μl of distilled water on a piece of Parafilm using a mechanical micropipette. Then pull up the droplet with the micropipette without taking up any air. Use a fine tip black permanent marker to mark the meniscus of the water on the micropipette. Blow the water out. Micropipettes are siliconized by drawing a solution of 5% methylchlorosilane in chloroform into the micropipette and expelling it. The pipette should be rinsed twice with water after being siliconized. We resiliconize used pipettes before each experiment. Siliconizing micropipettes helps to keep cells from sticking to the sides of the micropipette and being damaged.
7. Digestion medium should be mixed immediately before use. Stock solutions of 2.8 mM KH_2PO_4, 0.5 mM PIPES (acid, pH

not adjusted), and 2.5 mM $CaCl_2$ are prepared using distilled water and stored in 1 ml aliquots in 1.5 ml microcentrifuge tubes at −20 °C. Aqueous 1% potassium dextran sulfate does not store well even frozen, so we store 0.01 g aliquots of potassium dextran sulfate powder in closed 1.5 ml microfuge tubes at −20 °C and add 1 ml of distilled water to make a 1% solution immediately before the digestion medium is prepared. However, before the 1% solution of potassium dextran sulfate is added to the digestion medium, set aside five μl to use in the bursting solution (*see* **Note 10**). This small reduction of the volume of 1% potassium dextran sulfate does not adversely affect the digestion medium. To prepare the digestion medium, we mix the 1 ml aliquots of 2.8 mM KH_2PO_4, 0.5 mM PIPES, 2.5 mM $CaCl_2$ and 1% potassium dextran sulfate (minus the 5 μl reserved for the bursting solution) with an additional 1 ml of distilled water, then add 0.64 g mannitol and 0.1 g PVP. Mix using a plastic stirrer until the sugar and PVP are dissolved, then adjust the pH to 4.1 using 0.1 N KOH and/or 0.1 N HCl. We have found that it is essential to avoid sodium ions in the digestion medium in order to preserve the viability of meiocytes, so it is necessary to adjust the pH with KOH, not NaOH, and not to substitute any sodium salts in the digestion medium. Furthermore, the water used in the digestion media should be distilled, not deionized for optimum meiocyte viability. We use potassium dextran sulfate from Calbiochem, but Calbiochem no longer offers this product. We have used Nacalai as an alternative supplier with some success for spreading SCs, but it seems stronger than the Calbiochem product and has interfered with MLH1 labeling on tomato SC spreads. A reduced concentration of the Nacalai product (*e.g.*, half or three-fourths) may work better in both the digestion and bursting media. Carbosynth and Karlan Research Products Corporation also offer potassium dextran sulfate, but we have not tried their products.

8. To desalt cytohelicase, make a 0.5% aqueous mixture (using deionized or distilled water), and pass it through a column packed with Sephadex G25. The enzyme is light-sensitive, so desalt in subdued light, and do not use a fraction collector with a UV-monitor that will inactivate the enzyme. If the cytohelicase is too concentrated, it will come down the column as finger-like projections rather than as a solid front. This can be solved by further diluting the cytohelicase mixture. Pectinase is desalted the same way. Freeze-dry the tan-colored salt-free eluants. Desalting gets rid of sodium and possibly other undesirable ions that can injure primary microsporocytes.

9. In a hood add 4 g of paraformaldehyde powder and 1 ml of 1 N NaOH to 95 ml of deionized or distilled water in a beaker on a stirring hot plate. The temperature of the solution should not rise above 60 °C. Once the solution clears, cool it to room temperature, and then add 1 ml of 0.05 M sodium borate to buffer the solution. With stirring in a hood, titrate the solution to a pH of 8.2–8.5 with 1 N HCl (and 1 N KOH or NaOH, if needed). This 4% solution of formaldehyde can be aliquoted into small bottles and stored indefinitely at −20 °C. The formaldehyde solution can be thawed and refrozen a few times without discernable loss of activity.
10. Bursting medium is made by mixing three solutions: 0.05% IGEPAL® CA-630 (Sigma, a nonionic, nondenaturing detergent previously known as Nonidet P-40), 0.1% potassium dextran sulfate, and 4% paraformaldehyde. Make 20 ml of 0.05% IGEPAL® CA-630 using distilled water and store 500 μl aliquots in 650 μl microcentrifuge tubes at −20 °C. While preparing the digestion medium, thaw one of these 500 μl aliquots. For the potassium dextran sulfate, mix the reserved 5 μl of the 1% solution made during the preparation of the digestion medium (*see* **Note 7**) with 45 μl of distilled water (=1/10 dilution). Transfer 5 μl of this diluted potassium dextran sulfate solution into the tube holding the thawed 500 μl of 0.05% IGEPAL. (The remaining 45 μl of diluted potassium dextran sulfate can be discarded.) Add 41 μl of 4% formaldehyde (from stock, *see* **Note 9**) to the IGEPAL–potassium dextran sulfate mixture and mix by inversion. Keep the bursting medium on ice before and during use in the spreading procedure.
11. Tetko nylon screen worked for this application, while nylon cloth purchased from a fabric store did not. The mesh size is not particularly important (*e.g.*, 100–400 μm). See [14] for mechanism of silver staining.
12. As long as the silver nitrate solution remains clear, it can be stored indefinitely in a closed bottle at room temperature. The silver solution will stain clothes and skin black, so avoid contact.
13. We affinity-purify antibodies against the recombinant protein that was used for immunization of the animals. To affinity-purify rabbit antibodies against MLH1, express and purify the recombinant N-terminal segment of the protein as described by [4] and covalently link the purified recombinant peptide to a resin using a kit (AminoLink Plus Immobilization kit from ThermoFisher Scientific). Affinity-purify the antibodies using either high salt or low pH according to instructions provided with Gentle Ag/Ab Binding and Elution Buffers

(ThermoFisher Scientific). In our experience, either method of elution is useful, although some antibodies may be easier to use and/or bind more specifically with less background in cytological preparation with one elution buffer compared to another. For rabbit MLH1 antibodies, we obtained the best results using low pH elution.

14. Excessive drying of DNA pellets can result in pellets that are difficult to get back in aqueous solution.
15. During anther dissection and SC spreading, small amounts of distilled water may need to be added periodically to keep up the liquid volume in the depression slide and to keep the digestion medium from getting too concentrated. This can be detected by white precipitate of mannitol on the edges of the depression. If this occurs, a drop of water can be added and mixed in.
16. After a tassel is removed from a plant, all steps from collecting anthers to bursting protoplasts should take place in a room with the temperature adjusted to 20 °C (68 °F) because maize protoplasts die at higher temperatures. We tried alternatives to cooling the room such as placing cool water or ice under the anthers during digestion, but all were unsuccessful at producing live cells and good SC spreads. Possibly the interior of a maize stalk is normally maintained at ~20 °C.
17. It is important to handle the anthers gently to avoid damaging and/or killing the microsporocytes. Excessive pressure on anthers damages cells by pushing cytoplasm and nucleoplasm through cytoplasmic channels between PMCs. This causes the artifact called cytomixis [15, 16] and, in more extreme cases, pressure will burst the cells. Anthers become dark at damage sites due to oxidation of phenolic compounds.
18. Use a small carpenter's level to see if the table top is flat. If not, the drop of protoplast suspension will not spread evenly and run downhill (more in one direction). This is undesirable. If the table top is not flat, use some modeling clay and make two pads across which a microscope slide is laid. Then place the level on the slide to see which side is higher or lower. Press the part of the slide that is too high until the slide is level. This leveled slide will act as a flat base to assure that the glow discharged slide it supports is level and drops of protoplast suspension spread evenly in all directions.
19. Some plastic pipette tips are very hydrophobic, and when a drop is expelled from the tip, the drop falls off the tip. We have found that yellow tips are more hydrophilic, and the expelled aqueous droplet remains attached to the end of the tip until touched off. This is desirable.
20. Allow at least 1 h drying time. We have had problems with SCs washing off the slide if less time is allowed to complete drying.

21. We have found that the effective time to use the maize protoplasts (even if the protoplasts still appear to be healthy) is about an hour after the 45 min digestion. After that, we have gotten few slides with usable SC spreads.
22. The slides can be silver stained immediately after air-drying, but there will be less background staining if the slides are stained the following day.
23. There will be 10 SCs in a complete set of SCs from maize. Each SC represents one bivalent. On good slides a mixture of complete SC sets and scattered SCs are observed. If only scattered SCs are visible, the protoplasts burst too early and may have been handled too roughly. There are several possible ways to solve this problem: Reduce the time and/or intensity of agitation before the bursting/protoplast mixture is placed on the slide. The time can be reduced between placing the bursting/protoplast mixture on the slide and nebulizing 4% formaldehyde on the slide in the hood. See if the diameter of the micropipette used to pick up the protoplasts is too small, that is, less than 0.13 mm in diameter, which can damage the protoplasts. Be sure the micropipette has been siliconized.

    If no SCs are observed, it is possible that PMCs at the wrong meiotic stage were used. SCs are only visible in zygotene, pachytene, and early diplotene. If neither SCs nor dark contracted nuclei are visible when you are sure pachytene microsporocytes were used (based on squash preparations and anther lengths), then you may have washed them off the slides. Be sure not to agitate the slides in the washes.

    If only darkly stained nuclei and/or SCs in tight uninterpretable bunches are observed, the primary microsporocytes may have been dead when they were added to the bursting medium. This can be due to the cells having been roughly handled sometime from harvesting tassels to cutting anthers in half and pressing protoplasts out of anther halves. Check to see that protoplasts in the digestion medium are spherical and clear (alive). If the protoplasts are alive, too much digestion medium may have been drawn into the micropipette with the protoplasts, that is, more than a total volume of 0.5 μl.
24. Removing the graphite helps reduce background under fluorescence.
25. Take care to prevent drying the slides between steps of the immunolabeling procedure.
26. This step helps quench any reactive aldehydes remaining after the fixation step and reduces background.
27. This step helps to block nonspecific binding of antibodies and reduces background.

28. The optimal concentration for each antibody must be determined empirically. We used anti-AFD1 (from rat serum) diluted 1:500 and anti-MLH1 (affinity-purified from rabbit serum) diluted 1:50. We usually dilute secondary antibodies 1:500.
29. Centrifugation helps to remove any large protein aggregates that may cause background.
30. To avoid bubbles under the plastic coverslip, use forceps to pick up the right side of the long axis of the coverslip, and carefully place the plastic coverslip on the drop of antibody suspension in the middle of the clear part of the microscope slide (with the frosted end to the left). Let the solution spread out a bit before gently lowering the coverslip from left to right, allowing any bubbles to pass to the edge of the slip and out.
31. We observe less fluorescent background when the slides are incubated upside down.
32. To remove the plastic coverslip from the slide, hold each slide upside down over a sink and add TBST to the overhanging edge of the plastic coverslip until the coverslip floats free and moves easily. Then, carefully remove the coverslip by grasping the coverslip with a pair of fine forceps and peeling the coverslip away from the slide rather like opening an upside down can of sardines.
33. As a precaution to avoid bleaching the fluorescent probes, we usually turn off the overhead lights during this step and use indirect lighting from lights mounted under cabinets in the same room and from the lighted room next door.
34. One way to avoid bubbles when adding a cover glass is to lay the slide flat with the frosting up and to the left. Holding a cover glass on the right side with fine forceps in the right hand, tilt the cover glass to touch the clear glass near the frosting and block leftward movement of the cover glass with fingers of the left hand. Release pressure on the forceps so the right side of the cover glass is supported by the lower tine of the forceps only. Slowly lower the right side of the cover glass to make contact the liquid on the slide, and slowly withdraw the forceps to let the cover glass rest on top of the liquid.
35. Scanning and imaging a slide generally takes an hour or more, even for an average slide, so leave the rest of the slides in the refrigerator to minimize any possible degradation in the fluorescent signal by unnecessary exposure to light and multiple warming and cooling cycles.
36. Be careful not to get oil on the edges of the cover glass because any oil that gets under the cover glass could ruin the slide.

37. We used a rotator with gentle agitation (40–50 rpm) to increase the effectiveness of washing, but the procedure will work without the rotator.
38. Slides can be prepared up to this point and stored at 4 °C or −20 °C before doing the hybridization steps. We have not determined the maximum amount of time the slides can be stored, but we have waited for a few days to a week with satisfactory results.
39. The relatively high concentration of herring sperm DNA is included in the hybridization mixture to protect knob DNA by blocking any chemical or surface feature that indiscriminately binds DNA.
40. Probes can be kept on ice for longer periods of time.
41. This time and temperature is critical, so efforts should be made to adhere to these values as closely as possible. We have found that the block can be as hot as 82 °C without adversely affecting the SC spreads. It is also important to limit the number of slides added to the block at any one time to prevent their excessively cooling the block.
42. This combination of temperature, SSC, and formamide represents 80% hybridization stringency [17].
43. Be sure to seat the slide properly in the stage clips to enable each SC spread to be found easily.

Acknowledgments

We thank the National Science Foundation for supporting our work and Associate Professor Song-Bin Chang in the Department of Life Sciences at National Cheng Kung University, Taiwan for selecting the PCR priming sequences for knob DNA amplification.

References

1. Stack S, Shearer LA, Lohmiller L, Anderson LK (2017) Meiotic crossing over in maize knob heterochromatin. Genetics 205:1101–1112
2. Carpenter ATC (1975) Electron microscopy of meiosis in *Drosophila melanogaster* females: II: the recombination nodule—a recombination-associated structure at pachytene? Proc Natl Acad Sci U S A 72:3186–3189
3. Anderson LK, Stack SM (2005) Recombination nodules in plants. Cytogenet Genome Res 109:198–204
4. Lhuissier FGP, Offenberg HH, Wittich PE, Vischer NOE, Heyting C (2007) The mismatch repair protein MLH1 marks a subset of strongly interfering crossovers in tomato. Plant Cell 19:862–876
5. Moses M (1968) Synaptinemal complex. Annu Rev Genet 2:363–412
6. Golubovskaya IN, Hamant O, Timofejeva L, Wang CJR, Braun D, Meeley R, Cande WZ (2006) Alleles of afd1 dissect Rec8 functions during meiotic prophase I. J Cell Sci 119 (3306):3315
7. Mercier R, Mezard C, Jenczewski E, Macaisne N, Grelon M (2015) The molecular biology of meiosis in plants. Annu Rev Plant Biol 66:5.1–5.31

8. Charlesworth B, Langley CH, Stephan W (1986) The evolution of restricted recombination and the accumulation of repeated sequences. Genetics 112(947):962
9. Longley AE (1939) Knob positions on corn chromosomes. J Agric Res 59:475–490
10. Ananiev EV, Phillips RL, Rines HW (1998) Complex structure of the knob DNA on maize chromosome 9: Retrotransposon invasion into heterochromatin. Genetics 149:2025–2037
11. Harlow E, Lane D (1988) Antibodies a laboratory manual. Cold Spring Harbor Laboratory, Cold Spring Harbor
12. Chang MT, Neuffer MG (1994) Chromosomal behavior during microsporogenesis. In: Freeling M, Walbot V (eds) The Maize Handbook. Springer-Verlag, New York, pp 460–475
13. Stack SM, Anderson LK (2009) Electron microscopic immunogold localization of recombination-related proteins in spreads of synaptonemal complexes from tomato microsporocytes. In: Keeney S (ed) *Meiosis Volume 2.* Cytological Methods, Methods in Molecular Biology. Humana Press, Inc., Totowa, NJ, pp 147–169
14. Herickhoff L, Stack S, Robertson J (1992) Staining plant cells with silver. III. The mechanism. Biotech Histochem 67:171–182
15. Tarkowska J (1965) Experimental analysis of the mechanism of cytomixis. I. Cytomixis in vegetative tissues. Acta Soc Bot Pol 34:27–44
16. Tarkowska J (1966) Experimental analysis of the mechanism of cytomixis. II. Cytomixis in the pollen mother cells of lily: Lilium candidum L. Acta Soc Bot Pol 35:25–40
17. Schwarzacher T, Heslop-Harrison P (2000) Practical in situ Hybridization. BIOS Scientific Publishers Ltd., New York, pp 1–203

Chapter 9

Cytological Techniques to Study Cytomixis in Plant Male Meiosis

Sergey R. Mursalimov, Yuriy V. Sidorchuk, and Elena V. Deineko

Abstract

In this chapter we describe cytological techniques to study cytomixis, a process of nuclear migration between plant cells, in squashed plant male meiocytes of *Nicotiana tabacum* and *Secale cereale*. To perform immunostaining or fluorescence in situ hybridization (FISH) on meiotic cells involved in cytomixis common protocols are modified. During preparation of specimens for subsequent cytological analysis, it is necessary not only to make DNA and proteins accessible to DNA probes and antibodies, but also to preserve cell cytoplasm. There are also some important modifications in the protocols applied for meiocytes of different plant species. Here we describe protocols for immunostaining and FISH in rigid tobacco male meiocytes with dense cytoplasm and thick callose wall, that tolerate hard squashing, and in soft rye male meiocytes, that are easily damaged upon squashing, both to study cytomixis.

Key words Cytomixis, Tobacco, Rye, Meiosis, Immunolocalization

1 Introduction

Cytomixis is a unique process of nuclear migration between plant cells through special intercellular channels that are considerably larger than plasmodesmata. Such migration of the nuclei is most frequently found in plant male meiosis but has been also observed in plant tissues such as apical meristems [1] or tapetum cells [2] while never in plant female meiosis or in any type of animal cells. So far, cytomixis is documented in male meiosis of over 400 higher plant species belonging to 84 families. Among the angiosperms, the family Poaceae contains the highest number of documented species (82) in the monocots and the family Fabaceae (48) in the dicots. The vast majority of the instances of cytomixis are described in male meiosis of angiosperms. However, nuclear migration or cytomictic channel formation is also found in gymnosperms, quillworts, horsetails, and ferns [3]. Thus, cytomixis is a widespread phenomenon in higher plants and likely every researcher who studies plant male meiosis has faced it.

Mónica Pradillo and Stefan Heckmann (eds.), *Plant Meiosis: Methods and Protocols*, Methods in Molecular Biology, vol. 2061, https://doi.org/10.1007/978-1-4939-9818-0_9,

As a rule, cytomixis is found in meiotic prophase I. A nucleus can migrate between cells as a whole, resulting in binucleated and enucleated cell formation, or in parts, leading to micronuclei formation [4, 5]. All nuclear components including chromatin, nucleolus, and nuclear matrix migrate between cells enclosed within the nuclear membrane showing no signs of damage [6]. The nuclear migration frequency is highly variable (from 0% to 100% of meiocytes per anther) and typically increases in case of plant genome reorganization (e.g., polyploidization) or upon environmental stress(es) [7, 8].

A general cytological picture of cytomixis was described in detail for various plants. Currently the reasons and consequences of such nuclear migration are discussed (for review *see* [9]). Obviously, the migration of nuclei has big influence on involved cells. Cells with surplus/missing chromatin can either be eliminated from meiotic divisions or continue meiosis in altered form. The future fate of such cells is still disputable. The most fascinating possible result of cytomixis is unreduced and aneuploid pollen formation [10, 11]. This hypothesis is not supported by solid proofs yet but we believe, sooner or later, it will.

There is no difference between cytological analysis of plant male meiotic cells involved in cytomixis and noncytomictic ("normal") ones embedded in, for example, polyethylene glycol (PEG) or epoxy resin using tissue sections. However, protocols for squashed male meiotic cell preparations for immunostaining or FISH are modified to meet certain requirements:

- *Cells must have well-preserved cytoplasm* to provide an adequate cytological picture, that is, distinct, easily visible cell boundaries to prove nuclei migration between cells. The boundaries also enable to reveal that micronuclei formed after cytomixis do not locate in intercellular space but instead in a recipient cell cytoplasm. In addition, well preserved cytoplasm indicates absence of mechanical cell damage during specimen preparation.
- *Cells must be sufficiently squashed to form monolayers while not disrupting cell-to-cell contacts.* Cell monolayer formation is likely the most challenging part of squashed cell preparation protocols. If cells do not form a monolayer on the slide they are not suitable for most types of light microscopic analysis except confocal microscopy. Cell monolayers are obtained in different ways, however most commonly cells covered with a coverslip are mechanically squashed. Typically the more is pressed on a coverslip the better cells adhere to the slide. Unlike tobacco male meiocytes with dense cytoplasm and thick callose wall, that tolerate hard squashing, in cereals such as rye similar hard mechanical treatments cannot be applied due to their soft meiocytes.

- *DNA and proteins must be accessible to DNA probes and antibodies.* Before immunostaining meiotic cells are squashed, frozen and treated with surfactant to make their DNA and proteins accessible. Before FISH cells are additionally treated with RNase and proteinase.

Note that there is significant difference between staining protocols of meiocytes of different plant species. The meiocytes having thick callose wall and dense cytoplasm are squashed rather hard and treated with enzymes for longer time periods or with a higher concentration of the enzyme mixture, while soft meiocytes are fixed longer or with higher fixative concentrations and cannot be squashed hard.

Here we describe protocols for immunostaining and FISH of tobacco and rye meiocytes based on squashed preparations. Tobacco meiocytes are rigid and can tolerate a significant mechanical impact during treatment. Rye meiocytes, like meiocytes of other cereals, are rather soft and do not tolerate hard squashing. If the soft meiocytes are not sufficiently fixed or squashed too much they are separated losing cell-to-cell contacts (Fig. 1c) and/or their cytoplasm is lost (Fig. 1d). Such cells cannot be used for subsequent analysis. The immunostaining and FISH protocols modified for both species presented here allow acquiring pictures of the cells involved in cytomixis that meet all requirements mentioned above (Fig. 1a, b, e).

2 Materials

2.1 Plant Materials and Cultivation

Tobacco (*Nicotiana tabacum* L. cv. Petit Havana SR1, $2n = 4x = 48$) and rye (*Secale cereale* L. cv. Onokhoiskaya, $2n = 14$) plants are grown in a hydroponic greenhouse with a photoperiod of 16/8 h (day/night) at a temperature of 22/18 °C (day/night).

2.2 Fixation of Anthers and Preparation of Slides

1. Acetocarmine (2%): dissolve 10 g of carmine in 500 ml of 45% glacial acetic acid and boil for 6–8 h in a flask with an attached reflux column. Filter through filter paper when the solution is still warm. Store in a dark bottle at 4 °C.
2. Formaldehyde fixative (4% or 8% freshly prepared): formaldehyde solution in 1× PBS, pH 7.4. For 4% take 1 ml and for 8% take 2 ml of 37% formaldehyde and bring the volume up to 9.25 ml with 1× PBS (*see* **Notes 1** and **2**).
3. Phosphate-buffered saline (PBS; 10×): 1.37 M NaCl, 27 mM KCl, 0.1 M Na_2HPO_4, 18 mM KH_2PO_4, pH 7.4. Dissolve 80 g of NaCl, 2 g of KCl, 14.4 g of Na_2HPO_4, and 2.4 g of KH_2PO_4 in 800 ml of dH_2O. Mix, adjust pH with HCl, and make up to 1 L with dH_2O. Autoclave and store at 4 °C.

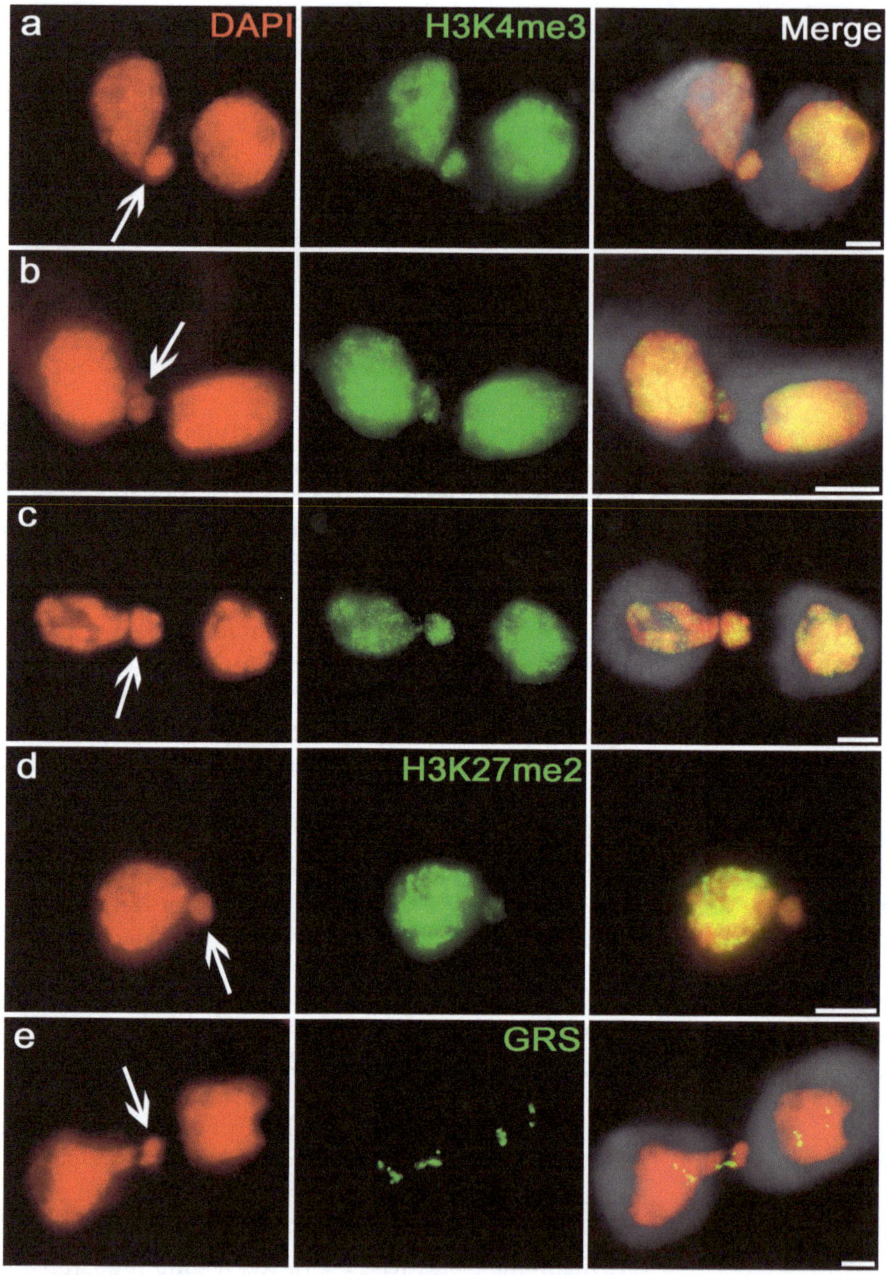

Fig. 1 Immunostaining (**a–d**) and FISH (**e**) in plant male meiocytes involved in cytomixis. (**a**, **e**) tobacco meiocytes and (**b–d**) rye meiocytes. In (**a**, **b**, **e**) cell structure is preserved and meiocytes are suitable for analysis. In (**c**) meiocytes lost their contact due to squashing and in (**d**) meiocyte lost its contacts and cytoplasm. Meiocytes in (**c**, **d**) are not suitable for analysis. The nuclei migrate from left to right. Arrows denote the migrating part of nuclei. Cytoplasm is grey in merged figures. Bar represents 5 μm

4. Enzyme mixture: 20% cellulose, 20% pectinase in 1× PBS, pH 7.4. Dissolve 2 g of cellulase and 2 g of pectinase in 10 ml of 1× PBS. Store 1 ml aliquots at −20 °C (*see* **Note 3**).
5. Poly-L-lysine slides.
6. Coverslips 24 × 40 mm.
7. Glass embryo dish.

2.3 Immunostaining

1. Wash buffer: 0.1% Tween 20 in 1× PBS, pH 7.4. Add 1 ml of Tween 20 to 1 L of 1× PBS and mix well. Store at 4 °C.
2. Blocking buffer: 3% bovine serum albumin (BSA) in wash buffer. Dissolve 0.3 g of BSA in 10 ml of wash buffer. Store at −20 °C.
3. Primary antibody solution: dilute primary antibody to detect a specific antigen in the sample (e.g., rabbit anti-H3K4me3; Agrisera, AS16 3190) 1:150 in 1× PBS containing 1% BSA and 0.1% Tween 20.
4. Secondary antibody solution: dilute fluorescently labeled secondary antibody to detect the primary antibody (e.g., anti-rabbit IgG-Atto 594; Sigma-Aldrich, 77,671) 1:150 in 1× PBS containing 1% BSA and 0.1% Tween 20.
5. Mounting medium: 0.5–1 μg/ml DAPI (4,6-diamidino-2-phenylindole, dihydrochloride) in antifade mounting. Store at −20 °C.
6. Moist chamber: a hermetic container (e.g., plastic box or 11 cm petri dish) with wet paper on the bottom and two rails in order to keep microscopic slides horizontally.
7. Dissection needles, fine tweezers, and a razor blade.
8. Filter paper.
9. Laboratory film Parafilm "M."
10. Incubator at 37 °C.
11. Shaking incubator for 1.5 ml tubes at 37 °C.
12. Horizontal or rotating shaker for 50 ml tubes.

2.4 Fluorescence In Situ Hybridization

1. DNA probe labeled with biotin- or digoxigenin-dUTP, 5–50 ng/μl (*see* **Note 4**).
2. Blocking DNA: autoclaved or sonicated salmon sperm DNA, 100-fold higher concentrated than the probe used (*see* **Note 5**).
3. Saline-sodium citrate buffer (SSC; 20×): 3 M NaCl, 300 mM $Na_3C_6H_5O_7$, pH 7.0. Dissolve 175.3 g of NaCl and 88.2 g of $Na_3C_6H_5O_7 \times 2H_2O$ in 800 ml of dH_2O. Mix, adjust pH with HCl, and make up to 1 L with dH_2O. Autoclave and store at 4 °C (*see* **Note 6**).

4. Proteinase buffer: 20 mM Tris–HCl, 2 mM $CaCl_2$. Add 20 μl of 1 M Tris–HCl (pH 7.5) and 2 μl of 1 M $CaCl_2$ to 978 μl of dH_2O.
5. $CaCl_2$ solution: 1 M $CaCl_2$ in dH_2O. Dissolve 1.47 g of $CaCl_2$ in 10 ml of dH_2O, autoclave and store aliquots at −20 °C.
6. Tris–HCl: 1 M Tris–HCl in dH_2O, pH 7.5. Dissolve 121.1 g of Tris base in 800 ml of dH_2O. Adjust pH to 7.5 by slowly adding approximately 6–7 ml of concentrated HCl. Bring final volume to 1 L with dH_2O. Autoclave and store at 4 °C.
7. Proteinase K solution: 1 mg/ml Proteinase K in proteinase buffer. Add 10 μl of commercial Proteinase K (20 mg/ml) to 190 μl of proteinase buffer.
8. FISH formaldehyde fixative (freshly prepared): 4% formaldehyde in 2× SSC, pH 7.0. Take 1 ml of 37% formaldehyde and bring the volume up to 9.25 ml with 2× SSC.
9. RNase A (1 mg/ml): add 10 μl of commercial RNase A (10 mg/ml) to 90 μl of 2× SSC.
10. Hybridization buffer (1.25×): mix in 1.5 ml plastic tube 500 μl of formamide, 200 μl of 50% (w/v) dextran sulfate, and 100 μl of 20× SSC. Heat the mixture at 90 °C for 3 min and mix well. Store at −20 °C.
11. FISH wash buffer: 0.1% Tween 20 in 4× SSC, pH 7.0. Add 1 ml Tween 20 to 1 L of 4× SSC and mix well. Store at 4 °C.
12. Fluorescently labeled avidin or anti-digoxigenin antibody solution: Fluorescently labeled avidin or anti-digoxigenin antibody (e.g., avidin-FITC; Sigma, A2901; or anti-digoxigenin-rhodamine; Roche, 11,207,750,910) 1:150 in 4× SSC containing 1% BSA and 0.1% Tween 20.
13. Alcohol series in 50 ml plastic tubes (70%, 90%, and 96%). Store at −20 °C.
14. Formamide solution (prepared immediately before use): 70% formamide in 2× SSC. Add 30 μl of 2× SSC to 70 μl of formamide.
15. Hot plate at 58–68 °C.
16. Coplin jars (or 50 ml plastic tubes).

3 Methods

3.1 Fixation of Tobacco Anthers and Preparation of Slides

1. Dissect 5–10 anthers from collected flowers buds of appropriate size (*see* **Note 7**). Cut the anthers in two parts with a razor blade to aid fixative access to meiocytes and place them in 5 ml of 4% formaldehyde fixative for 30 min on ice to inactivate all

enzymatic activity in the cells and thus to preserve DNA and proteins.

2. Wash anthers in 5 ml of 1× PBS 3× 15 min at RT to remove the fixative.
3. Squash anthers into suspension in 100 μl of 1× PBS in an embryo dish with a blunt end of a dissection needle. Use a 200 μl pipette tip to transfer the meiocyte suspension to a 1.5 ml tube containing 200 μl of enzyme mixture. Mix with a pipette and place the tube to a shaking incubator at 37 °C for 1 h at 700 rpm to digest callose cell wall.
4. Add 1 ml of 1× PBS to the tube and mix it with a pipette. Centrifuge the meiocytes at 6000 × *g* for 5 min to pellet the cells. Discard the supernatant. Resuspend the pellet by pipetting in 1 ml of 1× PBS and centrifuge again at 6000 × *g* for 5 min to remove the enzyme mixture.
5. Resuspend the pellet by pipetting in 20 μl of 1× PBS and transfer to a poly-L-lysine slide. Put a 24 × 40 mm coverslip on top, cover the slide with a piece of filter paper and press hard with your thumbs to adhere cells to the slide and to squash them. Do not let the coverslip slide!
6. Freeze the slides in liquid nitrogen for 10 s and remove the coverslip with a razor blade. Let the slide warm up to RT and immediately proceed with the next step. Do not let it dry.

3.2 Immunostaining of Tobacco Meiocytes

1. Add 40 μl of blocking buffer to the slide and cover with a 24 × 40 mm coverslip. Incubate for 1 h at 37 °C in a prewarmed moisture chamber to block nonspecific binding sites.
2. Remove the coverslip by gentle sliding (*see* **Note 8**) and hold the slide vertically with its bottom touching filter paper to remove most of the blocking buffer.
3. Add 40 μl of primary antibody solution to the slide and cover with a 24 × 40 mm coverslip. Incubate for 1 h at 37 °C in a prewarmed moisture chamber to detect specific antigen.
4. Remove the coverslip by gentle sliding (*see* **Note 8**), transfer the slide in a 50 ml plastic tube with 45 ml of wash buffer and wash it 3× 10 min at RT on a shaker (100–130 rpm) to remove unbound primary antibodies.
5. Add 40 μl of secondary antibody solution to the slide and cover with a 24 × 40 mm coverslip. Incubate for 1 h at 37 °C in a prewarmed moisture chamber in the dark to detect the target-bound primary antibodies.
6. Remove the coverslip and wash the slide in the dark as in **step 4** to remove unbound secondary antibodies.

7. Add 20 μl of mounting medium and cover with a 24 × 40 mm coverslip to stain DNA and protect fluorescently labeled molecules on the slide from fading. Cover the slide with a piece of filter paper and press hard with your thumbs. Do not let the coverslip slide!
8. Seal edges of the coverslip with nail polish to prevent drying and store the slide at 4 °C in the dark (*see* **Note 9**).

3.3 Fixation of Rye Anthers and Preparation of Slides

1. Dissect 5–10 anthers from collected flowers buds of appropriate size (*see* **Note 7**) and place them in 5 ml of 8% formaldehyde fixative for 60 min on ice (*see* **Note 10**) to inactivate all enzymatic activity in the cells and thus to preserve DNA and proteins.
2. Wash anthers in 5 ml of 1× PBS 3× 15 min at RT to remove the fixative.
3. Transfer anthers in a glass embryo dish with 200 μl of enzyme mixture. Do not use plastic tubes! Place the embryo dish in a prewarmed moisture chamber and incubate for 30 min at 37 °C to soften the anthers for the subsequent squashing.
4. Gently remove the enzyme mixture with a pipette and add 500 μl of deionized water for 5 min at RT. Repeat twice.
5. Transfer the anthers to a poly-L-lysine slide with tweezers. Gently squash them with a dissection needle in 20 μl of deionized water and spread specimen over the slide surface (*see* **Note 11**).
6. Dry for 5–10 min at RT for better adhesion and transfer to −20 °C for at least 10 min to enable ice crystal formation that will improve accessibility of target DNA and proteins.

3.4 Immunostaining of Rye Meiocytes

1. Add 50 μl of blocking buffer to the slide and cover with a 24 × 40 mm piece of Parafilm. Do not press! Incubate for 1 h at 37 °C in a prewarmed moisture chamber to block nonspecific binding sites.
2. Carefully remove Parafilm with tweezers. Hold the slide vertically with its bottom touching filter paper to remove most of the blocking buffer. Do not dry it completely!
3. Add 50 μl of primary antibody solution to the slide and cover with a 24 × 40 mm piece of Parafilm. Do not press! Incubate for 1 h at 37 °C in a prewarmed moisture chamber to detect specific antigen.
4. Carefully remove Parafilm with tweezers. Place the slide either into a 50 ml plastic tube or a Coplin jar each filled with wash buffer and wash 3× 15 min without shaking at RT to remove unbound primary antibodies.

5. Add 50 µl of secondary antibody solution to the slide and cover with a 24 × 40 mm piece of Parafilm. Do not press! Incubate for 1 h at 37 °C in a prewarmed moisture chamber in the dark to detect target-bound primary antibodies.
6. Wash in the dark as in **step 4** to remove unbound secondary antibodies.
7. Add 20 µl of mounting medium and cover with a 24 × 40 mm coverslip to stain DNA and protect fluorescently labeled molecules on the slide from fading. Do not press! Keep it overnight at 4 °C in the dark.
8. Seal edges of the coverslip with nail polish to prevent drying and store at 4 °C in the dark (*see* **Note 9**).

3.5 Fluorescence In Situ Hybridization on Tobacco and Rye Meiocytes

Steps 2–8 are only for tobacco meiocytes. Omit them for rye and continue from **step 9** onward.

1. Fix cells and prepare slides as described in Subheading 3.1 for tobacco or in Subheading 3.3 for rye (*see* **Note 12**).
2. Add 40 µl of RNase A to the slide and cover with a 24 × 40 mm coverslip. Incubate for 1 h at 37 °C in a prewarmed moisture chamber to degrade RNA.
3. Remove the coverslip by gentle sliding (*see* **Note 8**). Transfer the slide either into a 50 ml plastic tube or a Coplin jar each filled with 2× SSC and wash it 3× 5 min without shaking at RT to remove RNase A.
4. Add 40 µl of proteinase K solution to the slide and cover with a 24 × 40 mm coverslip. Incubate for 30 min at 42 °C in a prewarmed moisture chamber to remove RNase A residues and to digest partially dense cell cytoplasm.
5. Remove the coverslip by gentle sliding (*see* **Note 8**). Transfer the slide either into a 50 ml plastic tube or Coplin jar each filled with 2× SSC and wash it 3× 5 min without shaking at RT to remove proteinase K.
6. Transfer the slide either into a 50 ml plastic tube or Coplin jar each filled with FISH formaldehyde fixative for 10 min at RT to inactivate proteinase residues.
7. Transfer the slide either into a 50 ml plastic tube or a Coplin jar each filled with 2× SSC and wash it 3× 10 min without shaking at RT to remove the fixative.
8. Dehydrate the slide through an alcohol series at RT (70%, 90%, and 96% ethanol for 1 min each). Air-dry the slide to remove water and to prepare for formamide treatment.
9. Add 40 µl of 70% formamide to the slide, cover with a 24 × 40 mm coverslip and immediately transfer it to a hot

plate either at 68 °C for 5 min for tobacco or at 58 °C for 1 min for rye to denature cellular DNA.

10. Immediately remove the coverslip by gentle sliding (*see* **Note 8**) and dehydrate the slide through a precooled (−20 °C) alcohol series (70%, 90%, and 96% ethanol for 1 min each) to prevent renaturation of cellular DNA.
11. Air-dry the slide at RT.
12. Prepare hybridization mixture: add 1 μl of DNA probe, 2 μl of blocking DNA and 1 μl of dH_2O to 16 μl of 1.25× hybridization buffer (*see* **Note 13**), denature hybridization mixture at 95 °C for 5 min and immediately transfer the mixture on ice for 5 min to prevent renaturation of the DNA probe (*see* **Note 14**).
13. Add 20 μl of the ice-cold hybridization mixture to the slide and cover with a 24 × 40 mm coverslip. Incubate overnight at 37 °C in a prewarmed moisture chamber to enable hybridization of cellular and probe DNAs.
14. Remove the coverslip by gentle sliding (*see* **Note 8**) and wash the slide either in a 50 ml plastic tube or a Coplin jar to remove unbound DNA probe as follows:
    - 10 min in 2× SSC prewarmed to 42 °C
    - 3× 5 min in 0.1× SSC prewarmed to 42 °C
    - 10 min in 2× SSC prewarmed to 42 °C.
15. Add 40 μl of fluorescently labeled avidin or anti-digoxigenin antibody solution to the slide and cover either with a 24 × 40 mm coverslip for tobacco or with a 24 × 40 mm piece of Parafilm for rye. Incubate for 1 h at 37 °C in a prewarmed moisture chamber in the dark to detect hybridized DNA probes.
16. Remove the coverslip/Parafilm and wash the slide either in a 50 ml plastic tube or a Coplin jar each filled with FISH wash buffer 3 × 5 min in the dark at RT to remove unbound avidin or anti-digoxigenin antibody.
17. Add 20 μl of mounting medium and cover with a 24 × 40 mm coverslip to stain DNA and to protect fluorescently labeled molecules on the slide from fading. Keep overnight at 4 °C in the dark.
18. Seal edges of the coverslip with nail polish to prevent drying and store the slide at 4 °C in the dark (*see* **Note 9**).

4 Notes

1. Typically formaldehyde fixative freshly prepared from paraformaldehyde powder performs better than diluted from a 37% formaldehyde solution stabilized with methanol. However, for

our purposes both fixatives work equally well. Since preparation of a working dilution from a 37% formaldehyde solution is easy and quick we use this type of fixative.

2. Long term (more than an hour) formaldehyde fixation induces red autofluorescence of plant cell cytoplasm. In this case, use green fluorescence dyes for target retrieval.
3. The use of 1× citrate buffer (10×: 40 mM citric acid, 60 mM sodium citrate, pH 4.8) instead of 1× PBS, increases the efficiency of the enzyme mixture significantly. Thus, use it if you cannot remove the callose wall by suggested protocol.
4. There are various probe types and different ways to label and detect them. Establish which one works best for your purpose. In conventional protocols, probes are labeled with biotin- or digoxigenin-dUTP. The dUTPs are incorporated in a probe by PCR or Nick translation. After hybridization biotin- or digoxigenin labeled probes cannot be observed immediately, they are detected either by fluorescently labeled avidin or anti-digoxigenin antibodies, respectively. In Fig. 1e we used a probe labeled by biotin-dUTP via PCR with primers for tobacco genomic repetitive sequence (GRS) present in five tobacco chromosomes (*see* details in [12]).
5. Blocking unlabeled DNA blocks repetitive sequences prior to specific hybridization. The blocking DNA is fragmented by autoclaving or sonication. Desired size range of the blocking DNA is a complex of 100–600 bp fragments.
6. Pay attention to the form of salts you use (anhydrous/hydrated). For example, to prepare 20× SSC you use 88.2 g of dihydrated trisodium citrate ($Na_3C_6H_5O_7 \times 2H_2O$) while only 77.4 g of its anhydrous form.
7. To define the meiotic stage, take one anther from a flower bud (place the remaining anthers on wet filter paper), stain it with acetocarmine or any other chromosome dye and analyze under a microscope. Use the remaining anthers if you found the desired meiotic stage.
8. For gentle sliding, put your finger on the coverslip and carefully move it away horizontally or carefully shake the slide vertically. If it does not work, dip the slide in respective buffer for five to ten times (e.g., in a 50 ml plastic tube) until the coverslip falls off.
9. The slides after immunostaining or FISH can be stored at 4 °C in the dark for 1–2 months.
10. Ice cold fixation of cereal meiocytes should be avoided in case of cytoskeleton staining.
11. Deionized water improves cell adhesion to a slide and is used when squashing with a coverslip is not suitable.

12. For FISH both formaldehyde and acetic alcohol, three parts of absolute ethanol to one part of acetic acid fixative (v/v), fixatives can be used. After acetic alcohol fixation (1–24 h with subsequent replacement by 70% alcohol) material can be stored at 4 °C for at least 6 months. Contrariwise, formaldehyde fixation does not allow storing material but preserves cell structure better.
13. You can vary volumes of the components in the mixture: 1–3 μl of DNA probe, 1–3 μl of blocking DNA and 0–2 μl of dH_2O.- Concentrations of DNA probe and blocking DNA are interdependent (must be 1:100) and you have only 4 μl to adjust it. The final volume of the hybridization mixture is 20 μl with 16 μl of 1.25× hybridization buffer.
14. If you work with multiple samples at the same time do not prepare hybridization mixture for all samples in one single plastic tube since the mixture cannot be mixed well. Take an individual tube for each sample (20 μl per tube).

Acknowledgments

Rye seeds were kindly provided by Silkova O.G. (Institute of Cytology and Genetics, Russia). The work was supported by the Russian Foundation for Basic Research [16-34-60007 mol_a_dk] and Siberian Branch of the Russian Academy of Science under the program "Molecular genetic bases of regulation of genes expression, morphology, differentiation and cell reprogramming" [0324-2019-0042].

References

1. Guzicka M, Wozny A (2005) Cytomixis in shoot apex of Norway spruce [*Picea abies* (L.) karst.]. Trees 18:722–724
2. Kravets EA, Sidorchuk YV, Horyunova II et al (2016) Intra- and intertissular cytomictic interactions in the microsporogenesis of mono- and dicotyledonous plants. Cytol Genet 50:267–277
3. Mursalimov SR, Sidorchuk YV, Deineko EV (2013) New insights into cytomixis: specific cellular features and prevalence in higher plants. Planta 238:415–423
4. Sidorchuk YV, Novikovskaya AA, Deineko EV (2016) Cytomixis in the cereal (Gramineae) microsporogenesis. Protoplasma 253:291–298
5. Singhal VK, Kumar P (2008) Impact of cytomixis on meiosis, pollen viability and pollen size in wild populations of Himalayan poppy (*Meconopsis aculeata* Royle). J Biosci 33:371–380
6. Mursalimov S, Permyakova N, Deineko E et al (2015) Cytomixis doesn't induce obvious changes in chromatin modifications and programmed cell death in tobacco male meiocytes. Front Plant Sci 6:1–13
7. Barton DA, Cantrill LC, Law AMK et al (2014) Chilling to zero degrees disrupts pollen formation but not meiotic microtubule arrays in *Triticum aestivum* L. Plant Cell Environ 37:2781–2794
8. Reis A, Sousa S, Viccini L (2016) High frequency of cytomixis observed at zygotene in tetraploid *Lippia alba*. Plant Syst Evol 302:121–127
9. Mursalimov S, Deineko E (2018) Cytomixis in plants: facts and doubts. Protoplasma 255:719–731

10. Fakhri Z, Mirzaghaderi G, Ahmadian S, Mason AS (2016) Unreduced gamete formation in wheat x *Aegilops* spp. hybrids is genotype specific and prevented by shared homologous subgenomes. Plant Cell Rep 35:1143–1154
11. Farooq U, Saggoo MIS (2014) Male meiosis and behaviour of sex chromosomes in different populations of *Rumex acetosa* L. from the Western Himalayas, India. Plant Syst Evol 300:287–294
12. Mursalimov S, Deineko E (2017) Cytomixis in tobacco microsporogenesis: are there any genome parts predisposed to migration? Protoplasma 254:1379–1384

Chapter 10

Analysis of Meiosis in Nonmodel Tropical Plants: The Case of *Carica papaya* Linn

José Mora-Calderón, Kalani Scott-Moraga, and Pablo Bolaños-Villegas

Abstract

To develop plants that are more tolerant to drought, marginal soil fertility, and diseases and that satisfy demands for high yield, new cultivars of the tropical fruit papaya (*Carica papaya* L.) are needed. Nonetheless, in many cases, these traits are available in only wild relatives found throughout Latin America. Understanding meiotic progression may facilitate the introgression of desirable traits into commercial cultivars that maintain high fertility. In this protocol, we describe a practical and simple method to effectively isolate male meiocytes in order to document the behavior of papaya meiotic chromosomes.

Key words Meiosis, Nonmodel organism, Papaya, Plant chromosomes, Tropical agriculture

1 Introduction

Papaya (*Carica papaya* L.) is an economically important fruit tree grown in tropical and subtropical regions (Fig. 1a) [1]. It has a small diploid genome of approximately 372 Mbps [2], of which 51% correspond to repetitive sequences [3]. The genome of papaya is comprised by nine pairs of chromosomes ($2n = 18$) with a genetic length of approximately 1068.6 centimorgans (cM), a distance per cM of about 348 Kbps, and approximately 12 linkage groups [1, 4]. The length of these nine chromosomes during metaphase I is very similar (between 1 and 4 μm); however, chromosomes 1, 2, 3, 4, 5, 7, and 8 are metacentric, while 6 and 9 are submetacentric [4]. It is possible to identify individual chromosomes during pachytene by performing Fluorescent in situ Hybridization (FISH) with the assistance of 5S rRNA probes, as during pachytene meiotic chromosomes appear about ten times larger than during metaphase I [4, 5]. Papaya chromosomes are mostly euchromatic, except for the Y chromosome (chromosome 1), as inferred by intense staining with 4′,6-diamidino-2-phenylindole (DAPI) [5, 6]. A single 45S rDNA locus is located in the middle of chromosome 4, while at least thirteen 5S rDNA loci are located

Mónica Pradillo and Stefan Heckmann (eds.), *Plant Meiosis: Methods and Protocols*, Methods in Molecular Biology, vol. 2061, https://doi.org/10.1007/978-1-4939-9818-0_10,

Fig. 1 Morphology of plants and flowers of *Carica papaya* hybrid Pococí. (**a**) view or commercial farm in Northern Costa Rica, plants are approximately 4 m tall. (**b**) hermaphrodite flower with open anthers (orange). (**c**) female flower without anthers. (**d**) 5-month-old plant displaying an axial flower raceme. The longest flower buds were approximately 4 cm long. *Scale bars*, **a**–**c**, 4 cm

on chromosomes 1, 3, 5, 8, and 9. Usually 5S rDNA loci are stained brightly by DAPI [5]. Chromosome 8 contains two 5S rDNA loci, chromosome 5 contains two large and two small loci; while chromosome 3 contains only four small loci [5]. At least two small 5S rDNA loci are detected in the male-specific region of the Y chromosome (MSY), while 5S rDNA loci are not detected on the X chromosome [5].

The use of FISH during pachytene also allows to map most linkage groups to specific chromosomes [4]. For instance linkage groups 2 and 3 correspond to chromosomes 4 and 6, linkage groups 7 and 12 correspond to chromosome 7, groups 8, 1 and 10 correspond to chromosome 9, while groups 9 and 11 correspond to chromosome 2 [4]. Chromosome 1 is the sex chromosome [4]. Linkage groups in papaya show suppression of recombination and are believed to be centromeric; however, the end of linkage groups 2–5 show recombination hot spots [1]. These hotspots are believed to be gene-rich regions, in which beneficial allele combinations may have been concentrated by natural selection during the evolution of papaya [1]. These regions are important since they may harbor yet-unidentified QTLs for crucial traits such as sugar content, fruit size, shape, weight and disease tolerance [1].

Papaya is trioecious, with three basic gender types: female, male, and hermaphrodite [5]. Gender in papaya is genetically controlled by a sex-linked region that behaves like an XY sex chromosome, with maleness and hermaphroditism being determined by different regions within the Y chromosome: Yh (HSY) in hermaphrodites, and Y (MSY) in males [7]. Both HSY and MSY are approximately 8.1 Mb in length, and recombination with the X chromosome is suppressed [7]. Moreover, the MSY region may span up to 13% of chromosome 1 and include the centromere [8]. The corresponding region of the X chromosome is only 3.5 Mb [7]. Any combination of the Y and Yh chromosomes (YY, YYh, or YhYh) is inviable, and the embryos abort 25–50 days after pollination, implying that the Y chromosome types are both missing an essential gene that is functional in the X chromosome [7]. In cultivated papaya, most plants segregate at a rate of two-thirds hermaphrodite and one-third female plants, although dioecious varieties exist [7] (Fig. 1b, c). The presence of natural populations in Mexico and Central America and the cultivation in Mexico and Belize before the arrival of the Spaniards suggest a Mesoamerican origin [7].

Papaya is the most economically important fruit crop of the family *Caricaceae*, which consists of 6 genera and 35 species, and is grown for its edible fruit and the production of the papain proteolytic enzyme [9, 10]. Introgression of new traits from wild relatives has been proposed in papaya, most notably for tolerance to the mosaic virus and the fungus *Phytophthora palmivora* and for improving flesh thickness, flesh consistency and skin texture [11]. Other traits of interest are photosynthetic rate, water use and nutrient uptake [12]. Wild relatives such as *Vasconcellea longiflora*, *V. monoica*, *V. goudotiana*, *V. quercifolia*, *V. sphaerocarpa*, *V. palandensis*, and *Jacaratia spinosa* appear to share the same number of chromosomes with *C. papaya* [11, 13, 14]; however, defects in meiotic chromosome segregation including formation of micronuclei have been observed in several species including *V. cauliflora*, *V. cundinamarcensis*, and *V. sphaerocarpa*, suggesting the occurrence of hybridization events [14]. Intergeneric and interspecific hybrids have been reported [11, 14]. However, the information reported is usually sparse, and molecular characterizations of their genomes are lacking. No reports exist about female meiosis in papaya, either. Therefore, the study of meiotic behavior in papaya and the development of practical protocols may facilitate the screening of fertility patterns in new hybrids and species, thus allowing for quicker selection times.

The purpose of this protocol is to facilitate male meiotic chromosome analysis in papaya to in order to select varieties and species in which meiotic behavior and fertility are optimal for plant breeding. Eventually, this protocol may be used as well to map the location of beneficial loci [1], or to determine chiasmata formation rates in response to changes in meiotic recombination [15], either in papaya or in other tropical/subtropical crops such as cacao or coffee.

2 Materials

Hermaphrodite plants from papaya hybrid Pococí were grown at the Fabio Baudrit Agricultural Research Station at La Garita (San José district), Alajuela province, Costa Rica. The location is 010°00′18″N 084°15′56″W and 840 m above sea level. Average temperature is 22 °C.

1. Fixative buffer: ethanol (99.5%), 50% v/v, formaldehyde (36.5–38%), 10% v/v, glacial acetic acid, 5% v/v. Store at 4 °C.
2. Chromosome spreading solution: glacial acetic acid solution, 65% (v/v). Store at 4 °C.
3. Chromosome stain: Fluoroshield with DAPI. Store at 4 °C.
4. Glass slides: poly-L-lysine–coated glass slides.
5. Coverslips: 22 × 22 mm.
6. Tubes: 1.5-ml microcentrifuge tubes, and 15 ml centrifugation tubes.
7. Tube racks: 80-place tube racks.
8. Tube boxes: 32/8-place tube boxes.

9. Syringes: 26G × 1/2.
10. Scalpel and blades: Shandon stainless-steel scalpel blade handle, surgical carbon blades.
11. Forceps.
12. Spatula.
13. Dry bath.
14. Microscopes: for meiocyte observation, an epifluorescence microscope coupled with a CCD camera. For flower dissection, a stereomicroscope with a built-in camera and tablet display.

3 Methods

Hermaphrodite flowers are sampled from 5-month-old hermaphrodite plants (Fig. 1b). Be careful not to collect female flowers as not all flowers are necessarily hermaphrodite (Fig. 1c). Flowers are collected, measured, and fixed on site, in the morning preferably between 9:00 and 10:00. Flowers are stored at 4 °C. All solutions are prepared with ultrapure water (15–18 MΩ-cm), and molecular biology-grade reagents are used fresh. All procedures are performed at room temperature. All waste products must be disposed of following safety regulations.

3.1 Preparation of Materials Prior to Collection in the Field Plots

All tools must be carefully prepared before leaving the laboratory, as distance to the plant breeding plots makes it inconvenient to return once flower collection has begun.

1. Mix the reagents for the fixative buffer and dispense in 15-ml microcentrifuge tubes. Precool at 4 °C for at least 15 min before going to the field.
2. Place the precooled microcentrifuge tubes filled with fixative inside a box.
3. Place a scalpel and blades inside the box.
4. Place a notebook and pen inside the box.

3.2 Collection of Samples in the Field

Collection of fresh material is a must; you must proceed as quick as possible to keep the samples cool.

1. Once in the field, collect flower buds (Fig. 1d). Excise with a scalpel and knife. Measure length and place in a microcentrifuge tube (*see* Table 1). Mark the corresponding length on top of the tube and place buds of similar length in the same tube. Collect at least ten buds of the same length from at least three different plants. Place the tubes inside a tube rack or a tube box by using forceps. Write down the date and contents with tape on the side or on top of the rack or box. Annotate the same in the notebook.

Table 1
Flower bud length and corresponding meiotic stage in *Carica papaya* cv. Pococí

Flower bud length (mm)	Stage
1.5–2.5	Prophase
2.5–3	Prometaphase and metaphase I
4	Anaphase I and prophase II
7	Late anaphase II and telophase II
37	Pollen mitosis I

2. Once inside the lab, place the racks inside the refrigerator to cool at 4 °C for at least 15 min. Change the fixative three times for at least 5 min. Allow to fixate for at least 8 h for 4 °C.

3.3 Flower Dissection and Isolation of Meiocytes

Meiocytes are fragile and must always be kept moist. Proper handling ensures the isolation of high-quality chromosome spreads. Proceed with extreme caution.

1. The next day after collection, turn on the dry block and set it at 45 °C.
2. Drop 20–40 μm of 65% acetic acid solution on a slide, place flowers on the slide and slit the base with a needle to release anthers. Under an Olympus SZX 61 stereoscope, remove petals and sepals and puncture the anthers with a syringe to release the meiocytes. Keep the slide moist with acetic acid solution, 65% v/v, always. Use a spatula to spread the meiocytes out evenly. Meiocytes appear as grains of sand to the naked eye.
3. Place the slide on the dry block previously set at 45 °C. Add 20–40 μl of acetic acid solution, 65% v/v, allow to evaporate a bit, then drop 20–40 or even 100 μl of cold fixative (kept at –20 °C) from a height of about 25 cm. Remove the slide from the dry block carefully.

3.4 Staining of Meiocytes

Meiocytes in papaya are large; therefore, a variety of stains may be employed to image chromosomes. The stain 4′,6-diamidino-2-phenylindole (DAPI) is a fluorescent dye that gives off a consistent signal with little bleaching over time, therefore it is the one used in this protocol (Fig. 2).

1. Add 10–20 μl Fluoroshield with DAPI to the slide, place a coverslip on top very slowly to prevent bubble formation, allow DAPI to spread out evenly and seal the slide with regular nail polish. Keep in the dark to allow the slide to seal and prevent exposure to the light. Store at 4 °C if necessary.

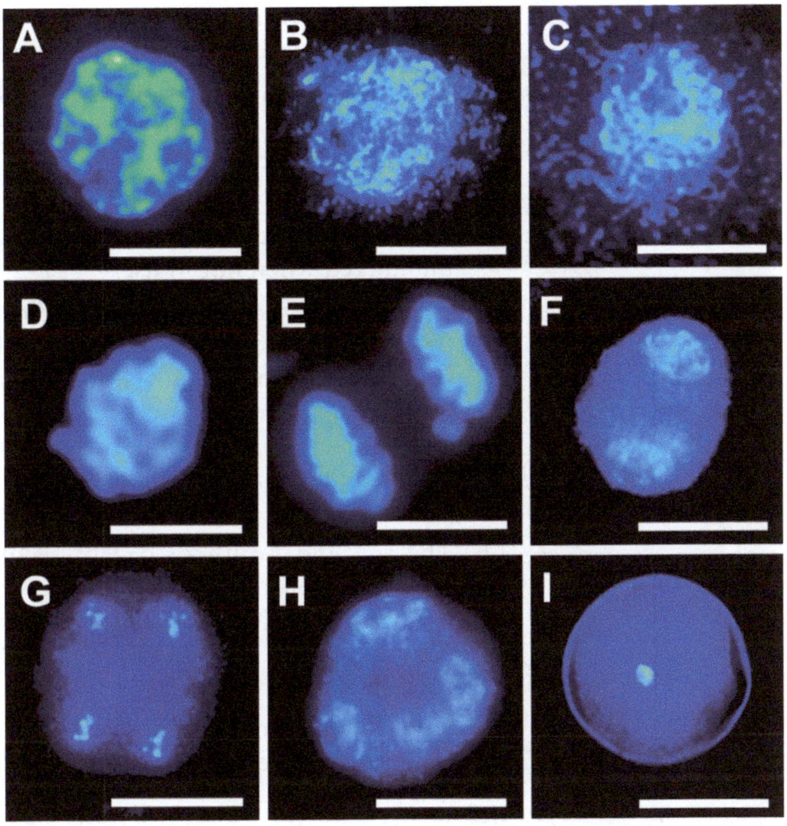

Fig. 2 Male meiosis in *C. papaya* cv. Pococí. (**a**) interphase; (**b**) leptotene; (**c**) pachytene; (**d**) metaphase I; (**e**) anaphase I; (**f**) prophase II, (**g**) late anaphase II; (**h**) telophase II; pollen mitosis (**i**) Cells were stained with DAPI. *Scale bars*, **a–e**, **j–l**, 20 μm, **f**, **g**, and **i**, 10 μm

2. Slides are analyzed under an epifluorescence microscope (Olympus BX53) with a DAPI filter under 100× magnification. At least three photos per cell are taken (Olympus Q Color 5 camera) and saved as .tiff files without any compression.
3. All files must show the collection date, approximate flower bud length, and the name of the person who saved the file.
4. In order to study chromosome behavior images are processed by using Adobe Photoshop CS5 (Adobe Systems). Images are analyzed at 300 dpi by using curve, level and exposure adjustments. No other image manipulations such as erasing are performed.

4 Notes

1. Costa Rica is a tropical country with high relative humidity. All our microscopy equipment is confined within a room and

protected with dehumidifiers that operate 24/7. Microscope internal dehumidifying pills are replaced yearly.

2. Collection of plant material is restricted to the early morning because during the rainy season (May to November), it rains heavily after noon.
3. Thunderstorms are frequent, when in use all laboratory equipment is connected to industrial voltage surge suppressors and unplugged when not in use.
4. Not all flowers from hermaphrodite papaya plants (as genotyped by PCR) necessarily develop as hermaphrodite flowers; a very few are female. Thus, flowers must be dissected on site to verify that flower buds contain anthers. This is a must. The study of female meiosis is better suited by collecting from a female individual from a dedicated breeding line used as a pollen acceptor, such as the Costa Rican L1 line.
5. The age and health of papaya seedlings is important. Five-month-old plants yield healthy flowers with many meiocytes. The yield of meiocytes in older plants (2–3 years old) is limited. Also, the height of plants is usually <2 m, which allows for easy collection of flowers.
6. Unlike in the model plant *Arabidopsis thaliana* [16], in papaya, no enzyme digestion is required for the release of meiocytes from anthers.
7. Most meiotic events are observed in flower buds 1.5–4 mm long (Fig. 2a–h).
8. Observation of meiocytes at stages late anaphase II; and telophase II (Fig. 2g–h) suggests that tetrad formation is normal in hybrid Pococí.
9. We did not observe tricellular pollen in any of our samples. Such pollen may be found in flower buds longer than 37 mm.
10. The actual morphology of individual chromosomes at metaphase I cannot be resolved by use of our current CCD camera. More powerful equipment may be required.
11. It is possible to store the slides in the dark at 4 °C for up to 5 months. Before observation under the microscope, the slides need to be placed at room temperature to remove unwanted humidity condensation.

Acknowledgments

Our work was supported by the Vicerrectoría de Investigación intramural grants # B5A13 and B0185, and we particularly wish to thank Vicechancellor Fernando García-Santamaría. Our Dell Precision Tower 7820 computer system was kindly donated by the Spanish Agency for International Development Cooperation (Agencia Española de Cooperación Internacional para el Desarrollo). We thank previous students Stefano Albertazzi, Sergio Castro-Pacheco (now a student at the Erasmus Mundus Master in Plant Breeding program), and Romano Porras (now at Universität Hohenheim, Stuttgart, Germany) for their valuable work. We also thank Eric Mora-Newcomer and Walter Barrantes-Santamaría for providing flower samples from their papaya breeding program. Pablo is a young member affiliate of the World Academy of Sciences (TWAS/UNESCO) and a member of the American Society of Plant Biologists.

References

1. Yu Q, Tong E, Skelton RL et al (2009) A physical map of the papaya genome with integrated genetic map and genome sequence. BMC Genomics 10:371
2. Chen C, Yu Q, Hou S et al (2007) Construction of a sequence-tagged high-density genetic map of papaya for comparative structural and evolutionary genomics in Brassicales. Genetics 177:2481–2491
3. VanBuren R, Ming R (2013) Dynamic transposable element accumulation in the nascent sex chromosomes of papaya. Mob Genet Elem 3:e23462
4. Zhang W, Jiang J (2014) Molecular cytogenetics of papaya. In: Ming R, Moore PH (eds) Genetics and genomics of papaya, plant genetics and genomics: crops and models 10. Springer Science+Business Media, New York, pp 157–168
5. Zhang W, Wai CM, Ming R et al (2010) Integration of genetic and cytological maps and development of a pachytene chromosome-based karyotype in papaya. Trop Plant Biol 3:166–170
6. Wai CM, Moore PH, Paull RE et al (2012) An integrated cytogenetic and physical map reveals unevenly distributed recombination spots along the papaya sex chromosomes. Chromosome Res 20:753–767
7. VanBuren R, Zeng F, Chen C et al (2015) Origin and domestication of papaya Yh chromosome. Genome Res 25:524–533
8. Zhang W, Wang X, Yu Q et al (2008) DNA methylation and heterochromatinization in the male-specific region of the primitive Y chromosome of papaya. Genome Res 18:1938–1943
9. Abreu IS, Carvalho CR, Soares FAF (2015) Early sex discrimination in Carica papaya by nuclei FISH. Euphytica 206:667–676
10. Liao Z, Yu Q, Ming R (2017) Development of male-specific markers and identification of sex reversal mutants in papaya. Euphytica 213:53
11. Narducci Da Silva E, Neto MF, Pereira TNS et al (2012) Meiotic behavior of wild Caricaceae species potentially suitable for papaya improvement. Crop Breed Appl Biotechnol 12:52–59
12. Jiménez VM, Mora-Newcomer E, Gutiérrez-Soto MV (2014) Biology of the papaya plant. In: Ming R, Moore PH (eds) Genetics and genomics of papaya, plant genetics and genomics: crops and models 10. Springer Science +Business Media, New York, pp 17–33
13. Corrêa DJP, Pereira TNS, Neto MF et al (2010) Meiotic behavior of Carica papaya and Vasconcellea monoica. *Caryologia* 63:229–236
14. Caetano C, Lagos T, Sandoval C et al (2008) Citogenética de especies de Vasconcellea (Caricaceae). Acta Agronómica 57:241–245
15. Kurzbauer M-T, Pradillo M, Kerzendorfer C et al (2018) Arabidopsis thaliana FANCD2 promotes meiotic crossover formation. Plant Cell 30:415–428
16. Bolaños-Villegas P, Yang X, Makaroff CA et al (2014) Protocol for the preparation of Arabidopsis meiotic chromosome spreads and fluorescent in situ hybridization. The Plant Journal 4:e1102

Chapter 11

Analytical Methodology of Meiosis in Autopolyploid and Allopolyploid Plants

Pilar Prieto and Tomás Naranjo

Abstract

Meiosis is the cellular process responsible for producing gametes with half the genetic content of the parent cells. Integral parts of the process in most diploid organisms include the recognition, pairing, synapsis, and recombination of homologous chromosomes, which are prerequisites for balanced segregation of half-bivalents during meiosis I. In polyploids, the presence of more than two sets of chromosomes adds to the basic meiotic program of their diploid progenitors the possibility of interactions between more than two chromosomes and the formation of multivalents, which has implications on chromosome segregations and fertility. The mode of how chromosomes behave in meiosis in competitive situations has been the aim of many studies in polyploid species, some of which are considered here. But polyploids are also of interest in the study of meiosis because some of them tolerate the loss of chromosome segments or complete chromosomes as well as the addition of chromosomes from related species. Deletions allow to assess the effect of specific chromosome segments on meiotic behavior. Introgression lines are excellent materials to monitor the behavior of a given chromosome in the genetic background of the recipient species. We focus on this approach here as based on studies carried out in bread wheat, which is commonly used as a model species for meiosis studies. In addition to highlighting the relevance of the use of materials derived from polyploids in the study of meiosis, cytogenetics tools such as fluorescence in situ hybridization and the immunolabeling of proteins interacting with DNA are also emphasized.

Key words Centromeres, Chromosome dynamics, Chromosome pairing, In situ hybridization, Synapsis, Recombination, Telomeres

1 Introduction

1.1 The Meiotic Program in Polyploids

Polyploid organisms are those containing more than two sets of chromosomes. Polyploidy is pervasive and represents a major mechanism of speciation in plants. Some estimates suggest that between 47% and 70% of angiosperms are polyploids [1] and many lineages show evidence of ancient genome duplications [2]. Polyploids have been broadly classified into two types: autopolyploids, which have three or more copies of the same genome, and allopolyploids, which originated after interspecific hybridization between related diploid progenitors. Thus, allopolyploids

Mónica Pradillo and Stefan Heckmann (eds.), *Plant Meiosis: Methods and Protocols*, Methods in Molecular Biology, vol. 2061, https://doi.org/10.1007/978-1-4939-9818-0_11, © Springer Science+Business Media, LLC, part of Springer Nature 2020

contain two or more distinct subgenomes, whose chromosomes, called homoeologs, preserve some degree of affinity because of their common origin. The polyploid condition creates a competitive situation at the start of meiosis, in which, each chromosome has two or more potential partners to interact physically in order to repair the double-strand DNA breaks (DSBs) produced in early leptotene. Three or more related chromosomes, homologs in autopolyploids and homologs and homoeologs in allopolyploids, usually interact simultaneously, rather with a single partner, and synapse to form complex synaptic multivalents in zygotene and complete homologous/homoeologous recombination.

Although competitive chromosome interactions represent the specific meiotic feature of polyploids, meiosis in these organisms takes place following most of the steps that make up the basic meiotic program of the diploid progenitor species. That is, physical interactions leading to chromosome sorting are preceded of the formation of programed DSBs catalyzed by the topoisomerase-like conserved protein, Spo11, in conjunction with a number of additional proteins [3, 4]. After a DSB occurs, sections of DNA around the 5′ ends of the break are cut away in a process called resection. An overhanging 3′ end of the broken DNA molecule then associate with the recombinase RAD51 and/or the meiosis-specific recombinase DMC1 to form a nucleoprotein filament, which invades a similar or identical DNA molecule that is not broken [5]. During leptotene, chromosomes decondense and elongate, and form the axial element, a protein-rich backbone that will keep the sister chromatids together until the second division. Proteins of the axis, such as ASY1-ASY3 in *Arabidopsis* or PAIR2-PAIR3 in rice, promote the search of a repair template in the homologous chromosome [6–9]. Thus, these early steps of the DSBs repairing are involved in the identification of the homologous partner and are indispensable for homologous alignment and the assembly of the synaptonemal complex (SC) in zygotene in many organisms [10–13]. Homologs are fully synapsed at pachytene. The SC maintains homologs in close juxtaposition along their length and serves as a scaffold to recruit factors of the recombinational repairing machinery [14]. In plants and animals, most of DSBs are repaired following a noncrossover (NCO) pathway, only a minor fraction (5%) give rise to crossovers (COs) [15]. Thus, the production of DSBs, chromosome sorting, SC assembly and COs formation share similar molecular mechanisms in diploids and polyploids.

1.2 Chromosome Interactions in Early Meiosis in Autopolyploids

Chromosome interactions leading to pairing, synapsis, and CO production in competitive situations are essentially different from interactions in diploids. Assuming that the premeiotic arrangement of homologs is not quite different in polyploids and diploids, that is, they occupy separated nuclear territories, chromosomes have to move at the start of meiosis to find a partner. This movement can be based both in the telomere migration to form the bouquet and in the unfolding of chromatin that increase largely the length of the chromosomes [16, 17]. Once the chromosomes have come close enough to interact, a choice has to be made between potential partners when more than two are available. In some polyploids, axial cores of each homologous set are aligned in bundles along much of their length [18–23]. In others, pairing is restricted to the distal regions while the rest of the chromosomes remain largely distant [24–26]. Differences between these two types of presynaptic alignment fall in the number and distribution of autonomous pairing sites (APSs), sites where pairing can be initiated between any two of three or more homologous chromosomes. Multiple APSs are scattered along the chromosome axes in the extensive presynaptic alignment and two APSs located at both chromosome ends are mainly formed in the distal alignment. Once the first APS is activated, activation of a second APS between two chromosomes, one of which is involved in the first APS, and synapsis extension give rise to a paring partner switch (PPS) in the SC multivalent formed. The number of PPSs in SC multivalents provides an estimates of the number of APSs, APSs = 1.5 PPSs +1 [18]. Multivalents formed in autopolyploids in prophase I show stretches of SC that normally involve two homologs at any location of the chromosomes. SC multivalents formed by homologous sets after distal alignment or extensive pairing differ in the number of PPSs, one PPS is mainly formed in the first case and many in the second.

Variation of the meiotic program introduced by autopolyploidy concerns mainly the type, distal or complete, of presynaptic alignment, the number and distribution of synapsis initiation points, the number of partner exchanges and its evolution through zygotene and pachytene, and the existence or not of preferences in the partner election. All these aspects of meiotic prophase I of autopolyploids can be addressed when individual chromosomes can be identified. To date, the behavior of individual chromosomes in competitive situations in prophase I has been carried out in polyploids of species such as *Crepis capillaris* where homologous sets differ in their length [20, 22] or in primary trisomics as those of rye (*Secale cereale* L.) [27]. In the absence of protocols leading to identify individual chromosomes, inferences on pairing and synapsis have been made from metaphase I observations [28]. Nevertheless, a reduction of multivalents have been observed both trough prophase I and from prophase I to metaphase I in different polyploids, regardless individual chromosomes were identified ([29] and references therein). Some SC multivalents formed in zygotene are transformed into bivalents in pachytene, but those persisting during pachytene give rise to bivalents at metaphase when not all SC are stabilized by COs formation. The frequency of multivalents at metaphase I has been found to be lower in natural autopolyploids of different species, sourced many generations ago, than in new colchicine induced autopolyploids [30, 31]. Such a result suggests the occurrence of partial diploidization of the meiotic behavior, which may be a common feature of autopolyploid species evolution. Identification of the diploidization mechanism requires extending the analysis to pairing and synapsis.

1.3 Chromosome Interactions in Early Meiosis in Allopolyploids

Allopolyploids possess homoeologous subgenomes from two or more evolutionary related progenitors. Such subgenomes differentiated by accumulation of changes both in the DNA sequence, which produce homoeoalleles, and in the chromosome architecture and gene order, but their chromosomes retain some degree of genetic affinity. Such a genetic affinity confers homoeologous chromosomes the possibility of competing with homologues in interactions leading to pairing, SC assembly, and COs. However, homoeologous subgenome diversification suggests differences in the meiotic behavior between autopolyploids and allopolyploids. Multivalents are expected to be formed more often in autopolyploids than in allopolyploids. This was confirmed in a survey consisting of 171 reports on neopolyploids [32], which yielded a higher frequency of multivalents at diakinesis and metaphase I in autopolyploids (28.8%) than in allopolyploids (8.3%). These multivalent frequencies suggest partial diploidization of many autopolyploids and complete diploidization of most allopolyploids. Exclusive bivalent formation at metaphase I is vital to ensure regular homologous disjunction at anaphase I and reproductive stability in allopolyploids. This diploid-like behavior has been demonstrated to be the result of the emergence of genetic regulatory systems in allopolyploids such as wheat, cotton, *Avena sativa*, *Festuca arundinacea*, and *Brassica napus*, in amphidiploids related to the diploid species *Lolium perenne*, *L. multiflorum*, and *L. rigidum*, and in polyploids in the genera *Aegilops*, *Hordeum*, *Nicotiana*, and *Coffea* [33].

Common bread wheat, *Triticum aestivum*, is an allohexaploid species with $2n = 6x = 42$ chromosomes (genome formula AABBDD) that is considered a reference model in the analysis of the meiotic behavior in allopolyploids [34]. Genomes A and D are derived from *T. urartu* and *Aegilops tauschii*, respectively, while *Ae. speltoides* or a closely related species to *Ae. speltoides* is, most likely, the progenitor of the B genome [35]. Hexaploid wheat arose after two hybridization events, the first hybridization originated the tetraploid wheat (AABB) *T. turgidum* and the second one the hexaploid wheat *T. aestivum*. Relevance of bread wheat in the analysis of meiotic pairing is reinforced by the importance of this species as a crop and because its regulatory system of meiotic pairing is effective both in wheat itself and in interspecific hybrids, which has implications in the transfer to wheat of alien genes controlling important agronomical traits.

Tetraploid and hexaploid wheat species form at metaphase I as many bivalents as homologous pairs have, 14 and 21 bivalents, respectively. This meiotic behavior of polyploid wheat species is controlled by the major homoeologous pairing suppressor *Ph1* located on the long arm of chromosome 5B [36–38]. *T. aestivum* carries another homoeologous pairing suppressor, called *Ph2*, which is less efficient than *Ph1* and locates on the short arm of chromosome 3D [39, 40]. Meiotic phenotype of mutants of *Ph1* is characterized by the presence of multivalents at metaphase I involving homologous and homoeologous chromosomes, as well as by a reduction in the number of ring bivalents due to the increase of the number of univalents and rod bivalents, relative to the wild type. The meiotic phenotype at metaphase I of mutants of *Ph2* is similar to that of the wild type [34]. Despite the regular formation of homologous bivalents at metaphase I in the wild type genotypes, 28% of chromosomes of hexaploid wheat, and 39% of tetraploid wheat, are involved in the formation of SC multivalents during zygotene. A similar SC multivalent frequency is produced in the *Ph* mutants. While SC multivalents are corrected and transformed into pairs of bivalents in the course of pachytene in the wild type and in the *ph2b* mutant, synapsis correction is not completed in the mutants of *Ph1*, *ph1b* in hexaploid wheat and *ph1c* in tetraploid wheat. In these mutants, chiasmata are formed between homologous and between homoeologous chromosomes in some SC multivalents, which persist until metaphase I ([34], and references therein).

Unequivocal meiotic phenotypes of wild type and all mutants of *Ph* genes are found when interspecific hybrids of wheat and related species are produced. Wild type, *ph2b*, and *ph1b* hybrids of wheat and related species, such as rye, *Ae. longissima*, and *Ae. sharonensis*, show low (<2), intermediate [5–11], and high [12–18] chiasmata at metaphase I. The number of chiasmata between wheat and related species chromosomes increases with their degree of closeness, it is higher in wheat–*Ae. longissima* and wheat–*Ae. sharonensis* hybrids than in wheat–rye hybrids [34]. *Ae. speltoides* is an exception, the three types of hybrids with wheat show a high frequency of chiasmata, since this species carries genes that suppress the effect of *Ph1* [34].

The mode of action of the *Ph1* locus has been the focus of numerous studies since its discovery 60 years ago. A number of hypothesis (reviewed in [34]) were advances to explain how *Ph1* restricts recombination to homologous chromosomes. *Ph1* seems to exert its action on two different steps of the process of bivalent formation: the first impedes consolidation of COs that could be initiated between homoeologous chromosomes; the second promotes the correction of SC multivalents, which are transformed into bivalents. The fact that the number (19.5) of ring bivalents at metaphase I is higher than the number (17.8) of SC bivalents at late zygotene suggests that some corrected SC bivalents form chiasmata in resynapsed stretches, which implies that the recombination machinery can be loaded after multivalent correction at pachytene. The lack of an accurate identification of *Ph1* itself has represented a great obstacle in the clarification of its mode of action. The *Ph1* locus is confined to a region of 3 Mb in the middle of chromosome arm 5BL [41]. This region contains 36 genes and the *Ph1* locus was assigned to a subregion carrying a subtelomeric heterochromatin segment inserted into an array of seven *CDK2*-like copies [42]. Included in the heterochromatin segment is the *TaZYP4-B2* gene. This gene was proposed to be *Ph1* as supported by the high level of homoeologous pairing found in hybrids of two *Tazyp4-B2* mutants of wheat with *Ae. variabilis* [43]. An alternative approach based on gene silencing identified another gene located in a 450-Kb segment of the *Ph1* region, which was termed *C-Ph1* (candidate *Ph1*). Silencing of *C-Ph1* results the same meiotic phenotype as the *ph1b* mutant [44].

The *Ph2* gene has not yet been identified. This gene resides in a segment of 80 Mb of 3DS, which is deleted in the *ph2a* mutant; *ph2b* was suggested to be a point mutation [45]. The mismatch repair *TaMSH7* gene, which shows highest similarity to *MSH7* of *Arabidopsis* and maize, is one of the genes located in this 80 Mb subregion and was postulated a possible candidate for *Ph2* [46]. Some role of *MSH7* on recombination in cereals was suggested after the reduction of fertility caused by RNAi-induced silencing of this gene in barley [47]. Other minor suppressor genes located on 3AS, 2DL, and 4D as well as promoters of meiotic pairing located on group 2, 3, and 5 chromosomes have been reported (reviewed in [34]).

Brassica napus is an allotetraploid species ($2n = 4x = 38$, genomes AACC) in which a homoeologous pairing suppressor gene, termed *PrBn*, has been identified [48]. The diploid progenitors of *B. napus* are *B. rapa* ($2n = 18$, AA) and *B. oleracea* ($2n = 20$, CC), which underwent recurrent hybridizations to originate the current tetraploid. *B. napus* shows 19 bivalents at metaphase I, but haploids from different varieties show a low (~10 univalents) or high (~4 univalents) level of homoeologous pairing at metaphase I, which are controlled by different alleles of *PrBn* [49]. This locus maps within an interval of 10–20 cM in the linkage group C9. Thus, *PrBn* resembles *Ph2* of wheat, since its effect on meiosis is manifested in the allohaploid genotypes. SC analysis of euploid plants of varieties with different *PrBn* alleles revealed no difference in their synaptic behavior. Although some multivalents appeared at pachytene, most homologous pairs formed SC bivalents [50]. Allohaploids with different genotype showed also a similar synaptic pattern, characterized by the presence of incomplete synapsis between homoeologous chromosomes. Differences between genotypes concerned the number of class I COs and were only manifested in the allohaploids [50].

2 Plant Materials Derived from Polyploids Used in Meiotic Studies

2.1 Introgression Lines

In addition to natural and synthetic polyploids and interspecific hybrids mentioned above, other materials, such as introgression lines derived from allopolyploids, emerged in the last two decades as powerful tools in the study of meiosis in polyploids. The addition of a pair of alien chromosomes to the full genome complement of a crop species is frequently used as a first step for accessing genetic variation from the secondary gene pool. However, introgression lines can be also used in the study of interspecific genetic interactions, in the chromosomal location of genetic markers and in the study of chromosome structure and behavior in somatic and meiotic cells.

Many studies on meiosis, both in diploids and in polyploids, are focused on the isolation of meiotic mutants, identification of the mutant phenotype and molecular characterization of the affected meiotic step. An alternative approach consists in the study of the behavior of chromosomes introgressed in the genetic background of a different species. Such an approach can be used in polyploid plants such as bread wheat, which tolerates the introgression of chromosomes from related species. Introgression lines of a number of Triticeae species into wheat are available. The study of meiosis in the complete set of introgression lines of rye into wheat has shown that individual chromosomes of rye behave different when the rye genetic background is changed by that of wheat. Differences in the meiotic behavior or rye chromosomes concern chromatin remodeling produced at leptotene, synapsis, and chiasma formation [51–53].

The identification of a specific chromosome or chromosome segments from one species in another genetic pool strongly depends on the phylogenetic proximity between both species. In the case of rye, introgressed chromosomes or segments are identified in fluorescence in situ hybridization (FISH) studies using rye specific repetitive DNA probes, which label almost completely the chromosomes (Fig. 1). Alternatively, total genomic DNA can be used as a probe in genomic in situ hybridization (GISH) experiments (Fig. 1) [54]. In the case of barley, where two sets of cultivated (*Hordeum vulgare*) and wild (*H. chilense*) barley additions were developed in wheat [55–57], GISH protocols has been used in the analysis of meiosis and identification of chromosome rearrangements [58, 59]. Double monosomic wheat–barley addition lines containing one chromosome from *H. vulgare* and the other from *H. chilense*, belonging or not to the same homoeologous group have been also developed to study chromosome recognition and pairing during meiosis in the absence of a homologous partner [59]. Observations confirmed that *Hordeum* homoeologous chromosomes can recognize each other and associate

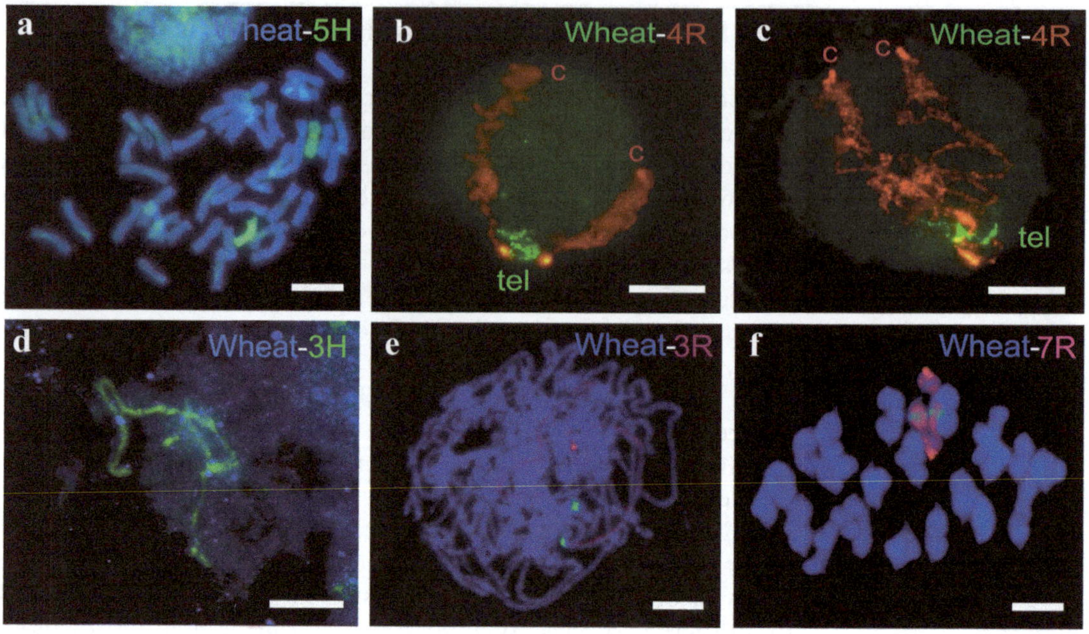

Fig. 1 Painting of alien chromosomes in mitotic and meiotic cells of wheat–barley and wheat–rye addition lines. Total genomic DNA from *H. chilense* (green in **a**, **d**) and DNA probes pUCM600 (for most rye chromatin, red), pSc74 (for subtelomeric heterochromatin, red in **b**, **c**; green in **e**, **f**), pAWRC1 (for rye centromeres, red in **b**, **c**, **e**, **f**), and pAtT4 (for wheat and rye telomeres, green in **b**, **c**, **e**) were used. **(a)** Somatic metaphase of a *T. aestivum*–*H. chilense* disomic addition line for chromosome 5H. **(b)** Nucleus at the leptotene–zygotene transition, rye chromosomes (4R) occupy separate territories, but their ends converge in the telomere cluster. **(c)** Nucleus at the leptotene–zygotene transition with less packaged rye chromatin (4R) than **(a)**. **(d)** Nucleus at zygotene with a pair of *H. chilense* chromosomes (3H) almost fully associated. **(e)** Complete synapsis of rye chromosomes (3R) at pachytene. **(f)** Cell at metaphase I showing a ring bivalent formed by chromosome pair 7R. *tel* telomeres; *c* centromeres. Scale bars represent 10 μm in all panels

correctly in pairs in prophase I, even in the presence of the *Ph1* locus of wheat, but no CO was formed as they appeared as univalents at metaphase I [59]. Sister chromatids segregation and misdivision of univalents at anaphase I was also analyzed in double monosomic additions [60, 61]. Chromosome segregation can occur independently of whether or not related chromosomes associate in pairs in early meiosis in the wheat background.

Telosomic addition lines derived from wheat–rye standard disomic additions are also available and have been used to assess the effect of the *Ph1* locus on chromosome association and synapsis [17] or the effect of colchicine on meiotic chromosome behavior [51]. Some deletion lines lacking a terminal segment of the alien chromosome added to wheat has been also isolated from wheat–rye or wheat–barley additions [61, 62]. Such lines are informative on the role of subtelomeres on chromosome recognition/pairing at early meiosis.

In polyploids, the replacement of a chromosome by its homoeolog is called chromosome substitution. The development of substitution lines involves the replacement of a pair of chromosomes in one variety or species, the recipient, by the homologous pair from another variety or species, the donor. They can be easily obtained after crossing between addition lines and monosomics of the recipient species. Substitution lines are available in different polyploids, but their development may differ according to the ploidy level. In addition to their possible use in the study of meiosis, a considerable amount of genetic information can be obtained from the genomes concerned. In fact, an important advance in understanding the genetic architecture of hexaploid wheat was based in compensation-substitution tests using nullitetrasomic lines. These tests demonstrated the existence of seven homoeologous groups with three chromosomes each [63]. Later, two series of D-genome chromosome substitution lines in tetraploid durum wheat were developed by replacement of one chromosome from the A or B genome by its homoeolog from the D-genome [64]. Examples of substitution lines in hexaploid wheat involving chromosomes from other species are (1) a set of disomic substitutions (DS) of S-genome chromosomes from *Ae. speltoides* for the wheat B-genome chromosomes, which were studied to unzip the effect of such chromosome substitution during meiosis [65], (2) the substitution lines in hexaploid wheat for all barley chromosomes except 1H and 5H [66–68], and (3) the wheat–rye substitution lines [69].

Other alien introgressions in wheat are chromosome translocations. Robertsonian translocations are the result of fusion of two misdivision products originated in the same meiotic cell. Intergenomic translocations can be also originated by meiotic recombination between homoeologous chromosomes. Some wheat–rye translocations have been developed for breeding purposes to introgress genetic traits, such as resistance to diseases, from rye into wheat [70–72]. Others have been used in the identification of CO-rich regions and their effect on homologous interactions at early meiosis [73]. A list of wheat lines carrying rye genetic introgressions for breeding purposes is available on line [74]. Translocations between chromosomes of wheat, *H. chilense* and *H. vulgare* have been obtained from plants carrying chromosomes of these three species, which can be used in breeding related programs and in studies on meiosis [75].

Chromosome rearrangements such as inversions have been also identified in wheat–rye additions. They provide valuable information about chiasma location and chromosome dynamics in the search of the homologous partner, especially when homologous regions are located in opposite poles of the nucleus as in heterozygotes for large paracentric inversions [76, 77].

2.2 Deletion Lines

Finally, polyploids such as bread wheat tolerate the presence of deletions even in homozygous condition. The deletion stocks of wheat obtained after chromosome truncation induced by the action of gametocidal chromosomes [78] are widely used in genetic and genomic studies. They represent also an excellent material in studies on meiosis. For example, they showed that the CO events occur mainly in the distal regions of the chromosomes, with a gradient from the centromeres to the telomeres, as well as that recombination frequency can increase in intercalary regions when the chromosome truncation forces the first CO to move toward the centromere regions [79]. Nevertheless, recombination is highly restricted in the proximity of the centromere.

3 Cytogenetic Tools in the Study of Meiosis in Polyploids and Their Derivatives

3.1 Chromosome Content Screening in Somatic Cells

Cytological procedures such as in situ hybridization, which are capable of identifying alien chromatin introgressed in a given polyploid, are extensively used to confirm the presence of the alien genetic material in the introgression line, or in derivatives obtained in breeding related programs based on marker-assisted selection. Routinely, GISH and FISH are accomplished on chromosomes recovered either from rapidly dividing root tips that have a high mitotic index (Fig. 1). Root tip sampling can be performed on seedlings that are only a few days old, which does not destroy the individual. The root tips of more mature plants can also be sampled; the plant is removed from its pot, the ends of the roots are clipped, and ample water is applied. In a few days, newly grown root tips emerge that can be used for cytological analysis. A variety of treatments, including colchicine, cold, hydroxyquinoline, and nitrous oxide, have been used to arrest somatic cells in metaphase. Such treatments, in addition to accumulate a large number of metaphase-arrested cells, facilitate chromatin condensation and visualization of the morphology of the chromosomes [80–85]. After this treatment, root tips are fixed and stored at 4 °C, even for months, until performing the FISH or GISH experiment. Most of the protocols are focused on somatic chromosome spreads, but equipment, chemicals and solutions are quite similar for studying meiosis in FISH experiments.

3.2 Sampling of Meiotic Cells

The sampling of cells at prophase I or metaphase I requires growing plants to near the flowering stage and some knowledge of the best sampling time for each species. Anthers are the preferred source of meiocytes because of their abundance compared to ovaries. Here we summarize the whole procedure for studying meiosis in polyploid cereals, although this protocol can be adapted in other species. The outline of the procedure is represented in Fig. 2. An important issue is the identification of the maturity stage when a plant is entering meiosis. Previously to anther excision, the

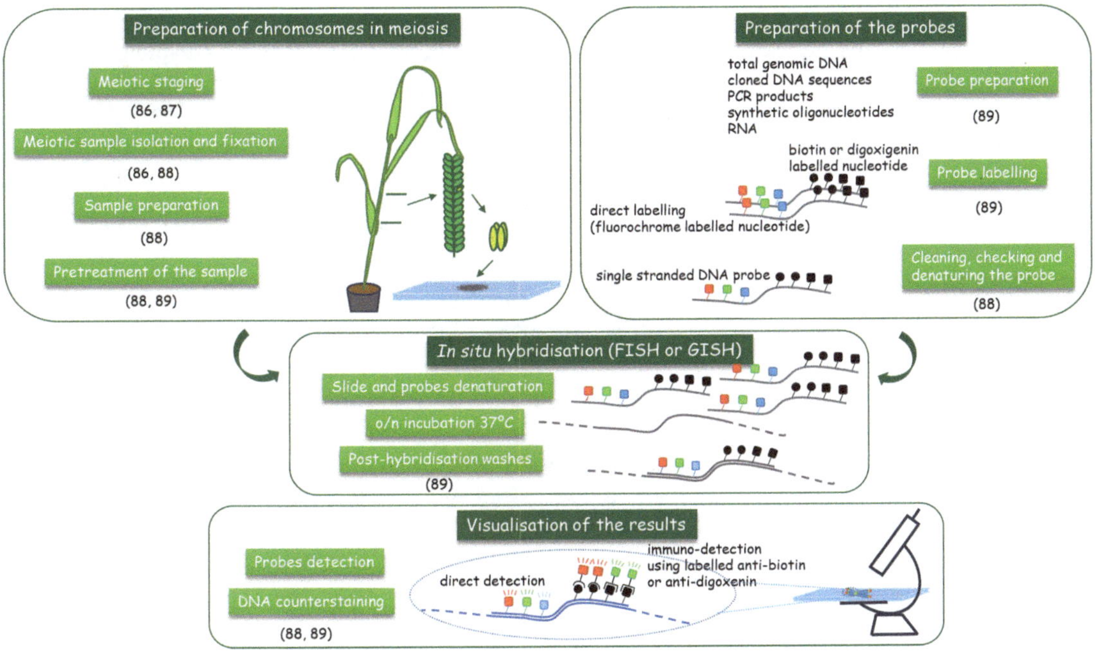

Fig. 2 Summary of the procedure for the development of in situ hybridization experiments in polyploid plants. Wheat is taken as an example. Numbers in brackets correspond to the references cited in the references list

developmental stage of each main shoot or tiller should be precisely estimated. Taking wheat as an example, meiosis occurs when the immature spike is still inside the tiller (Fig. 2). External features of the tiller help to recognize when the spike is in meiosis. Briefly, the flag leaf should be emerged and the spike enclosed within the sheaths of the leaves just in the internode between the leaves that precede the flag leaf. The spike length inside the tiller can be estimated by feeling gently with the fingers its bottom and top end. The immature spike is considered to be in meiosis when its length is around 5 cm. Anthers are fixed without pretreatment and stored at 4 °C. Materials, reagents, and equipment for anthers sampling and staging of PMCs (pollen mother cells) in FISH and GISH experiments have been described [86–89]. Given that meiosis may be affected by temperature, humidity, or day length, considerations to grow polyploid plants to fruitfully isolate meiocytes for FISH or GISH experiments should be also taken into account [86].

3.3 Probes and Equipment Used for GISH and FISH Analyses

Detailed protocols for GISH and multicolor FISH analysis, including materials and the necessary equipment for plant cytogenetic analysis, have been extensively described. DNA or RNA can be used to obtain the probe for in situ hybridization experiments. Total genomic DNA, cloned DNA sequences, PCR products and synthetic oligonucleotides are commonly used as probes [88] (Table 1). Molecular information concerning the DNA sequence,

Table 1
List of DNA probes and antibodies used for targeting specific meiotic events in polyploid species

DNA probes (FISH and GISH)			
Target	**DNA probes**	**Application in meiosis**	**References**
Alien genetic introgressions	Total genomic DNA	Homologous chromosome pairing and segregation	[51–53, 59, 60]
Centromeres	CCS1 CentO CentC CRB Bilby	Chromosome dynamics, role in chromosome recognition, association and segregation	[90–97]
Telomeres	pAtT4	Chromosome dynamics, role in chromosome recognition, association and segregation	[53, 98, 99]
Subtelomeres	HvT01 pSc119.2 pSc74	Role of subtelomeric sequences in homologous recognition and pairing	[59, 61, 62]
Several DNA sequences	Synthetic oligonucleotides	Diverse specific meiotic events	[100–102]

Antibodies (immunolocalization)			
Target	**Antibody**	**Application in meiosis**	**References**
Centromeres	Anti-CEN-H3	Chromosome dynamics, role of functional centromeric sequences in chromosome associations and segregation	[103, 104]
Synaptonemal complex:			
Axial element proteins	Anti-ASY1	Synapsis	[6, 105, 106]
Central element proteins	Anti-ZYP1	Formation of synaptonemal complex	[106–108]
Crossing overs	Anti-MLH1 Anti-RAD51	Crossing-over Early recombination nodules	[109, 110] [109, 111]
Cohesins and cohesin-associated proteins	Anti-SMC1 Anti-SMC3 Anti-SCC3 Anti-REC8/SYN1	Sister chromatid cohesion Sister chromatid cohesion Sister chromatid cohesion Chromosome condensation and pairing	[109] [109] [109, 110] [109]

number of copies, genomic organization and species specificity of the probes is of interest in the design of the in situ hybridization experiment. Most of this information is often known, however additional DNA in situ hybridization experiments may be performed in order to unravel long-range evolutionary changes involving the probe and its interspersion with other sequences.

Technical advances in DNA synthesis have allowed a large-scale de novo production of thousands of independent oligonucleotides (oligos). These massively synthesized oligos have been successfully labeled to be used as FISH probes and have paved the way for the development of chromosome-specific painting probes in plants, which also have the potential to be used in meiosis studies in polyploids [100–102].

RNA in situ hybridization gives information about the location and amount of expression of particular genes. Detailed information on probes (total genomic DNA, cloned DNA sequences, PCR fragments, and RNA) preparation and in situ hybridization protocols, equipment, supplies, chemical and solutions has been described [88, 89].

3.4 Antibodies and Equipment in Immunolocalization

The use of immunolocalization of chromosome-associated proteins, alone or combined with FISH and GISH, is an indispensable tool that has greatly contributed to the study of meiosis in plants [105, 112]. The immunolocalization technique is based on the principles of antigen-antibody reactions to find highly specific molecules in a cell or tissue. The common reagent for immunolocalization are polyclonal or monoclonal antibodies, both produced in animals such as rabbits, goats, rats, mice, among others by specific procedures [113]. There are many commercially available antibodies to detect chromosomal proteins in plants. Some of them, for example, the anticentromeric H3 histone, are of universal use [103]. For detection of proteins by immunolocalization, cells must be fixed and permeabilized to facilitate the access of antibodies to the nucleus. Different procedures for immunodetection of proteins associated to chromosomes to study meiosis in plants have been recently reviewed [87, 109].

3.5 Identification of Alien Chromatin in a Polyploid Background in Somatic and Meiotic Cells by FISH or GISH

In situ hybridization is routinely used in the identification of alien chromatin introgressed in a given polyploid. *GISH*, which was first applied to plant tissues in the latest eighties [114], and FISH are well-established methodologies to study meiosis. Highly efficient protocols for cytological identification of specific sequences in plants have been deeply described [88, 89]. These protocols were carried out in somatic or meiotic chromosome spreads, but can be also adapted with little modifications to 3D vibratome plant tissues, which allows the analysis in intact cells [115].

GISH is used to differentiate chromosomes from different genomes in interspecific hybrids and allopolyploids [114, 116].

Homologous and homoeologous chromosome associations can be identified using GISH in meiosis across a wide range of Triticeae species (Fig. 1) [59, 77]. Protocols referred above can be somehow customized with few modifications according to the types of cells and plant species concerned. For example, in *Brassica*, GISH is routinely performed to study genome stability and interspecific chromosome associations during metaphase I in hybrids between *Brassica* and related species [117]. In banana (*Musa* sp.), GISH has been revealed very useful to investigate meiotic chromosome pairing of different chromosome in the allotetraploid (AABB) of *M. acuminata* and *M. balbisiana* [118].

In wheat–rye introgression lines, FISH has been used to visualize the complete rye chromosomes through all meiotic stages (Fig. 1) [62].

3.6 FISH Analysis of Chromosome Dynamics in Meiosis

Multicolor FISH is performed as a routine tool to study meiosis in numerous plant cytogenetics laboratories due to the development of versatile probe-labeling reagents and techniques, signal detection systems, optical filter technology and digital image analysis systems for pseudocoloring and merging of images. This methodology is usually applied to visualize chromosome structures such as centromeres or telomeres, which play a relevant role on chromosome dynamics during meiosis. Many efforts have been focused on centromeres and telomeres behavior, especially in hexaploid wheat [51–53, 90–92, 119, 120]. A list of useful DNA probes and antibodies to target chromosome dynamics during meiosis and specific meiotic events in polyploids is shown in Table 1.

In most plants, centromeres are usually composed of a small DNA repeat that is tandemly arranged thousands of times at each primary constriction [93]. A cereal centromeric sequence (CCS1 sequence), identified within a clone (Hi-10) from *Brachypodium sylvaticum*, hybridizes to the centromeres of wheat and some other Triticeae species [121]. This CCS1 and the rye centromere-specific repetitive sequence pAWRC.1 [94] have been used in the analysis of centromere dynamics during early meiosis [90–92, 119, 120] as well as to investigate the neocentric activity of 5RL chromosome in wheat [122] or the centromeric structure of wheat–rye 1BL.1RS translocation lines [123]. Synthetic oligonucleotides can replace DNA probes such as CCS1 or pAWRC.1 in the identification of centromeres in FISH analysis in somatic cells [100, 101] but it is possible to use them also for meiosis studies. In fact, oligo-based probes have been employed to monitor pairing of a specific homoeologous chromosome pair throughout meiosis in *Cucumis hystrix* × cucumber hybrids [102].

In other species, a particular type of retrotransposon is present in centromeres, such as the tandem repeats of rice (CentO, previously named RCS2) and maize (CentC) centromeres, which are interspersed with a particular family of retrotransposons called

Centromeric Retrotransposons of Rice (CRR) and Maize (CRM) [95, 96, 124, 125]. A centromere-specific retrotransposon in *Brassica* (CRB) has been also detected by in situ hybridization as a major component of all centromeres in three diploid *Brassica* species and their allotetraploid relatives [97].

Telomeres are highly conserved structures among plants. An *Arabidopsis* telomere clone has been used for in situ hybridization to chromosomes of several monocot and dicot species. In all of them, the sequences hybridize exclusively to the ends of chromosomes indicating a high level of conservation of the telomeric repeats [98, 99]. This telomeric repeat has been used to investigate telomeres dynamics during meiosis in polyploids [53].

In species with large genomes, as hexaploid wheat, chromosomes enter in meiosis arranged in the Rabl orientation, that is, centromeres are concentrated in one pole of the nucleus and telomeres fanned out the opposite hemisphere. This arrangement facilitates a codirectional orientation of homologous chromosome segments [126]. Then, telomeres undergo an oriented migration and associate in a bouquet [127]. The bouquet configuration brings homologues together, which is considered an initial step in the search and recognition of the homologous partner. In hexaploid wheat, telomere convergence is affected by the *Ph1* locus [128]. Concomitant with telomere clustering, a conformational change in the chromatin is triggered, which facilitates physical interaction between homologs and the initiation of pairing and synapsis in distal positions [53, 129]. Later, pairing and synapsis extend from both ends to the chromosome center.

Since the DNA sequence of the telomeres is largely conserved, it remains poorly understood how chromosomes identify their homologous partners to correctly associate in pairs. Further FISH and GISH experiments may shed light on homologous recognition. Some works are focused on telomeres-associated sequences (TAS). These sequences are located on the ends of the chromosomes, can be visualized using FISH and their role on chromosome recognition and pairing is being assessed [61]. TAS are organized in long subtelomeric arrays of hypervariable tandemly repeat units. In rye, TAS are highly repeated and form subtelomeric heterochromatin chromomeres of variable size [130], which reveals the position of the adjacent telomere in the prophase I nucleus [53]. The probability of inclusion of the rye telomeres in the telomere cluster was found to be chromosome conformation dependent in wheat–rye additions, which has implication in the subsequent development of synapsis. The behavior of the polymorphic nature of subtelomeres is an exciting area for study, but also presents a difficult challenge from the technical perspective. Gene-rich regions, less evolutionary conserved than telomeres, are also distally located and represent hot spots of recombination [131]. The presence of repeated and single copy DNA sequences in subtelomeric regions most likely confer an increased difficulty when it comes to discerning what type of sequence play a critical role on chromosome recognition. The subtelomeric repeat sequences are much more divergent than evolutionary conserved telomeric sequences and they have been cloned and used as molecular markers to cap for example genetic maps in hexaploid wheat [132].

Bacterial artificial chromosomes probes (BAC-FISH) represent an additional tool in FISH experiments aimed to study the role of chromosomal regions different from centromeres or telomeres on chromosome pairing in meiosis of polyploids and interspecific hybrids. For example, BAC-FISH has allowed tracing the meiotic behavior of specific chromosome pairs in allotriploid hybrids between cultivated potato, *Solanum tuberosum*, and its wild relatives *S. commersonii* [133]. BAC-FISH has contributed also to predict, from meiotic homoeologous recombination, the possibility to transfer some resistant traits coded by genes of the RR subgenome of allotetraploid *Brassicoraphanus* (RRCC, $2n = 4x = 36$) into *Brassica napus* (AACC, $2n = 4x = 38$) [134].

3.7 Immuno-FISH

Immuno-FISH combines immunolabeling with FISH in order to visualize simultaneously the nuclear distribution of proteins and specific DNA sequences on the chromosomes (Fig. 3). This methodology is particularly interesting in studies of meiosis because morphogenesis and function of meiotic proteins are associated to stage-specific conformational changes of chromatin.

Different procedures for immunodetection of chromosome associated proteins in plant meiosis have been reviewed [87, 109]. Several proteins are associated with the DNA of the

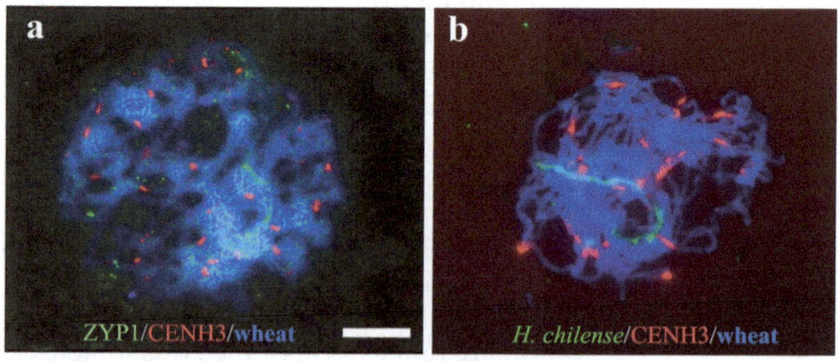

Fig. 3 Immunolocalization and immuno-FISH experiments on wheat and a wheat–barley addition line, respectively, during pachytene. (**a**) Simultaneous immunolocalization of ZYP1 (green) and CENH3 (red) on a wheat nucleus. ZYP1 formed long thick tracts along the synapsed homologous chromosomes. (**b**) Immuno-FISH on a *T. aestivum–H. chilense* addition line using the CENH3 antibody to detect centromeres (red) and total genomic DNA from *H. chilense* as a probe to label the synapsed barley chromosomes (green). Scale bar represents 10 μm

meiotic chromosomes; centromeric proteins associate constitutively others transiently. Centromeric histone H3 (CENH3) is extensively used for labeling of centromeres and also in studies on meiosis in polyploids [104]. Some proteins involved in the synaptonemal complex assembly during zygotene can be also immunolabeled. For example, ASY1 was reported as an axial element (AE)-associated protein in plants that is loaded onto the unsynapsed chromosome axes before synapsis [105]. ZYP1 is a protein that polymerizes between the unsynapsed axial cores of homologous chromosomes during zygotene until synapsis is complete at pachytene [107]. Antibodies raised against the *Arabidopsis* ASY1 and ZYP1 have been used to detect their homologous proteins in barley and rye, where they conserve the same localization in the synaptonemal complex [106, 135]. ASY1, ZYP1, and CENH3 have been combined with the telomeric repeat DNA probe in immune-FISH experiments to analyze centromere dynamics and the timing of chromosome synapsis in wheat prophase I [104, 108]. In addition, dynamics during plant meiosis of other proteins such as cohesins SCC3 and SYN1, histone HTR12 (the histone H3 variant of *A. thaliana*) or recombination proteins REC8, RAD51, and MLH1 has been deeply studied by immunolabeling [109, 110]. All of these proteins are useful tools to unzip chromosome dynamics during meiosis in polyploids (Table 1) [108, 111, 136].

3.8 Vibratome Sectioning for 3D Meiosis Study in Intact Cells in FISH and Immune-FISH Experiments

It is worthy to mention here, that most interesting process of plant cell biology and development such as meiosis, occur in cells deep within tissues of the plant. These cells are difficult to image, even when confocal microscopy is used. Several studies on wheat and maize using FISH have all shown the importance of examining well-preserved material, to be confident of both the 3-D preservation of delicate nuclear structure and the cell types that are being analyzed [137]. In squashed preparations of anthers, cell types could be misclassified, or polarized centromeres and telomere clusters produced during early meiosis can easily be displaced or brought together artificially. Several procedures have been determined for undertaking 3D in situ approaches on rice and *Arabidopsis*. In wheat, FISH was applied on three-dimensionally preserved tissue sections derived from intact plant structures such as florets [115]. This method combines vibratome sectioning with in situ hybridization and confocal microscopy, does not require embedding of tissue wax or resin, and can be applied to fixed or unfixed tissues. Vibratome tissue sectioning produces relatively thick sections (20–50 μm) which can be imaged to reveal the three-dimensional structure of the underlying tissues. It has the advantage of preserving well the 3D structure, so that subcellular structures can be reliably imaged. Furthermore, reliable identification of cell types often requires an accurate assessment of the tissue context, which is lost when cytological squash preparations are made. For example meiocytes can be clearly identified at all stages of meiosis and can be imaged in the context of their surrounding maternal tissue. FISH can be used to localize centromeres, telomeres and subtelomeric regions, and total genomic DNA can be used as probes to visualize introgressed chromosomes or chromosome segments. Vibratome sections can be used in other labeling experiments, such as RNA-FISH to detect transcripts, in antibody immunofluorescence labeling, in labeling of nascent transcripts by BrUTP incorporation in unfixed tissue, and in combined antibody and BrUTP labeling with FISH (e.g., [138]). Vibratome sectioning, which has been used to study meiosis in florets of wheat and related species [90, 91, 137] and in polyploid rice (*Oryza sativa* L.) [139], is a reliable tool for investigating meiotic events in plants, over a wide range of genome and taxa.

3.9 Diagnostic Meiotic Configurations

Meiotic configurations of polyploids may include the presence of univalents, bivalents and multivalents. In the absence of chromosome translocations, the types of multivalents that can be found in polyploids depend on the ploidy degree: trivalents are expected in triploids; trivalents and quadrivalents in tetraploids; trivalents, quadrivalents, pentavalents, and hexavalents in hexaploids; and so on. Triploids show a very low fertility, since they form genetically balanced gametes (n or $2n$) with a probability of $(1/2)^{n-1}$, where n is the haploid chromosome number. Thus, this ploidy level is barely transmitted by sexual reproduction. In contrast, tetraploids produce diploid gametes ($2n$) with a relatively high frequency, which ensures the transmission of the tetraploid condition. Evolutionarily, tetraploids represent the first step in the formation of polyploids species either in autopolyploids or allopolyploids. By this reason we will concern mainly meiotic configurations of tetraploids. Studies of pairing and synapsis have been usually carried out in nuclei at early and mead prophase I in the absence of chromosome-specific markers. This methodology makes it difficult to establish whether there are differences between chromosomes or not. The existence of a genome reference sequence in many species offers the possibility of generating chromosome specific molecular tools to identify individual chromosomes in prophase I. This is the case of a recent report showing synapsis of individual chromosomes in polyploid *Solanum* species [140]. Oligonucleotide-based probes were developed to label four different potato chromosomes. Each tetrasome of cultivated potato (*S. tuberosum*) formed a quadrivalent in 66–78% of cells at pachytene. However, in the hexaploid *S. demissum*, the six chromosomes of each homologous set formed three bivalents in 79–98% of cells at pachytene, which suggest a more evolved pairing genetic control in this species. In the meantime similar genomic tools are designed in other species, information about specific chromosomes can be obtained from metaphase I observations.

Sybenga [28] developed a model that allow to estimate the frequency of the two modes of synapsis, one quadrivalent or two bivalents, involving four homologous chromosomes with two distal pairing initiation sites per chromosome, as well as the frequency of chiasmate bonds in each chromosome arm in both types of SC configurations. The model concerns metacentric chromosomes with undistinguishable arms in organisms with exclusive localization of chiasmata in distal regions. Under this assumption, the following metaphase I configurations can be produced: ring quadrivalents (frequency rq); chain quadrivalent (frequency cq); trivalent plus univalents (frequency t); ring bivalent (frequency r); open bivalent (frequency o); and univalent pair (frequency u). All multivalents are obviously derived from the SC quadrivalent, ring bivalents are derived from two SC bivalents and the open bivalents and univalents pairs can be derived from both SC configurations. The frequency f of quadrivalents resulted to be:

$$f = \frac{(t + 2cq + 4rq)2}{16rq}$$

Frequencies a' and b' of chiasmate bond of the two chromosome arms in the SC quadrivalent type can be deduced from the following equations:

$$rq = fa'2b'2$$

$$cq = 2\mathrm{f}(a'b'^2 + a'^2b' - 2a'2b'^2)$$

Frequencies a and b of chiasmate bond of the two arms in the two bivalents synaptic type can be derived from a' and b' and the mean frequencies A and B of association of each chromosome arm obtained from all metaphase I configurations.

$$f \times a' + (1 - f) \times a = A$$

$$f \times b' + (1 - f) \times b = B$$

In tetraploid *Tradescantia virginiana* ($2n = 4x = 24$) f was estimated to be 0.658, close to 0.667 (2/3), the expected frequency of SC quadrivalents in random pairing. The means of chiasmate association of the two arms were $A = 0.946$ and $B = 0.683$. The estimates of a', b', a, and b were: $a' = 0.950$, $b' = 0.605$, $a = 0.938$, and $b = 0.830$.

This model was later refined and adapted to individual rye chromosomes 1R, 2R, and 5R, where interstitial chiasmata in addition to distal chiasmata were formed and distinction between chromosome arms was possible [141]. The frequencies estimated for one quadrivalent and two bivalents differed from those expected for random pairing in the three chromosomes. An increased bivalent pairing was found, which was not produced by preferential pairing.

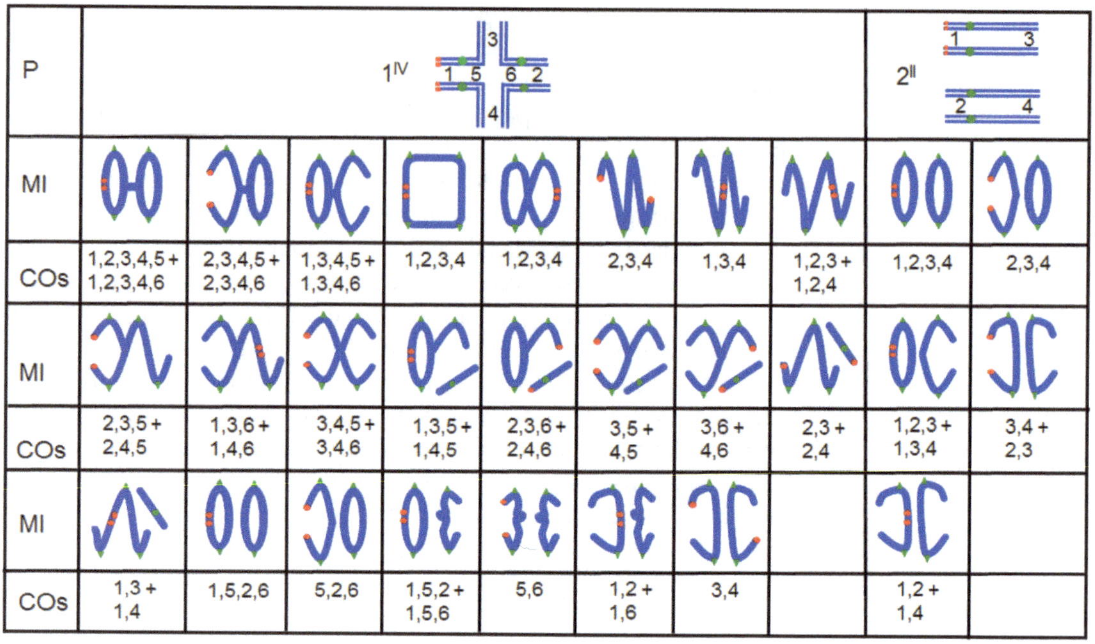

Fig. 4 Configurations at pachytene and metaphase I of four homologs showing homomorphic pairing. Numbers indicate chromosome segments where COs locate

Either spatial separation of the four homologs in two groups of two chromosomes in premeiotic cells, or correction of SC multivalents, or both, were suggested to be responsible of the increase in the frequency of two bivalents. Estimates of chiasma frequency of each chromosome arm were different after quadrivalent and bivalent pairing. This model may be extended to other tetraploids.

The occurrence of preferential pairing can be investigated when a set of homologous chromosomes is polymorphic for a chromosome marker. This is the case of the homologous set indicated in Figs. 4 and 5. The tetrasome consists of four submetacentrics, two with and two without a distal marker in the short arm. Such a situation can be generated when chromosome duplication is produced in a diploid heterozygote for the chromosome marker. Because of the difference in length between arms, the pairing partner switch in the quadrivalent is assumed to locate in the long arm. In such a situation, interstitial chiasmata in addition to distal chiasmata can be formed in this arm. Two pairing types are expected: (1) homomorphic pairing (Fig. 4), when the short arms carrying the marker, as well as those unlabeled, synapse either in the mode of one quadrivalent or two bivalents; (2) heteromorphic pairing (Fig. 5), when synapsis occurs between arms carrying and lacking the marker. Configurations at metaphase I assuming a minimum of one COs per chromosome pair are shown in both pairing types (Figs. 4 and 5). Markers positioning identifies each pairing type, which allows to assess whether chromosomes pair or

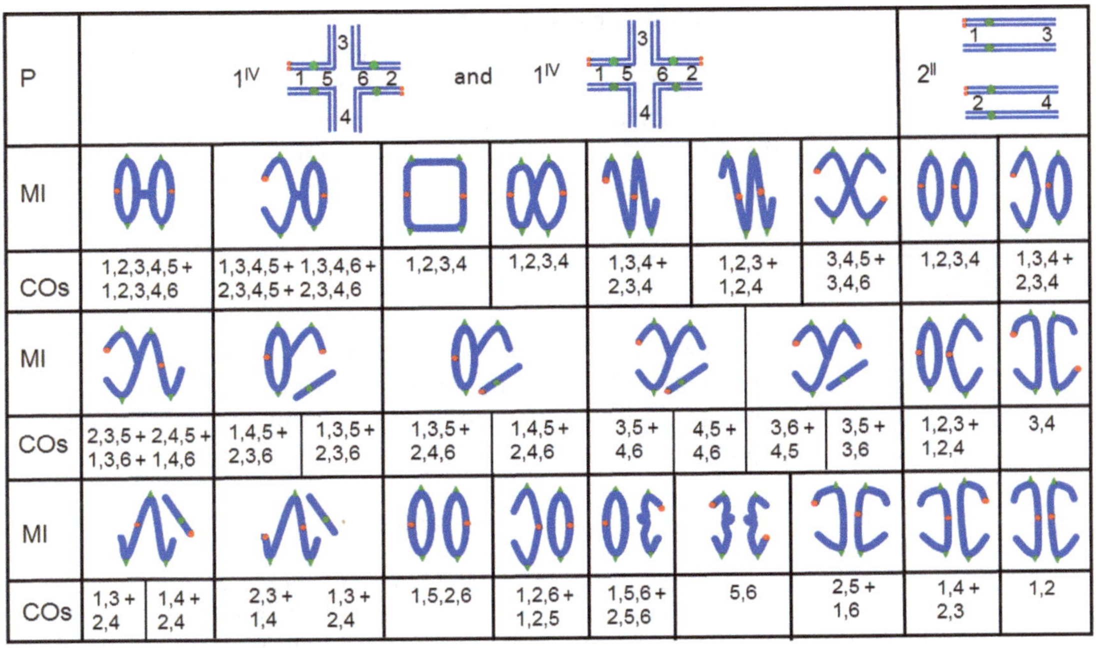

Fig. 5 Configurations at pachytene and metaphase I of four homologs showing heteromorphic pairing. Numbers indicate chromosome segments where COs locate

not at random. Metaphase I configurations provide also information of the frequency of two COs, intercalary and distally located, respectively. Such situation is apparent in trivalents and quadrivalents with COs formed in the interstitial segment 5 or 6 of Figs. 4 and 5. Of course, estimates of *f*, *a*, *b*, *a′*, and *b′* can be derived as indicated above. Similar models are possible for chromosomes with a different conformation or carrying markers in both arms.

4 Concluding Remarks

We present here a summary of achievements reached in the study of meiosis in polyploids. We mainly emphasize the development of materials and methodologies that have favored the most relevant advances, which can be applied in future research. The main feature of the meiotic chromosome behavior in polyploids concerns the presence of more than two potential partners and their physical interactions leading to pairing, synapsis and recombination. Although the mode of how chromosome sorting is produced remains still poorly understood, polyploids, especially allopolyploids, can shed light on this issue when interactions between alternative combinations of potential partners are identified. Introgression lines can help to obtain such chromosome constructions and to identify the different pairs that can be generated. Advances produced on chromosome painting technologies can help also in this attempt. More difficult may be this analysis in autopolyploids, given that all chromosome sets they have are homologous. Perhaps, future research works should be focused on the identification of individual sets of homologs and the comparative analysis of their behavior especially in the first meiotic division.

Acknowledgments

This work was supported by grants AGL2015-67349-P and AGL2015-64833-R from Dirección General de Investigación Científica y Técnica, Ministerio de Economía y Competitividad of Spain.

References

1. Ramsey J, Schemske DW (1998) Pathways, mechanisms, and rates of polyploid formation in flowering plants. Annu Rev Ecol Syst 29:467–501
2. Renny-Byfield S, Wendel JF (2014) Doubling down on genomes: polyploidy and crop plants. Am J Bot 101:1711–1725. https://doi.org/10.3732/ajb.1400119
3. Keeney S, Giroux CN, Kleckner N (1997) Meiosis-specific DNA double-strand breaks are catalysed by Spo11, a member of a widely conserved protein family. Cell 88:375–384
4. Neale MJ, Keeney S (2006) Clarifying the mechanics of DNA strand exchange in meiotic recombination. Nature 442:153–158. https://doi.org/10.1038/nature04885
5. Hunter N, Kleckner N (2001) The single-end invasion: an asymmetric intermediate at the double-strand break to double-Holliday junction transition of meiotic recombination. Cell 106:59–70
6. Caryl AP, Armstrong SJ, Jones GH, Franklin FCH (2000) A homologue of the yeast *HOP1* gene is inactivated in the *Arabidopsis* meiotic mutant asy1. Chromosoma 109:62–71
7. Nonomura KI, Nakano M, Eiguchi M, Suzuki T, Kurata N (2006) PAIR2 is essential for homologous chromosome synapsis in rice meiosis I. J Cell Sci 119:217–225. https://doi.org/10.1242/jcs.02736
8. Wang KJ, Wang M, Tang D, Shen Y, Qin BX, Li M et al (2011) PAIR3, an axis-associated protein, is essential for the recruitment of recombination elements onto meiotic chromosomes in rice. Mol Biol Cell 22:12–19. https://doi.org/10.1091/mbc.E10-08-0667
9. Ferdous M, Higgins JD, Osman K, Lambing C, Roitinger E, Mechtler K et al (2012) Inter-homolog crossing-over and synapsis in *Arabidopsis* meiosis are dependent on the chromosome axis protein AtASY3. PLoS Genet 8:e1002507. https://doi.org/10.1371/journal.pgen.1002507
10. Roeder GS (1997) Meiotic chromosomes: it takes two to tango. Genes Dev 11:2600–2621
11. Baudat F, Manova K, Yuen JP, Jasin M, Keeney S (2000) Chromosome synapsis defects and sexually dimorphic meiotic progression in mice lacking Spo11. Mol Cell 6:989–998
12. Romanienko PJ, Camerini-Otero RD (2000) The mouse *Spo11* gene is required for meiotic chromosome synapsis. Mol Cell 6:975–987
13. Page SL, Hawley RS (2004) The genetics and molecular biology of the synaptonemal complex. Annu Rev Cell Dev Biol 20:525–558. https://doi.org/10.1146/annurev.cellbio.19.111301.155141
14. Zickler D, Kleckner N (2015) Recombination, pairing, and synapsis of homologs during meiosis. Cold Spring Harb Perspect Biol 7:a016626. https://doi.org/10.1101/cshperspect.a016626
15. Higgins JD, Osman K, Jones GH, Franklin FCH (2014) Factors underlying restricted crossover localization in barley meiosis. Annu Rev Genet 48:29–47. https://doi.org/10.1146/annurev-genet-120213-092509
16. Bass HW, Marshall WF, Sedat JW, Agard DA, Cande WZ (1997) Telomeres cluster de novo before the initiation of synapsis: a three-

dimensional spatial analysis of telomere positions before and during meiotic prophase. J Cell Biol 137:5–18

17. Maestra B, De Jong JH, Shepherd K, Naranjo T (2002) Chromosome arrangement and behaviour of two rye telosomes at the onset of meiosis in disomic wheat-5RL addition lines with and without the *Ph1* locus. Chromosom Res 10:655–667
18. Loidl J, Jones GH (1986) Synaptonemal complex spreading in *Allium*. I Triploid A sphaerocephalon. Chromosoma 93:420–428
19. Loidl J (1986) Synaptonemal complex spreading in *Allium*. II. Tetraploid *A. vineale*. Can J Genet Cytol 28:754–761
20. Vincent JE, Jones GH (1993) Meiosis in autopolyploid *Crepis capillaris*. I. Triploids and trisomics; implications for models of chromosome pairing. Chromosoma 102:195–206
21. Thomas HM, Thomas BJ (1993) Meiosis in triploid *Lolium*. I Synaptonemal complex formation and chromosome configurations at metaphase I in aneuploidy autotriploid L. multiflorum. Genome 37:181–189
22. Jones GH, Vincent JE (1994) Meiosis in autopolyploid *Crepis capillaris*. II Autotetraploids. Genome 37:497–505
23. Stack S, Roelofs D (1996) Localized chiasmata and meiotic nodules in the tetraploid onion *Allium porrum*. Genome 39:770–783
24. Gillies CB, Kuspira J, Bhambhani RN (1987) Genetic and cytogenetic analyses of the a genome of *Triticum monococcum*. IV Synaptonemal complex formation in autotetraploids. Genome 29:309–318
25. Chatterjee R, Jenkins G (1993) Meiotic chromosome interactions in inbred autotetraploid rye (*Secale cereale*). Genome 36:131–138
26. Santos JL, Cuadrado MC, Diez M, Romero C, Cuñado N, Naranjo T et al (1995) Further insights on chromosomal pairing of autopolyploids: a triploid and tetraploids of rye. Chromosoma 104:298–307
27. Díez M, Santos JL, Cuñado N, Naranjo T (2001) Meiosis in primary trisomics of rye: considerations for models of chromosome pairing. Chromosom Res 9:13–23
28. Sybenga J (1975) Meiotic configurations. Springer, Berlin
29. Jenkins G, White J (1990) Elimination of synaptonemal complex irregularities in a *Lolium* hybrid. Heredity 64:45–53
30. Santos JL, Alfaro D, Sanchez-Moran E, Armstrong SJ, Franklin FCH, Jones GH (2003) Partial diploidization of meiosis in autotetraploid *Arabidopsis thaliana*. Genetics 165:1533–1540
31. Wu JH, Datson PM, Manako KI, Murray BG (2014) Meiotic chromosome pairing behaviour of natural tetraploids and induced autotetraploids of *Actinidia chinensis*. Theor Appl Genet 127:549–557
32. Ramsey J, Schemske DW (2002) Neopolyploidy in flowering plants. Annu Rev Ecol Syst 33:589–639. https://doi.org/10.1146/annurev.ecolsys.33.010802.150437
33. Jenczewski E, Alix K (2004) From diploids to allopolyploids: the emergence of efficient pairing control genes in plants. Crit Rev Plant Sci 23:21–45. https://doi.org/10.1080/07352680490273239
34. Naranjo T, Benavente E (2015) The mode and regulation of chromosome pairing in wheat-alien hybrids (Ph genes, an updated view). In: Molnár-Láng M, Ceoloni C, Doležel J (eds) Alien introgression in wheat. Cytogenetics, molecular biology and genomics. Springer, Cham, pp 133–162
35. Feldman M, Levy AA (2015) Origin and evolution of wheat and related triticeae species. In: Molnár-Láng M, Ceoloni C, Doležel J (eds) Alien introgression in wheat. Cytogenetics. Molecular biology and genomics. Springer, Cham, pp 21–76
36. Sears ER, Okamoto M (1958) Intergenomic chromosome relationship in hexaploid wheat. In: Proc 10th int Congr genetics. University of Toronto Press, Toronto, pp 258–259
37. Riley R, Chapman V (1958) Genetic control of the cytologically diploid behaviour of hexaploid wheat. Nature 182:713–715
38. Riley R, Kempana C (1963) The homoeologous nature of the non-homologous meiotic pairing in *Triticum aestivum* deficient for chromosome V (5B). Heredity 18:287–306
39. Mello-Sampayo T (1971) Genetic regulation of meiotic chromosome pairing by chromosome 3D of *Triticum aestivum*. Nat New Biol 230:22–23
40. Prieto P, Moore G, Reader S (2005) Control of conformation changes associated with homologue recognition during meiosis. Theor Appl Genet 111:505–510. https://doi.org/10.1007/s00122-005-2040-6
41. Gill KS, Gill BS, Endo TR, Mukai Y (1993) Fine physical mapping of *Ph1*, a chromosome pairing regulator gene in polyploid wheat. Genetics 134:1231–1236
42. Griffiths S, Sharp R, Foot TN, Bertin I, Wanous M, Reader S, Colas I, Moore G (2006) Molecular characterization of *Ph1* as a major chromosome pairing locus in

polyploid wheat. Nature 439:749–752. https://doi.org/10.1038/nature04434

43. Rey MD, Martín AM, Higgins J, Swarbreck D, Uauy C, Shaw P, Moore G (2017) Exploiting the ZIP4 homologue within the wheat Phl locus has identified two lines exhibiting homoeologous crossover in wheat-wild relative hybrids. Mol Breeding 37:95. https://doi.org/10.1007/s11032-017-0700-2
44. Bhullar R, Nagarajan R, Bennypaul H, Sidhu GK, Sidhu G, Rustgi S, von Wettstein D, Gill KS (2014) Silencing of a metaphase I specific gene present in the *Ph1* locus results in phenotype similar to that of the *Ph1* mutations. Proc Natl Acad Sci U S A 111:14187–14192. https://doi.org/10.1073/pnas.1416241111
45. Sutton T, Whitford R, Baumann U, Dong C, Able JA, Langridge P (2003) The *Ph2* pairing homoeologous locus of wheat (*Triticum aestivum*): identification of candidate meiotic genes using a comparative genetics approach. Plant J 36:443–456
46. Dong C, Whitford R, Langridge P (2002) A DNA mismatch repair gene links to the *Ph2* locus in wheat. Genome 45:116–124
47. Lloyd AH, Milligan AS, Langridge P, Able JA (2007) *TaMSH7*: a cereal mismatch repair gene that affects fertility in transgenic barley (*Hordeum vulgare* L.). BMC Plant Biol 7:67. https://doi.org/10.1186/1471-2229-7-67
48. Grandont L, Jenczewski E, Lloyd A (2013) Meiosis and its deviations in polyploid plants. Cytogenet Genome Res 140:171–184. https://doi.org/10.1159/000351730
49. Cifuentes M, Eber F, Lucas MO, Lode M, Chèvre AM, Jenczewski E (2010) Repeated polyploidy drove different levels of crossover suppression between homoeologous chromosomes in *Brassica napus* allohaploids. Plant Cell 22:2265–2276. https://doi.org/10.1105/tpc.109.072991
50. Grandont L, Cuñado N, Coriton O, Huteau V, Eber F, Chèvre AM et al (2014) Homoeologous chromosome sorting and progression of meiotic recombination in *Brassica napus*: ploidy does matter! Plant Cell 26:1448–1463. https://doi.org/10.1105/tpc.114.122788
51. Corredor E, Naranjo T (2007) Effect of colchicine and telocentric chromosome conformation on centromere and telomere dynamics at meiotic prophase I in wheat-rye additions. Chromosom Res 15:231–245. https://doi.org/10.1007/s10577-006-1117-7
52. Naranjo T (2014) Dynamics of Rye telomeres in a wheat background during early meiosis. Cytogenet Genome Res 143:60–68
53. Naranjo T (2018) Variable patterning of chromatin remodeling, telomere positioning, synapsis, and chiasma formation of individual rye chromosomes in meiosis of wheat-rye additions. Front Plant Sci 9:880. https://doi.org/10.3389/fpls.2018.00880
54. Schwarzacher T, AnamthawatJonsson K, Harrison GE et al (1992) Genomic *in situ* hybridization to identify alien chromosomes and chromosome segments in wheat. Theor Appl Genet 84:778–786
55. Islam AKMR, Shepherd KW, Sparrow DHB (1978) Production and characterization of wheat-barley addition lines. In: Proc 5th Int Wheat Genet Symp, New Delhi, pp 365–371
56. Islam AKMR, Shepherd KW, Sparrow DHB (1981) Isolation and characterization of euplasmic wheat-barley chromosome addition lines. Heredity 46:161–174. https://doi.org/10.1038/hdy.1981.24
57. Miller TE, Reader SM, Chapman V (1982) The addition of Hordeum chilense chromosomes to wheat. In: Induced variability in plant breeding. Proc Int Symp Eucarpia, Pudoc, Wageningen, pp 79–81
58. Rey MD, Calderón MC, Prieto P (2015) The use of the *ph1b* mutant to induce recombination between the chromosomes of wheat and barley. Front Plant Sci 6:160. https://doi.org/10.3389/fpls.2015.00160
59. Calderón MC, Rey MD, Martín A, Prieto P (2018) Homoeologous chromosomes from two *Hordeum* species can recognize and associate during meiosis in wheat in the presence of the *Ph1* locus. Front Plant Sci 9:585. https://doi.org/10.3389/fpls.2018.00585
60. Friebe B, Zhang P, Linc G, Gill BS (2005) Robertsonian translocations in wheat arise by centric misdivision of univalents at anaphase I and rejoining of broken centromeres during interkinesis of meiosis II. Cytogenet Genome Res 109:293–297. https://doi.org/10.1159/000082412
61. Calderón MC, Rey MD, Cabrera A, Prieto P (2014) The subtelomeric region is important for chromosome recognition and pairing during meiosis. Sci Rep 4:6488. https://doi.org/10.1038/srep06488
62. Naranjo T, Valenzuela NT, Perera E (2010) Chiasma frequency is region specific and chromosome conformation dependent in a rye chromosome added to wheat. Cytogenet Genome Res 129:133–142. https://doi.org/10.1159/000314029
63. Sears ER (1966) Nullisomic-tetrasomic combinations in hexaploid wheat. In: Lewis DR

(ed) Chromosome manipulation and plant genetics. Oliver and Boyd, London, pp 29–47
64. Joppa LR, Williams ND (1977) D-genome substitution-monosomics of durum-wheat. Crop Sci 17:772–776
65. Friebe B, Qi LL, Liu C et al (2011) Genetic compensation abilities of *Aegilops speltoides* chromosomes for homoeologous B-genome chromosomes of polyploid wheat in disomic S(B) chromosome substitution lines. Cytogenet Genome Res 134:144–150
66. Islam AKMR, Shepherd KW (1992) Substituting ability of individual barley chromosomes for wheat chromosomes. 1. Substitutions involving barley chromosomes 1, 3 and 6. Plant Breed 109:141–150. https://doi.org/10.1111/j.1439-0523.1992.tb00164.x
67. Islam AKMR, Shepherd KW (1995) Substitution of barley chromosome 4 for group 4 homoelogous of wheat. In: Li ZS, Xin ZY (eds) Proc 8th int wheat genet symp, Beijing, China, 20–25 July 1993. China Agricultural Scientech Press, Beijing, pp 141–144
68. Ya-Ping Y, Xiao C, Si-He X, Islam AKMR, Zhi-Yong X (2003) Identification of wheat-barley 2H alien substitution lines. Acta Bot Sin 45:1096–1102
69. Silkova OG, Dobrovolskaya OB, Dubovets NI et al (2007) Production of wheat-rye substitution lines based on winter rye cultivars with karyotype identification by means of C-banding, GISH, and SSR markers. Russ J Genet 43:957. https://doi.org/10.1134/S1022795407080200
70. Friebe B, Jiang J, Raupp WJ et al (1996) Characterization of wheat-alien translocations conferring resistance to diseases and pests: current status. Euphytica 91:59–87
71. Villareal RL, Bañuelos O, Mujeeb-Kazi A, Rajaram S (1998) Agronomic performance of chromosomes 1B and T1BL.1RS near-isolines in the spring bread wheat Seri M82. Euphytica 103:195–202. https://doi.org/10.1023/a:1018392002909
72. Zhou Y, He ZH, Sui XX, Xia XC, Zhang XK, Zhang GS (2007) Genetic improvement of grain yield and associated traits in the northern China winter wheat region from 1960 to 2000. Crop Sci 47:245–453. https://doi.org/10.2135/cropsci2006.03.0175
73. Valenzuela NT, Perera E, Naranjo T (2013) Identifying crossover-rich regions and their effect on meiotic homologous interactions by partitioning chromosome arms of wheat and rye. Chromosom Res 21:433–445
74. Schlegel R (2016) Current list of wheats with rye and alien introgression. Vol. 05-16, 1–18. http://www.rye-gene-map.de/rye-introgression
75. Prieto P, Ramírez MC, Ballesteros J, Cabrera A (2001) Identification of intergenomic translocations involving wheat, *Hordeum vulgare* and *Hordeum chilense* chromosomes by FISH. Hereditas 135:171–174. https://doi.org/10.1111/j.1601-5223.2001.t01-1-00171.x
76. Lukaszewski AJ (2008) Unexpected behavior of an inverted rye chromosome arm in wheat. Chromosoma 117:569–578. https://doi.org/10.1007/s00412-008-0174-4
77. Valenzuela NT, Perera E, Naranjo T (2012) Dynamics of rye chromosome 1R regions with high and low crossover frequency in homology search and synapsis development. PLoS One 7:e36385. https://doi.org/10.1371/journal.pone.0036385
78. Endo TR, Gill BS (1996) The deletion stocks of common wheat. J Hered 87:295–307
79. Naranjo T (2015) Forcing the shift of the crossover site to proximal regions in wheat chromosomes. Theor Appl Genet 128:1855–1863. https://doi.org/10.1007/s00122-015-2552-7
80. Jamieson G, Evans IJ, Barnes SR (1986) An enzymic method of preparing plant chromosomes for *in situ* hybridization. Stain Technol 61:21–25
81. Murata M (1983) Staining air dried protoplasts for study of plant chromosomes. Stain Technol 58:101–106
82. Sallee PJ (1982) Prefixation and staining of the somatic chromosomes of corn. In: Sheridan WF (ed) Maize for biological research. University Press, University of North Dakota, Grand Forks, pp 119–120
83. Rayburn AL, Gill BS (1985) Use of biotin-labeled probes to map specific DNA sequences on wheat chromosomes. J Hered 76:78–81
84. De Carvalho CR, Saraiva LS (1993) An air drying technique for maize chromosomes without enzymatic maceration. Biotech & Histochem 68:142–145
85. Kato A (1999) Air drying method using nitrous oxide for chromosome counting in maize. Biotech & Histochem 74:160–166. https://doi.org/10.3109/10520299909047968
86. Draeger T, Moore G (2017) Short periods of high temperature during meiosis prevent normal meiotic progression and reduce grain number in hexaploid wheat (*Triticum aestivum* L.). Theor Appl Genet 30:1785–1800. https://doi.org/10.1007/s00122-017-2925-1

87. Sepsi A, Fábián A, Jäger K, Heslop-Harrison JS, Schwarzacher T (2018) ImmunoFISH: simultaneous visualisation of proteins and DNA sequences gives insight into meiotic processes in nuclei of grasses. Front Plant Sci 9:1193. https://doi.org/10.3389/fpls.2018.01193
88. Zhang P, Friebe B (2009) FISH on plant chromosomes. In: Liehr T (ed) Fluorescence *in situ* hybridization (FISH) application guide. Springer, Berlin, Heidelberg, pp 365–394
89. Xu SS, Liu Z, Zhang Q, Niu Z, Jan CC, Cai X (2016) Chromosome painting by GISH and multicolor FISH. In: Shahryar FK, MAK P (eds) Plant cytogenetics: methods and protocols, Methods Mol Biol, vol 1429, pp 7–21. https://doi.org/10.1007/978-1-4939-3622-9_2
90. Martinez-Perez E, Shaw PJ, Moore G (2000) Polyploidy induces centromere association. J Cell Biol 148:233–238
91. Martinez-Perez E, Shaw P, Moore G (2001) The Ph1 locus is needed to ensure specific somatic and meiotic centromere association. Nature 411:204–207
92. Martinez-Perez E, Shaw P, Aragon-Alcaide L, Moore G (2003) Chromosomes form into seven groups in hexaploid and tetraploid wheat as a prelude to meiosis. Plant J 36:21–29
93. Malik HS, Henikoff S (2009) Major evolutionary transitions in centromere complexity. Cell 138:1067–1082
94. Francki MG (2001) Identification of bilby, a diverged centromeric Ty1-copia retrotransposon family from cereal rye (Secale cereale L.). Genome 44:266–274
95. Cheng ZK, Dong F, Langdon T, Ouyang S, Buell R, Gu M, Blattner FR, Jiang J (2002) Functional rice centromeres are marked by a satellite repeat and a centromere-specific retrotransposon. Plant Cell 14:1691–1704. https://doi.org/10.1105/tpc.003079
96. Wu J, Yamagata H, Hayashi-Tsugane M, Hijishita S, Fujisawa M, Shibata M, Ito Y, Nakamura M, Sakaguchi M, Yoshihara R et al (2004) Composition and structure of the centromeric region of rice chromosome 8. Plant Cell 16:967–976
97. Lim KB, Yang TJ, Hwang YJ, Kim JS, Park JY, Kwon SJ et al (2007) Characterization of the centromere and peri-centromere retrotransposons in *Brassica rapa* and their distribution in related *Brassica* species. Plant J 49:173–183. https://doi.org/10.1111/j.1365-313X.2006.02952.x
98. Simpson PR, Newman MA, Davis DR, Ellis THN, Matthews PM, Lee D (1990) Identification of translocations in pea by *in situ* hybridization with chromosome-specific probes. Genome 33:745–749
99. Ganal MW, Broun P, Tanksley SD (1992) Genetic mapping of tandemly repeated telomeric DNA sequences in tomato (*Lycopersicon esculentum*). Genomics 14:444–448
100. Tang Z, Yang Z, Fu S (2014) Oligonucleotides replacing the roles of repetitive sequences pAs1, pSc119.2, pTa-535, pTa71, CCS1, and pAWRC.1 for FISH analysis. J Appl Genet 55:313–318. https://doi.org/10.1007/s13353-014-0215-z
101. Fu S, Chen L, Wang Y, Li M, Yang Z, Qiu L, Yan B, Ren Z, Tang Z (2015) Oligonucleotide probes for ND-FISH analysis to identify rye and wheat chromosomes. Sci Rep 5:10552. https://doi.org/10.1038/srep10552
102. Han Y, Zhang T, Thammapichai P, Weng Y, Jiang J (2015) Chromosome-specific painting in *Cucumis* species using bulked oligonucleotides. Genetics 200:771–779
103. Palmer DK, Oday K, Wener MH, Andrews BS, Margolis RL (1987) A 17-Kd centromere protein (CENP-A) copurifies with nucleosome core particles and with histones. J Cell Biol 104:805–815
104. Sepsi A, Higgins JD, Heslop-Harrison JS, Schwarzacher T (2017) CENH3 morphogenesis reveals dynamic centromere associations during synaptonemal complex formation and the progression through male meiosis in hexaploid wheat. Plant J 89:235–249. https://doi.org/10.1111/tpj.13379
105. Armstrong SJ, Caryl AP, Jones GH, Franklin FC (2002) Asy1, a protein required for meiotic chromosome synapsis, localizes to axis associated chromatin in *Arabidopsis* and *Brassica*. J Cell Sci 115:3645–3655
106. Mikhailova EI, Phillips D, Sosnikhina SP, Lovtsyus AV, Jones RN, Jenkins G (2006) Molecular assembly of meiotic proteins Asy1 and Zyp1 and pairing promiscuity in rye (*Secale cereale* L.) and its synaptic mutant sy10. Genetics 174:1247–1258
107. Higgins JD, Sanchez-Moran E, Armstrong SJ, Jones GH, Franklin FC (2005) The *Arabidopsis* synaptonemal complex protein ZYP1 is required for chromosome synapsis and normal fidelity of crossing over. Genes Dev 19:2488–2500
108. Khoo KH, Able AJ, Able JA (2012) The isolation and characterisation of the wheat

molecular ZIPper I homologue, TaZYP1. BMC Res Notes 5:106. https://doi.org/10.1186/1756-0500-5-106

109. Paula CMP, Techio VH (2014) Immunolocalization of chromosome-associated proteins in plants – principles and applications. Bot Stud 55:63. https://doi.org/10.1186/s40529-014-0063-5

110. Chelysheva L, Grandont L, Vrielynck N, le Guin S, Mercier R, Grelon M (2010) An easy protocol for studying chromatin and recombination protein dynamics during Arabidopsis thaliana meiosis: immunodetection of cohesins, histones and MLH1. Cytogenet Genome Res 129:143–153. https://doi.org/10.1159/000314096

111. Khoo KH, Jolly HR, Able JA (2008) The RAD51 gene family in bread wheat is highly conserved across eukaryotes, with RAD51A upregulated during early meiosis. Funct Plant Biol 35:1267–1277

112. Franklin AE, Mcelver J, Sunjevaric I, Rothstein R, Bowen B, Cande WZ (1999) Three-dimensional microscopy of the Rad51 recombination protein during meiotic prophase. Plant Cell 11:809–824

113. Boenisch T (2009) Antibodies. In: Kummar LG, Rudbeck L (eds) Immunohistochemical (IHC) staining methods. Dako, Carpinteria, CA, pp 1–9

114. Schwarzacher T, Leitch AR, Bennett MD, Heslop-Harrison JS (1989) *In situ* localization of parental genomes in a wide hybrid. Ann Bot 64:315–324

115. Prieto P, Moore G, Shaw P (2007) Fluorescence *in situ* hybridization on vibratome sections of plant tissues. Nat Protoc 2:1831–1838

116. Martín A, Rubiales D, Cabrera A (1999) A fertile amphiploid between a wild barley (*Hordeum chilense*) and crested wheatgrass (*Agropyron cristatum*). Int J Plant Sci 160:783–786

117. Ding L, Zhao ZG, Ge XH, Li ZY (2014) Different timing and spatial separation of parental chromosomes in intergeneric somatic hybrids between *Brassica napus* and *Orychophragmus violaceus.* Genet Mol Res 13:2611–2618

118. Jeridi M, Perrier X, Rodier-Goud M, Ferchichi A, D'Hont A, Bakry F (2012) Cytogenetic evidence of mixed disomic and polysomic inheritance in an allotetraploid (AABB) Musa genotype. Ann Bot 110:1593–1606. https://doi.org/10.1093/aob/mcs220

119. Naranjo T, Maestra B, Corredor E (2005) The search of the homologous partner at early meiosis in wheat. Chromosom Res 13:9

120. Corredor E, Lukaszewski AJ, Pachón P, Allen DC, Naranjo T (2007) Terminal regions of wheat chromosomes select their pairing partner in meiosis. Genetics 177:699–706. https://doi.org/10.1534/genetics.107.078121

121. Aragon-Alcaide L, Miller T, Schwarzacher T, Reader S, Moore G (1996) A cereal centromeric sequence. Chromosoma 105:261–268

122. Manzanero S, Puertas MJ, Jiménez G, Vega JM (2000) Neocentric activity of rye 5RL chromosome in wheat. Chromosom Res 8:543–554

123. Francki MG, Berzonsky WA, Ohm HW, Anderson JM (2002) Physical location of a HSP70 gene homologue on the centromere of chromosome 1B of wheat (*Triticum aestivum* L.). Theor Appl Genet 104:184–191

124. Dong F, Millar JT, Jackson SA, Wang FL, Ronald RC, Jiang J (1998) Centromeric regions consist of complex DNA. Proc Natl Acad Sci U S A 95:8135–8140

125. Ma J, Wing RA, Bennetzen JL, Jackson SA (2007) Plant centromere organization: a dynamic structure with conserved functions. Trends Genet 23:134–139

126. Fussell CP (1987) The Rabl orientation: a prelude to synapsis. In: Moens PB (ed) Meiosis. Academic Press, Orlando, pp 275–299

127. Koszul R, Kim KP, Prentiss M, Kleckner N, Kameoka S (2008) Meiotic chromosomes move by linkage to dynamic actin cables with transduction of force through the nuclear envelope. Cell 133:1188–1201. https://doi.org/10.1016/j.cell.2008.04.050

128. Richards DM, Greer E, Martin AC, Moore G, Shaw PJ et al (2012) Quantitative dynamics of telomere bouquet formation. PLoS Comput Biol 8:e1002812. https://doi.org/10.1371/journal.pcbi.1002812

129. Prieto P, Shaw P, Moore G (2004) Homologous recognition during meiosis is associated with a change in chromatin conformation. Nat Cell Biol 6:906–908

130. Vershinin AV, Schwarzacher T, Heslop-Harrison JS (1995) The large-scale genomic organization of repetitive DNA families at the telomeres of rye chromosomes. Plant Cell 7:1823–1833. https://doi.org/10.1105/tpc.7.11.1823

131. Linardopoulou EV, Williams EM, Fan Y, Friedman C, Young JM, Trask BJ (2005) Human subtelomeres are hot spots of interchromosomal recombination and segmental duplication. Nature 437:94–100. https://doi.org/10.1038/nature04029

132. Mao L, Devos KM, Zhu L, Gale MD (1997) Cloning and genetic mapping of wheat telomere-associated sequences. Mol Gen Genet 254:584–591

133. Gaiero P, Mazzella C, Vilaro F, Speranza P, de Jong H (2017) Pairing analysis and *in situ* hybridisation reveal autopolyploid-like behaviour in *Solanum commersonii* × *S. tuberosum* (potato) interspecific hybrids. Euphytica 213:137. https://doi.org/10.1007/s10681-017-1922-4

134. Zhan Z, Nwafor CC, Hou Z, Gong J, Zhu B, Jiang Y et al (2017) Cytological and morphological analysis of hybrids between Brassicoraphanus, and *Brassica napus* for introgression of clubroot resistant trait into *Brassica napus* L. PLoS One 12:e0177470. https://doi.org/10.1371/journal.pone.0177470

135. Higgins JD, Perry RM, Barakate A, Ramsay L, Waugh R, Halpin C, Armstrong SJ, Franklin FC (2012) Spatio-temporal asymmetry of the meiotic program underlies the predominantly distal distribution of meiotic crossovers in barley. Plant Cell 24:4096–4109

136. Colas I, Darrier B, Arrieta M, Mittmann SU, Ramsay L, Sourdille P, Waugh R (2017) Observation of extensive chromosome axis remodeling during the "diffuse-phase" of meiosis in large genome cereals. Front Plant Sci 8:1235. https://doi.org/10.3389/fpls.2017.01235

137. Aragon-Alcaide L, Beven A, Moore G et al (1998) The use of vibratome sections of cereal spikelets to study anther development and meiosis. Plant J 14:503–508

138. Thompson WF, Beven AF, Wells B, Shaw PJ (1997) Sites of DNA transcription are widely dispersed through the nucleolus in *Pisum sativum* and can comprise single genes. Plant J 12:571–581

139. Prieto P, Santos AP, Moore G, Shaw PJ (2004) Chromosomes associate premeiotically and in xylem vessel cells via their telomeres and centromeres in diploid rice (*Oryza sativa*). Chromosoma 112:300–307. https://doi.org/10.1007/s00412-004-0274-8

140. He L, Braz GT, Torres JA, Jiang J (2018) Chromosome painting in meiosis reveals pairing of specific chromosomes in polyploid *Solanum* species. Chromosoma 127:505–513. https://doi.org/10.1007/s00412-018-0682-9

141. Naranjo T, Orellana J (1984) Meiotic behaviour of chromosomes 1R, 2R and 5R in autotetraploid rye. Chromosoma 89:143–150

Chapter 12

Multicolored Fluorescent In Situ Hybridization to Assess Pairing Configurations at Metaphase I in *Brassica* Hybrids

Virginie Huteau and Olivier Coriton

Abstract

Genetic diversity can be introduced into polyploid crop species through meiotic recombination by exchanges between homologous or homoeologous chromosomes. Fluorescent in situ hybridization (FISH) enables the characterization of these homologous and homoeologous chromosome pairs during meiosis and identification of structural rearrangements during mitosis in metaphase I. In this chapter, we describe a protocol for the multicolored fluorescent labeling of chromosome spreads. This protocol allows the characterization of each A and C homoeologous subgenomes in a polyploid species using a genome-specific BAC combined with specific chromosome labeling BAC sequences.

Key words *Brassica* species, Chromosome, Cytogenetics, Meiosis, Mitosis, Fluorescence in situ hybridization

1 Introduction

Allopolyploidy, which results from the merger and duplication of divergent genomes, has played a major role in the evolution and diversification of flowering plants [1]. Many cultivated crop species are allopolyploid, including rapeseed, *Brassica napus* (AACC, $2n = 38$). This species was formed from few independent hybridization events between the diploid ancestors of *B. oleracea* (CC, $2n = 18$) and *B. rapa* (AA, $2n = 20$) and has undergone strong selection pressure leading to low intraspecific genetic diversity [2]. Moreover, all Brassica have experienced a triplication event 15–20 mya; therefore, each genomic region can be found up to three times per constitutive diploid genome and six times in *B. napus* [3, 4]. In order to introduce new genetic diversity and improve agricultural traits in an allopolyploid crop species, homologous and homoeologous chromosome pairing can be enhanced in hybrids by the production of resynthesized oilseed rape or by direct

Mónica Pradillo and Stefan Heckmann (eds.), *Plant Meiosis: Methods and Protocols*, Methods in Molecular Biology, vol. 2061, https://doi.org/10.1007/978-1-4939-9818-0_12,

crosses between the allopolyploid and its diploid ancestors [5–8]. Because of the high level of homology between the genomes of *B. napus* and the ones of its progenitors, we showed that crossover frequency and distribution is affected in allotriploid hybrids in spite of pairing mainly restricted to homologous chromosomes [9]. Similarly, homoeologous recombination frequently occurred between two closely related genomes in haploids [10] but also in synthetics [11, 12]. These methods have been successfully utilized to transfer disease resistance genes and quality traits between Brassica species [5, 7, 13]. However, this high level of homology between homoeologous genomes [14] leads to increased challenges to characterize precisely homologous and homoeologous chromosomes. Genotyping and sequencing data can reveal deleted and duplicated regions but the nature of the rearrangements (deletions vs. translocations) can only be determined physically on chromosomes using cytological observations during mitosis as well as their impacts on meiotic stability.

Thus, we developed a technique for fluorescent in situ hybridization (FISH) labeling of meiotic and mitotic chromosome spreads combining two types of BACs [12, 15]. A dispersed repeat sequence within a BAC (BoB014O06 from *B. oleracea*) is used as a probe to distinguish specifically C-genome chromosomes in *B. napus* [16] and BACs sequences specific were selected from *B. rapa* BAC libraries [17, 18]. Because of the high sequence homology between the diploid species, BACs probes from *B. rapa* also hybridized in *B. oleracea*. Therefore, BAC specific hybridize to one pair of chromosomes in each diploid species and to two homoeologous chromosome pairs in *B. napus* providing four FISH signals. Using this method, the structural variations can be distinguished physically by observation of pairing configuration during metaphase I. Here, we describe the preparation of slides (at meiosis and mitosis stages) and the subsequent hybridization procedures using bacterial artificial chromosome-fluorescent in situ hybridization (BAC-FISH) on *B. napus* chromosomes.

2 Materials

2.1 Plants

Brassica species plants grown in greenhouse, photoperiod 16 h per day and 8 h night; average temperatures of 18–22 °C.

For mitosis, harvest young roots of about 1–2 cm from young seedlings in pots or from seeds after germination in a petri dish (*see* **Note 1**).

For meiosis, excise entire inflorescences, outdoors or in the greenhouse, in cool or cloudy weather in the morning. Sample collection is preferably done on the main inflorescence axis, before the opening of the first floral bud (*see* **Note 2**).

2.2 Chromosome Preparation

For mitosis:

1. 8-hydroxyquinoline in blocking solution (290 mg/l) for 2 h at room temperature in the dark and 2 h at 4 °C. The 8-hydroxyquinoline is dissolved at 60 °C. The solution can be stored at 4 °C for 1 month.
2. Solution ethanol–acetic acid (3:1) for 12 h at 4 °C.
3. Ethanol 70%.
4. Citrate buffer 0.01 M pH 4.5: 1.47 g of tri sodium citrate and 1.05 g citric acid transfer to a graduated glass beaker. Add distilled water up to 400 ml. Adjust solution to final desired pH using HCl. Top up to 500 ml with distilled water. Make 50 ml aliquots. Store at −20 °C.
5. Digestion enzymatic mixture: 5% (w/v) cellulase from Trichoderma sp., 1% (w/v) pectolyase from *Aspergillus japonicus* in 0.01 M citrate buffer pH 4.5. Volume 5 ml of each enzyme and dissolve in 50 ml of 0.01 M citrate buffer pH 4.5. Aliquot by 1 ml and store at −20 °C. Place the roots in 250 μl enzyme mixture in during 30–35 min at 37 °C.
6. Ethanol (3 volumes)–acetic acid (1 volume).
7. DAPI solution (VECTASHIELD Antifade Mounting Medium with DAPI Vector Laboratories).
8. Stereomicroscope.
9. Microscope slides Superfrost.
10. Dissection needles, a hook (curved dissection nacelle).
11. Glass Pasteur pipette.
12. Incubator (37 °C).

For meiosis:

1. Carnoy's II (6 volumes ethanol, 3 volumes chloroform, and 1 volume acetic acid).
2. Ethanol 50% at 4 °C.
3. Belling's carmine: Dissolve 10 g carmine in 1 L of boiling 45% glacial acetic acid, let the solution simmer for 5 min. Filter into dark bottles and store at 4 °C. This solution can be stored for a long time.
4. Acetic acid 45%.
5. Watch glass, alcohol lamp.
6. Stereomicroscope.
7. Superfrost microscope slides (Thermo Scientific).
8. Dissection needles and fine forceps.
9. Liquid nitrogen.

2.3 BACs Labeling

1. Isolation of BAC DNA using Qiagen Plasmid Midi kit.
2. Bioprime DNA labeling (Invitrogen).
3. Modified Random prime for Alexa labeling.

2.4 Hybridization

1. 2× SSC: 50 ml of commercial 20× SSC (saline–sodium citrate) in 950 ml distilled water.
2. 4× SSC–Tween 20: 100 ml of commercial 20× SSC (saline–-sodium citrate), 2 ml Tween 20 (Merck Eurolab). Make up to 1 L with distilled water.
3. RNAse solution: 15 μl RNase 10 mg/ml (Sigma-Aldrich) + 50 μl TE 10–1 (Sigma-Aldrich). Store at −20 °C.
4. Pepsin solution: Dissolve 5 mg Pepsin (Sigma-Aldrich) in 50 ml HCl 0.01 M.
5. Paraformaldehyde 4%, 2 g paraformaldehyde in 40 ml distilled water. Heat while stirring to approximately 60 °C. Add 400 μl NaOH 1 N to 40 ml of solution. Top up to 50 ml with distilled water.
6. DNA single stranded from salmon tests 10 mg/ml.
7. SDS 20%: Dissolve 20 g SDS (sodium dodecyl sulfate) in 100 ml H_2O.
8. Dextran sulfate 50%: Dissolve 50 g dextran sulfate (Pharmacia) in 100 ml distilled water, stirring for 2–3 h. Store at −20 °C.
9. Formamide deionized 100%: Add 5 g Amberlite (Sigma-Aldrich) in 100 ml formamide, stirring for 1 h. Store −20 °C.
10. Formamide 70% with 2× SSC: 70 ml formamide +30 ml 2× SSC.
11. BSA 5%: 100 mg BSA (bovine serum albumin, Sigma-Aldrich) + 2 ml 4× SSC–Tween 20.
12. Hybridization mix: 25 μl formamide 100%, 10 μl dextran sulfate 50%, 5 μl 20× SSC, 0.6 μl 20% SDS (for 1 slide).

3 Methods

3.1 Chromosome Preparation

For mitosis:

1. Take the young roots out of the soil (1–2 cm) and put them in a tube containing 8-hydroxyquinoline in blocking solution (0.29 g/l) for 2 h at room temperature in the dark and 2 h at 4 °C.
2. Fix the roots in freshly ethanol–acetic acid (3:1) for 12 h at 4 °C in plugged tubes. The roots are stored in ethanol 70% at 4 °C.
3. Rinse the fixed roots for 2 × 10 min with distilled water at room temperature.

4. Replace the water with citrate buffer pH 4.5 at room temperature for 15 min.
5. Replace the water with 250 μl digestion enzyme mixture. Incubate for 30–35 min at 37 °C in a moist chamber.
6. Replace the digestion enzyme mixture with water for 30 min at room temperature.
7. Mounting meristems in a drop of ethanol–acetic acid [3-1], dilacerate the meristematic zone, spread on a slide (Superfrost microscope slides VWR). Air-dry the preparation.
8. DAPI solution (VECTASHIELD Antifade Mounting Medium with DAPI VH-1200 Vector laboratories), add 20 μl of DAPI solution per slide and cover with 22 × 32 coverslip.
9. Using a fluorescence microscope, select slides. Cells must be clear without too dense cytoplasm with well spread chromosomes (Fig. 1) (*see* **Note 3**).

For meiosis:

1. Fixation the inflorescences with Carnoy's II for 24 h at room temperature. Replace the fixative with the fresh fixative at the end of 8 h for chlorophyll extraction. The inflorescence can be stored in 50% ethanol 50% at 4 °C. The observations can be made from 2 days up to several months.
2. Place a drop of Belling's carmine on a slide and place a floral bud.
3. Under binocular loupe, remove the anthers and remove the other tissue with needles. Cut the stamens and extract the pollen mother cells (*see* **Note 4**).
4. Warm the slide briefly over a flam to accelerate the staining and to induce the cells swelling and cover with a 22 × 22 coverslip.
5. Squash with an absorbent paper and observe the slides under a microscope ×25.
6. If the cells are well dispersed and in metaphase I, apply a regression by adding a drop of 45% acetic acid under the coverslip, eliminate excess with absorbent paper. A very gentle heating may be necessary to remove cytoplasm as much as possible.
7. Observe under a microscope and press a last time. Incubate the slide in liquid nitrogen and remove the coverslip with razor blade. Air-dry (*see* **Note 5**).
8. To select the best slides for FISH, add a drop of DAPI, cover with a coverslip and observe under a fluorescent microscope (Figs. 2 and 3).

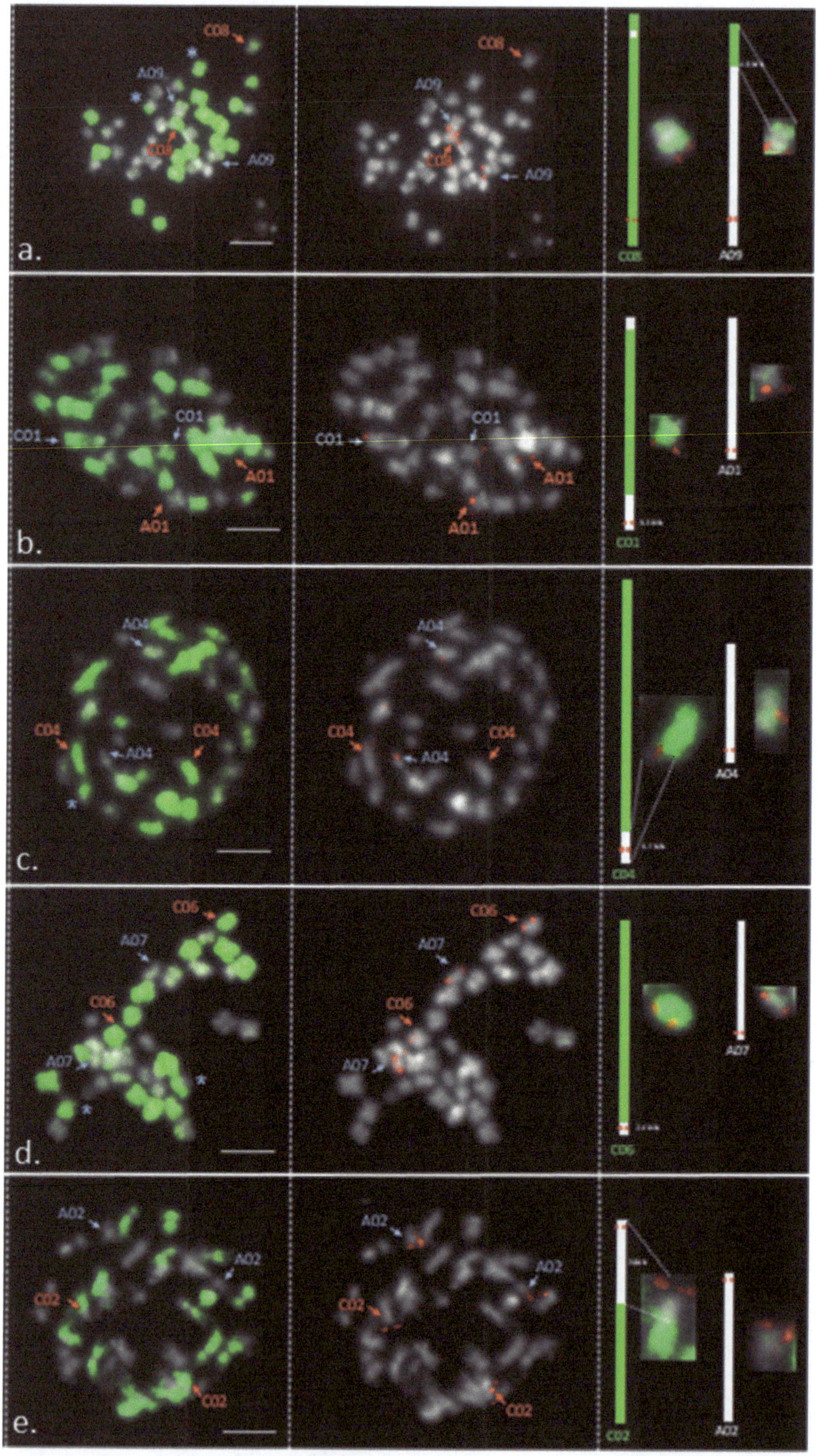

Fig. 1 Identification of nonreciprocal translocation in five resynthesized allotetraploid Brassica napus L. using BAC-FISH. BAC-FISH was carried out on somatic metaphase chromosomes using a BAC present within a

3.2 Hybridization

1. RNase Treatment: add 200 μl RNAse (100 ng/μl in 2× SSC), cover with a plastic coverslip and incubate at 37 °C for 1 h in a humid chamber (with 2× SSC buffer) (*see* **Note 6**).
2. Rinse the slides two times in 2× SSC for 3 min at 42 °C.
3. Treatment with Pepsin solution: add 100 μl Pepsin (0.05%), cover with a plastic coverslip and incubate for 10–20 min at 37 °C in a humid chamber (*see* **Note 7**).
4. Rinse the slides two times in 2× SSC for 3 min at 42 °C.
5. Treatment with paraformaldehyde: under the laboratory fume hood, immerse the slides in paraformaldehyde (4%) for 10 min at room temperature.
6. Wash in 2× SSC twice for 3 min.
7. Denature each slide: only for mitosis, under the laboratory fume hood, immerse the slides in formamide 70%, 2× SSC solution at 70 °C (water bath) for 2 min.
8. Dehydrate the slides in an ethanol series (70%, 90%, and 100%) at −20 °C, and air-dried at room temperature.
9. During **step 8**: prepare 50 μl of the hybridization mixture per slide: 40.6 μl Hybridization Mix, 100 ng BAC probe biotin, 100 ng BAC probes Alexa 488, 10 μg blocking DNA, make up to 50 μl with water.
10. Denature DNA probe mixture of the hybridization solution at 92 °C for 6 min. Chill rapidly and keep the denatured probe on glace.
11. Add the 50 μl of hybridization solution to the dry slides. Incubate in a moist chamber at 37 °C overnight.
12. For meiosis, apply the denatured probe on air-dried slide and denature DNA on each slide at 82 °C for 2 min in water bath. Incubate in a moist chamber at 37 °C overnight.

3.3 Wash

1. Wash slides twice by gently shaking in 2× SSC for 3 min at 42 °C.
2. Wash slides by gently shaking 50% formamide in 2× SSC for 3 min at 42 °C (*see* **Note 8**).

Fig. 1 (continued) deleted region (red) and the BAC Bob014006 clone that identifies all C chromosomes (green). Identification of a nonreciprocation: between A09 and C08 using BACs KBrB043F18 (in red) and Bob014006 (in green) clones (**a**); between A01 and C01 using BACs Bna191G20 (in red) and Bob014006 (in green) clones (**b**); between A04 and C04 using BACs KBrH009I04 (in red) and Bob014006 (in green) clones (**c**); between A07 and C06 using BACs KBrB021P15 (in red) and Bob014006 (in green) clones (**d**) and between A02 and C02 using BACs KBrB086G22 (in red) and Bob014006 (in green) clones (**e**). Additional translocations between A and C chromosomes are indicated by a star (∗). Chromosomes were counterstained with DAPI (blue). Bars represent 5 μm [12]

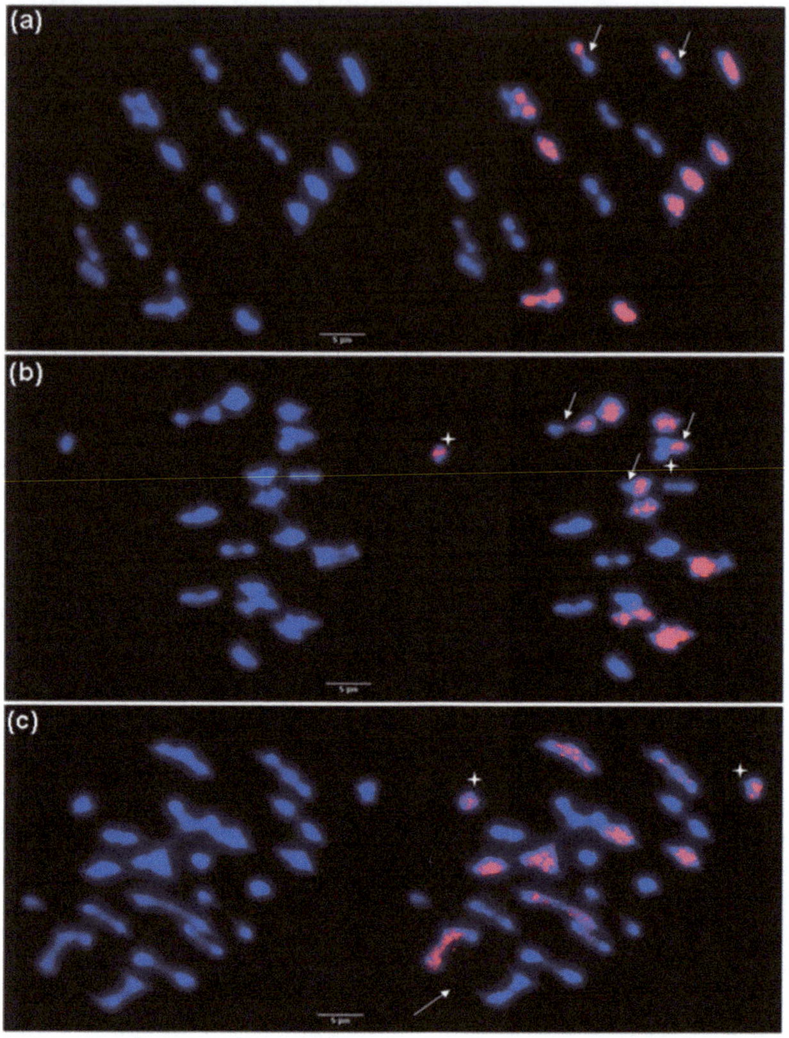

Fig. 2 A–C bivalents and multivalents at metaphase I of first meiosis of three euploid S0 synthetic B. *napus*. Unlabeled chromosomes (A genome) are in DAPI (blue), and Bob014006 (C-genome specific BAC clone) is stained in biotin (red). (**a**) Normal behavior on first S0 synthetic with 10 A–A pairs (in blue) and 9 C–C pairs (in red) (**b**) second S0 synthetic with two A–C bivalents (white arrow) and (**c**) third S0 synthetic with two univalents A chromosomes. Bars, 5 m [19]

3. Wash slides twice by gently shaking 2× SSC 3 min at 42 °C.
4. Wash slides by gently shaking in 4× SSC–Tween for 3 min at 42 °C.
5. Wash slides in 4× SSC–Tween for 3 min at 42 °C at room temperature.
6. Add 200 μl per slide blocking buffer (5% BSA) for 5 min at room temperature. Drain off the blocking buffer.

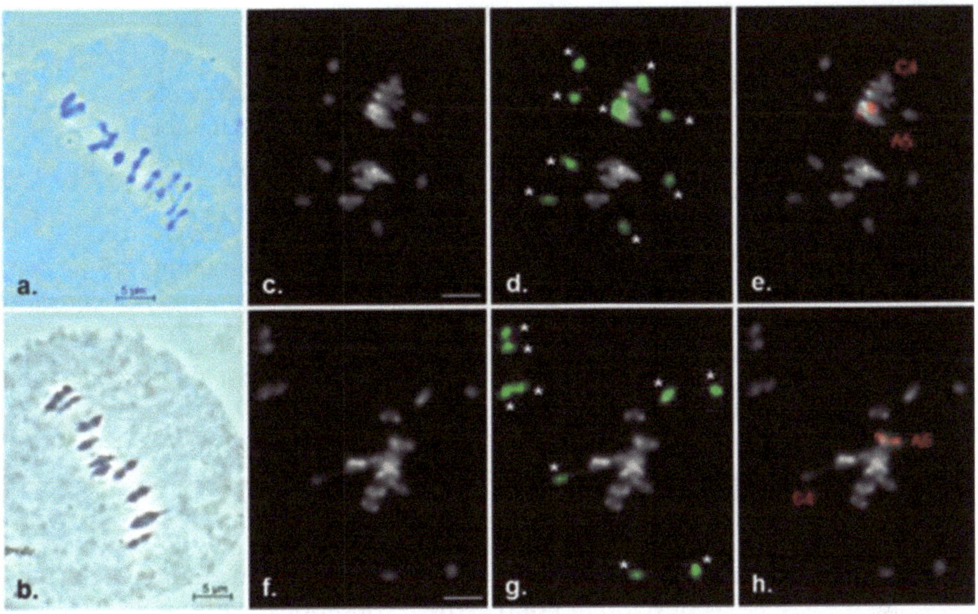

Fig. 3 Meiotic observations of AA and AAC F1 hybrids at metaphase I. (a, b) Pollen mother cells showing ten bivalents for (**a**) AA and (b) AA F1 hybrids. FISH analyses showing 10 A bivalents and 9 C univalents for (**c–e**) AAC and (**f–h**) AAC F1 hybrids. BAC-FISH was carried out using BACs Bob014006 and KBrH033J07 which identify all the C chromosomes (**d–g**, green) and the A05 and C04 chromosomes (**e–h**, in red), respectively. Univalents are indicated by a white star. Bars, 5 μm [9]

3.4 Detection

1. Prepare the detection solution anti-biotin fluorochrome diluted in BSA 5%. Biotinylated probes were immunodetected by Texas Red conjugated avidin antibodies (Vector Laboratories, Burlingame, USA).
2. Add 50 μl per slides and cover with a large plastic coverslip and incubate for 1 h at 37 °C in a humid chamber.
3. Wash slides twice by gently in 4× SSC–Tween for 3 min at 42 °C.
4. Add 200 μl per slide 5% BSA for 3 min at room temperature.
5. For detection BAC probes, signal amplification is essential for very weak signals.
6. Add the amplification solution: 5 μl anti-avidin biotin (0.5 mg/ml Vector BA300) dilute in 495 μl 5% BSA.
7. Add 50 μl per slides, cover with a large plastic coverslip and incubate for 1 h at 37 °C in a humid chamber.
8. Wash slides twice by gently in 4× SSC–Tween 3 min at 42 °C.
9. Prepare the detection solution by diluting avidin fluorochrome in BSA, 2.5 μl avidin Texas Red add 497.5 μl 5% BSA. Biotinylated probes were immunodetected by Texas Red

(conjugated avidin antibodies (Vector Laboratories, Burlingame, USA).

10. Add 50 µl per slides and cover with a large plastic coverslip and incubate for 1 h at 37 °C in a humid chamber.
11. Wash slides twice by gently in 4× SSC–Tween for 3 min at 42 °C.

3.5 Observation

1. Add 20 µl of DAPI in Vectashield containing 2.5 µg/ml 4′,6-diamidino-2-phenylindole (Vector Laboratories) per slides and cover with a 22 × 32 coverslip.
2. Observe the slides under a fluorescent microscope using the appropriate filters (Figs. 1, 2, and 3).

4 Notes

1. To collect root tips with a good mitotic index, the plants must develop in good growth conditions, without any stress.
2. In Brassica, bud size poorly correlated with pollen mother cell development stages but development stages are frequently homogenous among the various anthers of a single bud. Therefore, it is possible to select appropriate buds after acetocarmine staining of two anthers per bud. To do it, extract one anther from a fixed bud and place it on a slide. Add 30 µl of acetocarmine and crush the anther. Cover with a coverslip 22 × 22. Observe the slide under a bright field microscope. Three slides can be prepared from one bud.
3. Keep the slides in the fridge 4 °C.
4. Excise the entire inflorescences, preferably on the main axis of the inflorescence, before the opening of the first flower buds.
5. Keep the slides in the freezer −20 °C.
6. Slides pretreated well with RNAse. Because RNA can contribute significantly to the background signal, the pretreatment of the slides with RNAse is important.
7. Pepsin should be used to help remove excess cytoplasm covering chromosome preparations. Adjust time treatment to suit the preparations; if there is little cytoplasm, 10 min pepsin treatment, we increase the treatment time up to 20 min if too much cytoplasm.
8. Adjust percent stringency of DNA–DNA in situ hybridization as needed for washes if needed. Normally, in situ hybridization is performed with percent stringency ranging from 70% to 90%. In practice, the weakest background signal is obtained by hybridizing the probe and the target to a certain stringency, and subsequent posthybridization washes are performed at a

stringency of 2–5% higher. Factors that affect the stringency are temperature, formamide concentration, and salt concentration (buffer). If all these concentrations are increased, the stability of the double helix is lower.

Acknowledgments

We thank Anne-Marie Chèvre and Julie Ferreira-De-Carvalho for constructive reading of the manuscript.

References

1. Soltis DE, Visger CJ, Soltis PS (2014) The polyploidy revolution then ... and now: Stebbins revisited. Am J Bot 101(7):1057–1078
2. Nagaharu U (1935) Genome analysis in Brassica with special reference to the experimental formation of B. napus and its peculiar mode of fertilization. Jap J Bot 7:389–452
3. Schranz ME, Lysak MA, Mitchell-Olds T (2006) The ABC's of comparative genomics in the Brassicaceae: building blocks of crucifer genomes. Trends Plant Sci 11(11):535–542
4. Lysak MA, Cheung K, Kitschke M, Bures P (2007) Ancestral chromosomal blocks are triplicated in Brassiceae species with varying chromosome number and genome size. Plant Physiol 145(2):402–410
5. Leflon M, Brun H, Eber F et al (2007) Detection, introgression and localization of genes conferring specific resistance to Leptosphaeria maculans from Brassica rapa into B. napus. Theor Appl Genet 115:897–906
6. Fredua-Agyeman R, Coriton O, Huteau V, Parkin IA, Chèvre AM, Rahman H (2014) Molecular cytogenetic identification of B genome chromosomes linked to blackleg disease resistance in Brassica napus × B. carinata interspecific hybrids. Theor Appl Genet 127 (6):1305–1318
7. Stein A, Coriton O, Rousseau-Gueutin M, Samans B, Obermeier C, Parkin I, Chevre A-M, Snowdon R (2017) Mapping of homoeologous chromosome exchanges influencing quantitative trait variation in Brassica napus. Plant Biotechnol J 15:1478–1849
8. Mason AS, Chèvre AM (2016) Optimization of recombination in interspecific hybrids to introduce new genetic diversity into oilseed rape (Brassica napus L.). In: Mason AS (ed) Polyploidy and hybridization for crop improvement. Science Publishers, pp 431–444
9. Pelé A, Falque M, Trotoux G, Eber F, Nègre S, Gilet M, Huteau V, Lodé M, Jousseaume T, Dechaumet S, Morice J, Poncet C, Coriton O, Martin OC, Rousseau-Gueutin M, Chèvre AM (2017) Amplifying recombination genome-wide and reshaping crossover landscapes in Brassicas. PLoS Genet 13(5):e1006794
10. Nicolas SD, Leflon M, Monod H, Eber F, Coriton O, Huteau V, Chèvre AM, Jenczewski E (2009) Genetic regulation of meiotic crossovers between related genomes in Brassica napus haploids and hybrids. Plant Cell 21 (2):373–385
11. Gaeta RT, Pires JC, Iniguez-Luy F, Leon E, Osborn TC (2007) Genomic changes in resynthesized Brassica napus and their effect on gene expression and phenotype. Plant Cell 19(11):3403–3417
12. Rousseau-Gueutin M, Morice J, Coriton O, Huteau V, Trotoux G, Nègre S, Falentin C, Deniot G, Gilet M, Eber F, Pelé A, Vautrin S, Fourment J, Lodé M, Bergès H, Chèvre AM (2017) The impact of open pollination on the structural evolutionary dynamics, meiotic behavior and fertility of resynthesized allotetraploid Brassica napus L. G3 (Bethesda) 7 (2):705–717
13. Jenczewski E, Chèvre AM, Alix K (2013) Chromosomal and gene expression changes in *Brassica* allopolyploids. In: Chen ZJ, Birchler JA (eds) Polyploid and hybrid genomics. John Wiley & Sons, Inc., Oxford, pp 171–186
14. Chalhoub B, Denoeud F, Liu S, Parkin IA, Tang H, Wang X, Chiquet J, Belcram H, Tong C, Samans B, Corréa M, Da Silva C, Just J, Falentin C, Koh CS, Le Clainche I, Bernard M, Bento P, Noel B, Labadie K, Alberti A, Charles M, Arnaud D, Guo H, Daviaud C, Alamery S, Jabbari K, Zhao M, Edger PP, Chelaifa H, Tack D, Lassalle G,

Mestiri I, Schnel N, Le Paslier MC, Fan G, Renault V, Bayer PE, Golicz AA, Manoli S, Lee TH, Thi VH, Chalabi S, Hu Q, Fan C, Tollenaere R, Lu Y, Battail C, Shen J, Sidebottom CH, Wang X, Canaguier A, Chauveau A, Bérard A, Deniot G, Guan M, Liu Z, Sun F, Lim YP, Lyons E, Town CD, Bancroft I, Wang X, Meng J, Ma J, Pires JC, King GJ, Brunel D, Delourme R, Renard M, Aury JM, Adams KL, Batley J, Snowdon RJ, Tost J, Edwards D, Zhou Y, Hua W, Sharpe AG, Paterson AH, Guan C, Wincker P (2014) Plant genetics. Early allopolyploid evolution in the post-Neolithic Brassica napus oilseed genome. Science 345(6199):950–953

15. Suay L, Zhang D, Eber F, Jouy H, Lodé M, Huteau V, Coriton O, Szadkowski E, Leflon M, Martin OC, Falque M, Jenczewski E, Paillard S, Chèvre AM (2014) Crossover rate between homologous chromosomes and interference are regulated by the addition of specific unpaired chromosomes in Brassica. New Phytol 201(2):645–656
16. Howell EC, Barker GC, Jones GH, Kearsey MJ, King GJ, Kop EP et al (2002) Integration of the cytogenetic and genetic linkage maps of *Brassica oleracea*. Genetics 161(3):1225–1234
17. Mun JH, Kwon SJ, Yang TJ, Kim HS, Choi BS et al (2008) The first generation of a BAC-based physical map of Brassica rapa. BMC Genomics 9:280
18. Xiong Z, Gaeta RT, Pires JC (2011) Homoeologous shuffling and chromosome compensation maintain genome balance in resynthesized allopolyploid Brassica napus. Proc Natl Acad Sci U S A 108(19):7908–7913
19. Szadkowski E, Eber F, Huteau V, Lodé M, Huneau C, Belcram H, Coriton O, Manzanares-Dauleux MJ, Delourme R, King GJ, Chalhoub B, Jenczewski E, Chèvre AM (2010) The first meiosis of resynthesized Brassica napus, a genome blender. New Phytol 186(1):102–12

Chapter 13

Surface Spreading Technique in Plant Meiocytes for Analysis of Synaptonemal Complex by Electron Microscopy

Nieves Cuñado

Abstract

An improved method of preparing two-dimensional surface spreads of synaptonemal complexes (SCs) in higher plants for examination by electron microscopy is described. This protocol produces clear, well-spread preparations of SCs and unpaired axial cores from a range of meiotic prophase I stages (leptotene to pachytene) from meiocytes of different plant species. Synaptonemal complex (SC) analyses have been widely used in plant cytogenetic studies to address the process of meiotic chromosome synapses, because of the high-resolution allowed by electron microscopy. Although the real role of SC is still enigmatic, its presence and structural conservation in the vast majority of organisms reflect the importance of this protein structure in the meiotic process.

Key words Meiosis, Synaptonemal complex, Surface spreading, Silver staining, Electron microscopy

1 Introduction

Surface spreading techniques for making whole-mount preparations of synaptonemal complexes (SCs) have been revealed as powerful tools for providing direct and accurate observations of prophase I nuclei in plants. An optimized SC preparation protocol in pollen mother cell detailed here can be easily adapted to multiple plant species and it has been routinely applied in the characterization of their synapsis process [1–3] (Fig. 1). This protocol has been developed by carrying out some modifications to the surface spread methods described previously (*see*, e.g., [4–8]).

The synaptonemal complex (SC) is one of the hallmarks of meiotic prophase and an evolutionarily well-conserved structure as meiosis itself. They were first described more than 60 years ago [9, 10], and have since been found in almost all sexually reproducing eukaryotes analyzed thus far. SC is a multiprotein complex assembled between homologous chromosomes during prophase I of meiosis that is critical for the successful execution of meiosis (reviewed in [11–13]).

Mónica Pradillo and Stefan Heckmann (eds.), *Plant Meiosis: Methods and Protocols*, Methods in Molecular Biology, vol. 2061, https://doi.org/10.1007/978-1-4939-9818-0_13,

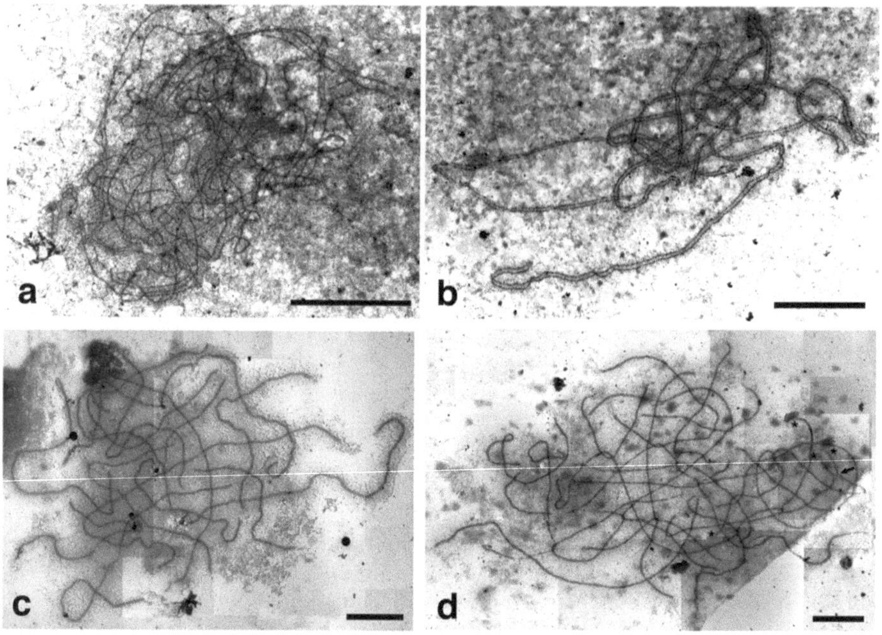

Fig. 1 Electron micrographs of silver stained meiotic prophase I nuclei of *Arabidopsis thaliana* (**a**, **b**) and allotetraploid *Brassica napus* (**c**, **d**). (**a**) Zygotene nucleus (**b**) Pachytene nucleus (**c**) Pachytene nucleus showing 19 bivalents (**d**) Pachytene nucleus showing 17 bivalents and one quadrivalent. Asterisks in (**d**) indicate the ends of the quadrivalent while one arrow indicates the synaptic partner switch. Bars represent 5 μm

Electron microscope (EM) analysis of whole-sectioned nuclei and spread preparations and along with immuno-EM and fluorescence imaging indicates that synapsed pachytene bivalents are highly organized in the following way. The chromatin rearrangements of meiotic prophase are accompanied by the assembly and disassembly of synaptonemal complexes (SCs); these are proteinaceous, zipper-like structures which keep homologous chromosomes (homologs) in close apposition along their entire length during meiotic prophase I. The chromatin from two sister chromatids of a meiotic prophase chromosome forms loops anchored on a common proteinaceous axis, called the axial element or axial core. As meiotic prophase proceeds and following coalignment when AEs have approached each other within 300 nm, the axial cores of homologous chromosomes become much more closely associated and are connected along their length by numerous transverse filaments (TFs), to form a robust structure of a synaptonemal complex (SC) This association, referred to as "synapsis," defines "zygotene" presence of complete SC defines "pachytene." At these stages, axial cores within the SC structure are referred to as lateral elements (LEs). Between the two LEs of an SC, a third longitudinal structure, the central element (CE), is formed on the TFs. Both outer protein axes (LEs) together with a third central element (CE), running in parallel, make up the tripartite structure of the SC, around 100 nm width.

In general, SC formation begins at the subdistal regions of the chromosomes, but additional interstitial synaptic initiation sites may also form, particularly in genomes with large chromosomes. Nevertheless, there is a considerable flexibility regarding this process and for instance, asynchronous of bivalent formation within a nucleus is common [1–3, 14].

The structure of the SC is amazingly consistent across species. Correspondingly, its components share a common underlying structural organization but the level of their primary amino acid sequence homology is very low, reflecting the fact that these molecules play primarily structural roles, rather than catalytic roles [13, 15]. In plants, mainly in Arabidopsis, rice, and barley, two axial proteins have been described: the HORMA domain–containing protein ASY1/PAIR2 and ASY3/PAIR3 [16–19]. The central-element proteins are poorly conserved at the sequence level, but they display the canonical structure of the yeast central-element ZIP1 protein, with a coiled-coil domain in the central region and a globular domain at each end. Plant *ZIP1* orthologs have been characterized in *Arabidopsis*, rice, maize, and wheat (*Triticum aestivum*) [20–24].

Fluorescence immunolabeling using antibodies against some of these specific SC proteins has allowed to determine their location on the chromosomes and at what stage of meiotic prophase I they are expressed. Moreover, in the last years this technique has represented a useful approach for the analysis of the synaptic process in different plant species both in wild type and in meiotic mutants. However, while using fluorescence microscopy has made important progress in the study of meiosis in plants, the analysis of the synaptonemal complex by silver staining using transmission electron microscopy still offers some advantages. While the resolution of fluorescence microscopy is restricted to ~200 nm due to the diffraction limit of light, electron microscopy allows for identifying cell substructures until a resolution of ~1 nm. In addition, the higher magnifications provide for the transmission electron microscope has permitted a more accurate knowledge of the initiation and progression of SC formation within individual bivalents (or multivalents) at early prophase I stages. For this reason, this technique is particularly useful for the analysis of complex patterns of synapsis in polyploids (Fig. 1c, d), haploids, hybrids, chromosomal mutants, etc. [25]. Besides, and not less important, silver staining allows all proteins involved in the SC structure to be dyed, even if they are not identified, and therefore, an advantage of this technique is the possibility of analyzing the patterns of synapsis in a wider range of plant species (wild type and mutants).

The protocol described in this chapter allows obtaining high-quality chromosome spreads as reflected in several articles in which it has been developed (Fig. 1) [1–3, 19, 26–33]. I include here slight modifications that should be applied to get better results depending on the plant materials (wild and cultivated species: *Gramineae*, *Brassica*, *Arabidopsis*, *Muscari*, *Rumex*, etc.).

2 Materials

2.1 Plant Material

1. Different dicots and monocots species can be used to analyze the synaptonemal complex by electron microscopy.
2. Plants from species like *Arabidopsis thaliana* and *Brassica* sp., are grown in a growth chambers under a 16-h light–8-h dark photoperiod, at 20 °C with 70% humidity.
3. Plans from Poaceae or Gramineae family like *Triticum*, *Aegilops*, and *Secale*, grown in field or greenhouse under natural conditions.
4. Likewise, plants from a variety of species like *Rumex* sp. (Polygonaceae) and *Muscari* sp. (Asparagaceae), grown in field or greenhouse under natural conditions.

2.2 Equipment and Materials

2.2.1 Collection of Inflorescences

1. Stereomicroscope or binocular loupe with cold light.
2. Plastic petri dishes (9 cm petri plate) with filter paper saturated with drinking water.
3. Watch glass.
4. Standard microscope slides.
5. 22 × 22 mm coverslips.
6. Dissection needles and fine tweezers.
7. Lancet or needle.
8. Alcohol burner.
9. Transmitted light microscopy.
10. Cooled metal plate (at 4–8 °C).

2.2.2 Synaptonemal Complex (SC) Preparations

1. Micropipettes and tips.
2. Cavity slides, single depression.
3. Metal rod to squash the anthers or buds.
4. High-quality, precleaned slides (glow discharged slides).
5. Overhead transparency (*see* Reagents).
6. Clear nail polish.
7. Fume cupboard.
8. Erlenmeyer flask.
9. Coplin jars.

10. For EM analysis, the slides need to be coated with a plastic film. Pour the plastic solution (made from 1% (w/v) triacetate dissolved in chloroform) (*see* Reagents) into a clean Coplin jar at room temperature. The clean slides are coated by rapidly dipping vertically in the plastic solution; remove and touch on the edge of the jar to drain off the excess fluid (*see* **Note 1**). Dry upright for about 10–15 min. Afterward, seal the plastic coating to the edges of the slide with regular clear nail polish. When dry, the slides could be stored in a slide box at 4 °C until necessary. The coated slides should be allowed to reach room temperature before use.
11. Crushed ice.
12. Cooled metal plate (at 4–8 °C).
13. Hot plate (30–40 °C).

2.2.3 Silver Staining

1. Flat square-tip tweezers; Fine-tip tweezers; Scissors.
2. Coplin jars.
3. Gloves.
4. Small glass bottle or tube with cap.
5. Small plastic box. The upper or lower half of a 24 × 60 mm coverslip box can be used.
6. Nylon mesh coverslips. We cut them out of nylon mesh into pieces (rectangles) of an approximate size of 24 × 50 mm (*see* **Note 2**).
7. Incubator (40–60 °C).
8. Moist chamber. We use crystal large petri dishes lined on the bottom with wet filter paper. Square wooden rods are placed on the paper to form a solid surface for slides to rest on.

2.2.4 Preparing for Electron Microscopy

1. Light microscope.
2. Object marker (Object marker for W 0.8 objective thread, for the circle-shaped marking of surface-spread nuclei on standard microscope slides with wipe-resistant red color – Zeiss).
3. Fine-tip tweezers; watchmaker's fine curved tweezers; scissors; scalpel handles; and blades.
4. Large staining jars containing distilled water.
5. Square 50 mesh copper TEM grids with asymmetrical center-marking (Fig. 2a) (*see* **Notes 3** and **4**).
6. Grid storage box with identification number (Fig. 2b).
7. Grid maps to record the locations of surface-spread nuclei (Fig. 2c) (*see* **Note 4**).
8. Storage box is made from heavy cardboard with lid with the bottom covered with filter paper.

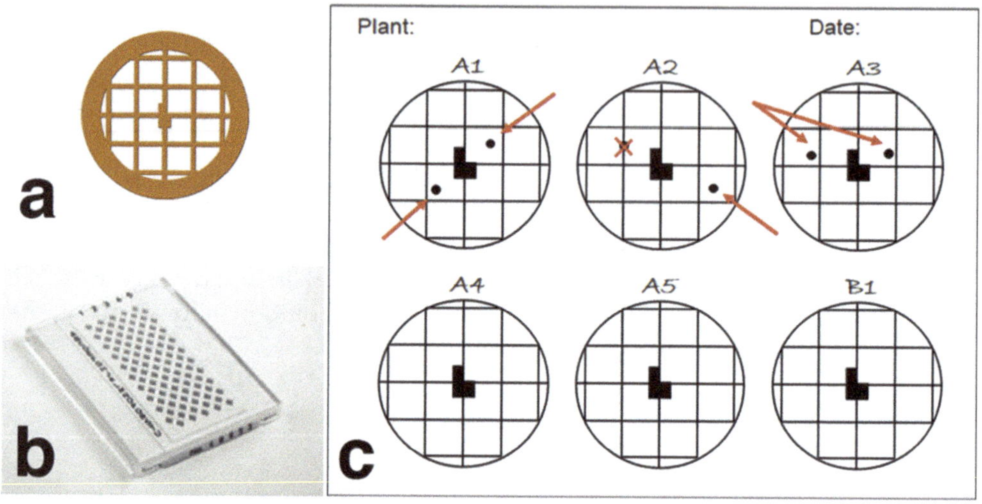

Fig. 2 Some materials for electron microscopy analysis (**a**) Square 50 mesh copper TEM grids with asymmetrical center-marking. (**b**) Grid storage box with identification number. (**c**) One example of a grid maps to record the locations of surface-spread nuclei

9. Fine brush to handle the grids (optional).
10. A4 Square Cut Folders cut into pieces with an approximate size of 40 × 100 mm.
11. Transmission electron microscope at 100 kV (Jeol 1010) and Megaview II camera and image acquisition software.
12. Image analyzing program (e.g., Adobe Photoshop, Image Tool)..

2.3 Reagents, Solutions, and Buffers

1. 1% acetocarmine solution: dissolve acetocarmine in 45% acetic acid in a flask. Boil for 8 h on a hot plate with an attached reflux column, and then cool rapidly by placing the flash on ice immediately. Store at room temperature until use.
2. 0.06 M sodium phosphate buffer, pH 7.4. Two stock solutions need to be prepared previously: 0.85 g Na_2HPO_4 (dibasic) in 100 ml distilled water and 1.34 g $2H_2O.NaH_2PO_4$ (monobasic) in 100 ml distilled water. Both solutions can be stored frozen. When necessary, mix 80 ml of Na_2HPO_4 solution with 20 ml of $2H_2O{\cdot}NaH_2PO_4$ solution. The pH of the final solution will be 7.4.
3. *Maceration solution* (for grasses): 0.5% (w/v) ClNa, 0.01 M EDTA, 6 mM sodium phosphate buffer, pH 7.5. Dissolve 0.5 g ClNa in 70 ml distilled water, 10 ml 0.1 M EDTA, and 10 ml 0.06 M sodium phosphate buffer. Adjust to pH 7.5 by adding 1 M NaOH or 0.1 M ClH. Finally, adjust the volume of the solution to 100 ml with distilled water. The solution can be stored in the fridge (at 4–8 °C) for up to 1 month.

4. *Digestion medium* (for the rest of species): 0.1 g snail gut enzyme "Cytohelicase" in 12.5 ml cold sterile distilled water, add 0.125 g polyvinylpyrrolidone and 0.19 g sucrose. The solution can be aliquoted and frozen.
5. *Swelling (spreading medium) solution*: 0.03% (v/v) Triton X-100 in 0.01 M EDTA, 6 mM sodium phosphate buffer, pH 7.5. In a bottle mix 10 ml 0.1 M EDTA, 10 ml 0.06 M sodium phosphate buffer and 80 ml distilled water. Check the pH to 7.5 (if necessary adjust the solution to pH 7.5 with 1 M NaOH or 0.1 M ClH by adding drop by drop). Finally, add 30 μl of Triton X-100 (*see* **Note 5**) with a cut pipette tip. Allow all Triton-X 100 to get out of the tip by pipetting up and down. Avoid foaming as much as possible. The solution can be stored in the fridge (at 4–8 °C) for up to 1 month.
6. 0.2 M sodium borate buffer: Dissolve 1.237 g boric acid (H_3BO_3) in 100 ml distilled water, adjust to pH 9.5 by adding 1 M NaOH. The solution can be stored in the fridge (at 4–8 °C) for up to 1 month.
7. *Fixative solution*: 4% (w/v) paraformaldehyde, 1.7% (w/v) sucrose in distilled water, pH 8.9. Prepare this solution in the fume hood with good protection. Put 80 ml of distilled water in a 250 ml beaker, adjust the pH to 9–9.5 with 0.2 M sodium borate buffer. Heat while stirring to approximately 60 °C, add 4 g of paraformaldehyde (PFA) powder, and stir constantly until the solution is clear. Then, allow to cool down before adding 1.7 g of sucrose and stir until completely dissolved. Adjust the pH to 8.9 with 1 M NaOH or 0.1 M ClH by adding drop by drop. Finally, adjust the volume of the solution to 100 ml with distilled water. The solution can be stored at 4–8 °C for up to 1 month. The pH must be checked before use.
8. To make the plastic coating, prepare a 1% (w/v) solution of overhead transparency in chloroform using the fume cupboard (*see* Subheading 2.2 and **Note 6**). Store solution in a refrigerator (at 4 °C) until preparing the slides. The solution, when necessary, should be brought up to room temperature before use.
9. For staining the slides: 40–50% (w/v) silver nitrate ($AgNO_3$) in distilled water. Solution should be prepared just before use.

3 Methods

3.1 Plant Material Preparation (Collect Buds or Anthers) (Staging)

This protocol is designed to be compatible with a wide range of plant species.

1. Inflorescences (buds or spikelet) prior to flower opening from healthy plants are cut, collected, and placed on a moist (wet) filter paper in a petri dish (*see* **Note 7**).

2. For Cruciferae and bulbs species, dissect out individual buds from an inflorescence, and order them by size in a petri dish containing damp filter paper, any bud with anthers that appear to be yellow can be discarded, as it is beyond the meiotic stages required for this technique.
3. Transfer separated buds to a watch glass or slide. Under the stereomicroscope, dissect out a single anther using fine forceps and needles. Stage one anther from a single bud. This can help to establish a possible link between the meiotic stage (prophase I, from leptotene to pachytene) and the bud size (*see* **Note 8**).
4. For Gramineae species, dissect all three anthers from each spikelet along the emerging spikes. Single anthers, with the correct size and color, are used to check the meiotic stage.
5. To determine the meiotic stage, in all cases transfer a single anther to one slide, with a drop of acetocarmine stain. Heat the slide containing the anther gently and evenly over a small flame of an alcohol burner until near-boiling. Apply a small (22×22 mm) coverslip to the material and disperse the material between glass slide and coverslip by tapping the coverslip gently with the flat back of a needle or lancet. Fold a filter paper in half around the slide, put it on a flat surface and then press the slide firmly using the thumb onto the filter paper above the area of the coverslip.
6. Examine under the light microscope to assess meiotic stage. If pollen mother cells are at required stage, that is, first meiotic prophase (leptotene–pachytene), save the rest of the anthers from that bud and the buds of similar size and color.
7. For grasses, repeat the control throughout the spike. In all cases, two sister anthers of the same floret at zygotene or pachytene are then kept.
8. The buds (Cruciferae family) or the anthers (cereals) with the appropriate meiotic stage are collected and placed on a wet filter paper in a petri dish. Keep them cold (in the refrigerator or put it over a cold metal plate) until necessary for making the SC preparations.

3.2 Chromosome (Nuclei) Preparation by Spreading

Unless otherwise indicated, all the buffers in this section should be precooled and kept on ice while in use.

1. Place solutions and the metal plate on ice while performing the surface spreading technique.
2. If required, unfreeze one aliquot of digestion medium in *crushed ice*.
3. Wipe with a piece of tissue paper and arrange the necessary excavated slides.

4. Add 5 μl of maceration or digestion medium (depending on plant species) in each of the cavity slides.
5. For cereals, using the forceps transfer four to six anthers to each of the excavated slides with 5 μl of *maceration solution* and tap out using a fine needle or a thin metal rod to produce a cell suspension. Remove debris as much as possible. Add another 5 μl of maceration solution to the cell suspension, mix, and place the slide on the cold metal plate or keep at room temperature for 4–6 min (*see* **Note 9**).
6. For others species (*Arabidopsis, Brassica, Muscari, Rumex* sp., etc.) transfer enough material for one preparation (typically 4–6 buds or their corresponding anthers) to a cavity slide with 5 μl *digestion medium* and suspended by using a thin metal rod. Add another 5 μl of enzyme solution to the cell suspension and mix. Remove debris as much as possible. Then incubate in the enzyme mixture at room temperature or onto the cold metal plate for 4–6 min (*see* **Notes 9** and **10**).
7. Afterward, add 10–15 μl of cold *Swelling solution*, mix well, and keep on cooled metal plate or at room temperature for 5–7 min (*see* **Note 11**).
8. Spreading is stopped by adding 20–30 μl of cold *Fixative solution* in drops around the suspension. Mix the suspension by gentle circular stirring with the metal rod and keep at room temperature for at least 10 min.
9. Then, the suspension is poured on the plastic-coated slide (*see* **Note 12**). Take good care not to damage the plastic support membrane. Adding more *Fixative solution* on the cavity slide will also help to transfer the last traces of the cell suspension.
10. Carefully spread out the suspension cells by tilting the slide in different directions. Then, allow the suspension to stand for another 15–20 min at room temperature.
11. Afterward, place the slides on a hot plate set to 30–35 °C in the fume cupboard for at least 24 h (or overnight) in order to let them dry completely.
12. Continue the protocol (silver staining) or store in a closed box and keep at room temperature (or at 4 °C for an extended period of time).

3.3 Silver Nitrate Staining

1. Preparations are washed one time for 1 min in a Coplin jar with distilled water and air-dried in a vertical position (*see* **Note 13**). The slides must be completely dry before staining.
2. Afterward, they are stained with an aqueous silver nitrate solution (33%). Prepare the staining solution just before use, by dissolving 1 g of $AgNO_3$ in 2 ml of distilled water (this is enough for 8–10 slides) (*see* **Note 14**).

3. Pour the silver nitrate solution into the little box suitable for dipping the patch of nylon mesh (about 24 mm × 50 mm) (*see* Subheading 2.2).
4. The piece of nylon cloth presoaked in the silver nitrate solution is placed on top of the slide. Make sure there are no bubbles under the nylon coverslip. Take good care not to damage the plastic support membrane (*see* **Note 15**).
5. The slides are placed in a prewarmed moist chamber and then incubated at 40–60 °C for 30–45 min until they turn into a caramel (golden brown) color (*see* **Note 16**).
6. Thereafter the nylon mesh is removed by immersing the slides in a large beaker filled with distilled water, the slides shook gently and then air-dried for at least 30 min or 1 h in upright position.

3.4 Transfer of Spreads to Grids (Electron Microscopy)

1. Check preparations for appropriate stages, spreading and silver staining of material by light microscopy using ×20 and ×40 objectives.
2. When suitable surface-spread nuclei are found, their locations are marked with a circle-shaped marking in the plastic coating around the SCs spreads using an object marker (Zeiss) containing water-resistant ink to guide the grid laying.
3. When all desired cell on a slide are located, crop a rectangle or square around the markers with a scalpel.
4. Breathe onto the slide and then float off the rectangle of plastic supporting membrane onto the surface of distilled water contained in a large square beaker/glass staining dish. This is achieved by gently lowering the slide at an angle, into the water, when the surface tension will pull the rectangle of plastic away from the slide.
5. Using watchmaker's forceps, place electron microscope grids very carefully at the marked circles and press down lightly so that it adheres to the plastic membrane (*see* **Note 17**).
6. Pick up the plastic membrane containing the grids immediately by touching the edge of the membrane vertically with a piece of cardboard (about 40 mm × 90 mm), dip and take everything out in smooth and fast movement. Then, turn it around and then the grids together with the cardboard paper and the plastic film are allowed to dry horizontally in a dust free box for 12–24 h.
7. Afterward, cut the supporting membrane carefully around the grids with a scalpel so it could be picked up with a fine forceps. Arrange the grids in row on a standard microscope slide and examine them under low power (×20) with LM.

8. Locate and map the exact location of SCs spreads in the grid (*see* Fig. 2c) and discard those that would not be suitable for EM analysis (*see* **Notes 4**, **18**, and **19**).
9. With precision, place the grids in the grids storage box with the correct identification number. Wrap the box with aluminum paper, write down the data, and store in the fridge (at 4–8 °C) until necessary.

3.5 Electron Microscopy and Image Processing

1. The located surface-spread nuclei are viewed in a transmission electron microscope at 80 KV. Specific microscopy instructions cannot be given, since they will differ depending on the brand, model, and configuration of the microscope used. Always follow the instructions of the technical assistant.
2. Take images following your microscope instructions and operation software guidance. Take as many images as necessary to make sure that the whole nucleus has been photographed (*see* **Notes 20** and **21**).
3. Check histograms of pixel brightness to ensure that all images are captured in the same conditions. Save images frequently in TIF format.
4. Electron micrographs are composed using *Adobe Photoshop* software.
5. After interpreting the different SC configurations (bivalents, multivalents) present in the fully traced nuclei (Fig. 3), the individual SC lengths, both absolute and relative, and the total SC complement length are measured (*see* **Note 22**).

4 Notes

1. In order to get a uniform thin plastic coated, the slides should be immersed and removed in a quick and continuous movement. If the movement is discontinuous, the plastic membrane can be irregular with thinner and thicker areas that can either hinder the passage of the electrons or cause the rupture of the membrane.
2. Nylon mesh coverslips are less expensive than glass, easier to handle, and are less damaging to plastic coating.
3. 50 Mesh grids offer a higher support to be able to withstand the vacuum chamber and electrons to pass through when examined under the electron microscope
4. Grids with an asymmetrical mark on the center offer a simple mechanism for identifying the exact position of SCs spreads, speeding up otherwise time consuming relocation of features at high magnification (Fig. 2a, c).

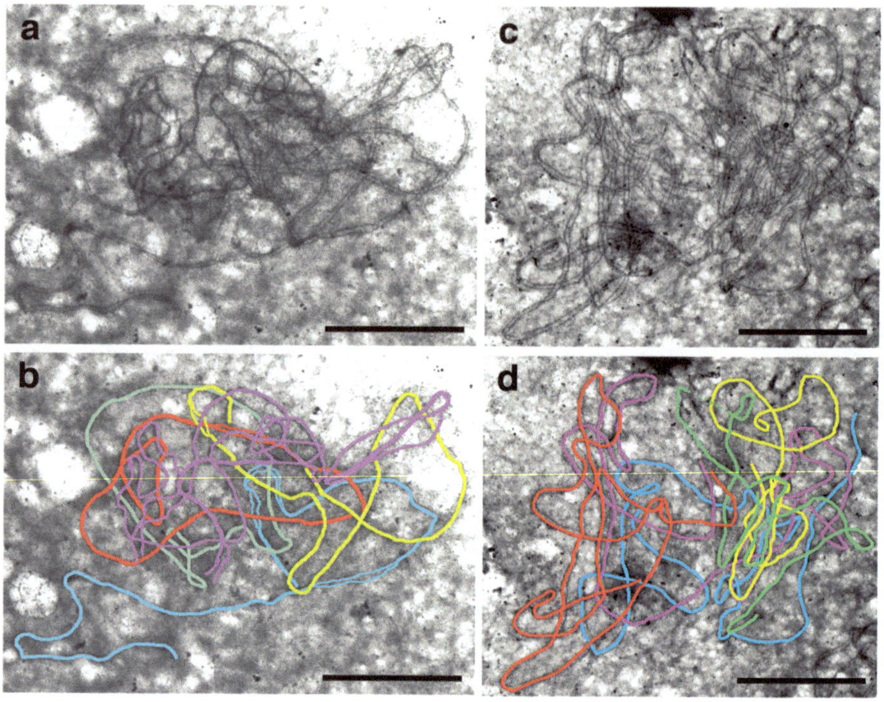

Fig. 3 Electron micrographs of silver stained meiotic prophase I nuclei of *Arabidopsis thaliana*. (**a**, **c**) and their corresponding drawing and interpretation (**b**, **d**). Late zygotene (**a**, **b**) and early pachytene (**c**, **d**). The five bivalents in each nucleus are drawn with different color
Bars represent 5 μm

5. Alternatively, Lipsol or a combination of Triton X-100 and Lipsol can be used. Likewise, the concentration of the detergent can be modified.
6. To prepare plastic solution, put 100 ml of chloroform in a 250 ml Erlenmeyer, add 1 g of overhead transparency and cover with Parafilm to prevent the chloroform evaporates. Keep at room temperature, shaking manually from time to time, until the complete dissolution. Then it can be used or stored in the refrigerator until necessary (for up to 1 week).
7. It is essential that plants are in perfect condition and, in general, the first inflorescences are the best to use because they produce large and synchronic anthers. This makes finding enough meiocytes at the suitable stage of meiosis much easier.
8. It is critical to check anthers or buds before making preparations to assess the correct stage (leptotene–pachytene). This is because their size is not absolutely reliable as an indicator of meiotic stage as it is subject to variation due to the environment, age of the plant and genotype in each plant material. For example, the link between meiotic stages and bud sizes is different in normal (wild type) and mutants in Arabidopsis.

9. Temperature and humidity are factors to be taken into account for suitable chromosome spreading as they affect how fast the cell suspension dries on to the slide. In a hot and dry environment, slides should be kept cold during the process, in order to avoid drying. If needed, add more solution and mix the cell suspension.
10. Representative times are given. This first step needs to be adjusted to the material and species used, by changing the time of incubation.
11. Representative times and concentrations are given, but in each case the conditions need to be established by trial and error. Different species and materials (i.e., cereals vs. cruciferous species, Arabidopsis wild-type vs. mutants, etc.) respond differently to spreading solution. Sometimes, it may be necessary to increase the time and/or the concentration of the detergent treatment. For example, in *Brassica napus* (Fig. 1c, d) [33] and *Atasy3* mutant [19] we achieved the best results by adding 0.05% Triton X-100 + 0.1% Lipsol in the swelling solution. On the contrary, in some cases, a reduction in time or concentration may be required, such as, for example, in early zygotene, in some meiotic mutants, or in haploids and hybrids, since their nuclei show a high percentage of unsynapsed axes that therefore can be broken more easily.
12. The coated slides should be allowed to *reach room temperature before* use.
13. If necessary, take the preps from the fridge and allow standing at room temperature before washing.
14. Silver nitrate solution stains skin, clothes, furniture, etc. even when very diluted. Wear gloves and avoid splashes. After use, all glass and plastic labware can be easily rinsed under the running tap water for several minutes.
15. Nylon mesh coverslips are easier to handle and less damaging to the plastic coated and material on the slide. The slips should be cut of nylon mesh to the size of a coverslip (24 mm × 50 mm), without extra material, to avoid them moving during subsequent handling, damaging the plastic coated.
16. The silver staining conditions (concentration, incubation time, and temperature) might need to be optimized for each plant material.
17. Electron microscope grids generally have a matte/dull and a glossy surface. Make sure to place the matte side (self-adhesive) down onto the plastic surface.
18. Using 50 mesh grids has a disadvantage because, unfortunately, several nuclei will probably be lost, since some of them may have been located under one of the square lines that form the grid.

19. Optionally, and only if there are few SC spreads in a preparation, the grids can be placed under the optical microscope with the help of a thin brush. In this case, cut the supporting membrane around the markers with a scalpel before placing the slide under the light microscope. Using a very fine brush, place the grids very carefully at the marked circles. Remove the slide from the microscope very carefully and breathe onto the slide with the grids and float off the rectangle of plastic supporting membrane as explained above.
20. Depending on the number and size of the chromosomes and the magnification it would be necessary to take between one and ten photos. For example, in *Triticum* and *Aegilops* species, with a low magnification (from 2000× to 4000×) frequently used, it would be necessary to take between six and ten photos. However, in *Arabidopsis* the magnification may be increased (e.g., from 6000× to 10,000×) and it would be necessary to take only one to four pictures (Fig. 1a, b and Fig. 3).
21. Be careful to overlap the photos enough so you can later assemble them into one.
22. Depending on the complexity of chromosome complement (i.e., nucleus), the SC complement can be analyzed in two ways. Individual SC can be traced out in the image resulting using Photoshop (Fig. 3) or from enlarged photographic prints on an acetate overlay. Afterward, the absolute lengths of synapsed and unsynapsed regions in each bivalent are measured manually using Image tool, NIH imaging, or a map measurer, respectively.

Acknowledgments

The author is very grateful for the contributions to synaptonemal complex research from former and current lab members, especially Juan Luís Santos and Mónica Pradillo. The author's research has been supported by several grants from the Ministerio de Ciencia, Innovación y Universidades of Spain and the Universidad Complutense de Madrid-Banco Santander of Spain (Grant number 910452).

References

1. Cuñado N, Callejas S, García MJ, Fernández A, Santos JL (1996) The pattern of zygotene and pachytene pairing in allotetraploid *Aegilops* species sharing the U genome. Theor Appl Genet 93:1152–1155
2. Cuñado N, García MJ, Callejas S, Fernández A, Santos JL (1996) The pattern of zygotene and pachytene pairing in allotetraploid Aegilops species sharing the D genome. Theor Appl Genet 93:1175–1179
3. López E, Pradillo M, Romero C, Santos JL, Cuñado N (2008) Pairing and synapsis in wild type *Arabidopsis thaliana*. Chromosom Res 16:701–708

4. Gillies CB (1981) Electron microscopy of spread maize pachytene synaptonemal complex. Chromosoma 83:571–591
5. Stack S (1982) Two-dimensional spreads of synaptonemal complexes from solanaceous plants. I the technique. Stain Technol 57:265–272
6. Albini SM, Jones GH, Wallace BM (1984) A method for preparing two-dimensional surface-spreads of synaptonemal complexes from plant meiocytes for light and electron microscopy. Exp Cell Res 152:280–285
7. Loidl J, Jones GH (1986) Synaptonemal complex spreading i*n Allium* I. Triploid A Sphaerocephalon. Chromosoma 93:420–428
8. Stack S, Herickhoff L, Sherman J, Anderson L (1991) Staining plant-cells with silver.1. The salt-nylon technique. Biotech Histochem 66:69–78. https://doi.org/10.3109/10520299109110553
9. Moses MJ (1956) Chromosomal structures in crayfish spermatocytea. J Biophys Biochem Cytol 2:215–218
10. Moses MJ (1958) The relation between the axial complex of meiotic prophase chromosomes and chromosome pairing in salamander (*Plethodon cinereus*). J Biophys Biochem Cytol 4:633–638
11. Von Wettstein D, Rasmussen SW, Helm PB (1984) The synaptonemal complex in genetic segregation. Annu Rev Genet 18:331–413
12. Heyting C (1996) Synaptonemal complexes: structure and function. Curr Opin Cell Biol 8:389–396
13. Zickler D, Kleckner N (1999) Meiotic chromosomes: integrating structure and function. Annu Rev Genet 33:603–754
14. Albini SM, Jones GH (1987) Synaptonemal complex spreading in *Allium cepa* and *A. fistulosum* I. The initiation and sequence of pairing. Chromosoma 95:329–338
15. Page SL, Hawley RS (2004) The genetics and molecular biology of the synaptonemal complex. Annu Rev Cell Dev Biol 20:525–558
16. Caryl AP, Armstrong SJ, Jones GH, Franklin FCH (2000) A homologue of the yeast HOP1 gene is inactivated in the Arabidopsis meiotic mutant asy1. Chromosoma 109:62–71
17. Armstrong SJ, Caryl AP, Jones GH, Franklin FC (2002) Asy1, a protein required for meiotic chromosome synapsis, localizes to axis-associated chromatin in *Arabidopsis* and *Brassica*. J Cell Sci 115:3645–3655
18. Wang KJ, Wang M, Tang D, Shen Y, Qin BX et al (2011) PAIR3, an axis associated protein, is essential for the recruitment of recombination elements onto meiotic chromosomes in rice. Mol Biol Cell 22:12–19
19. Ferdous M, Higgins JD, Osman K, Lambing C, Roitinger E, Mechtler K, Armstrong SJ, Perry R, Pradillo M, Cuñado N, Franklin FCH (2012) Inter-homolog recombination and synapsis in Arabidopsis meiosis is dependent on the chromosome axis protein AtASY3. PLoS Genet 8:e1002507
20. Higgins JD, Sánchez-Morán E, Armstrong SJ, Jones GH, Franklin FCH (2005) The Arabidopsis synaptonemal complex protein ZYP1 is required for chromosome synapsis and normal fidelity of crossing over. Genes Dev 19:2488–2500
21. Wang M, Wang KJ, Tang D, Wei CX, Li M et al (2010) The central element protein ZEP1 of the synaptonemal complex regulates the number of crossovers during meiosis in rice. Plant Cell 22:417–430
22. Golubovskaya IN, Wang CJR, Timofejeva L, Cande WZ (2011) Maize meiotic mutants with improper or non-homologous synapsis due to problems in pairing or synaptonemal complex formation. J Exp Bot 62:1533–1544
23. Khoo KHP, Able AJ, Able JA (2012) The isolation and characterisation of the wheat molecular ZIPper I homologue, *Ta* ZYP1. BMC Res Notes 5:106
24. Barakate A, Higgins JD, Vivera S, Stephens J, Perry RM, Ramsay L, Colas I, Oakey H, Waugh R, Franklin FCH, Armstrong SJ, Halpin C (2014) The synaptonemal complex protein ZYP1 is required for imposition of meiotic crossovers in barley. Plant Cell 26:729–740
25. Gillies CB (1989) Chromosome pairing and fertility in polyploids. In: Gillies CB (ed) Fertility and chromosome pairing: recent studies in plants and animals. CRC Press, Boca Raton, pp 137–176
26. Santos JL, Cuadrado MC, Díez M, Romero C, Cuñado N, Naranjo T, Martínez M (1995) Further insights on chromosomal pairing of autopolyploids: a triploid and tetraploids of rye. Chromosoma 104:298–307
27. Cuñado N, Santos JL (1999) On the diploidization mechanism of the genus *Aegilops*; meiotic behaviour of interspecific hybrids. Theor Appl Genet 99:1080–1086
28. Cuñado N, de la Herrán R, Santos JL, Ruiz Rejón C, Garrido-Ramos MA, Ruiz Rejón M (2000) The evolution of the ribosomal loci in the subgenus Leopoldia of the genus *Muscari* (Hyacinthaceae). Plant Syst Evol 221:245–252
29. Martínez M, Cuñado N, Carcelén N, Romero C (2001) The *Ph1* and *Ph2* loci play different roles in the synaptic behaviour of hexaploid

wheat *Triticum aestivum*. Theor Appl Genet 103:398–405

30. Cuñado N, Blázquez S, Melchor L, Pradillo M, Santos JL (2005) Understanding the cytological diploidization mechanism of polyploid wild wheats. Cytogenet Genome Res 109:205–209
31. Cuñado N, Navajas-Pérez R, de la Herrán R, Ruiz Rejón C, Ruiz Rejón M, Santos JL, Garrido-Ramos MA (2007) The evolution of sex chromosomes in the genus *Rumex* (Polygonaceae): identification of a new species with heteromorphic sex chromosomes. Chromosom Res 15:825–832
32. Pradillo M, López E, Romero C, Sánchez-Morán E, Cuñado N, Santos JL (2007) An analysis of univalent segregation in meiotic mutants of *Arabidopsis thaliana*: a possible role for synaptonemal complex. Genetics 175:505–511
33. Grandont L, Cuñado N, Coriton O, Huteau V, Eber F, Chèvre AM, Grelon M, Chelysheva L, Jenczewski E (2014) Homeologous chromosome sorting and progression of meiotic recombination in *Brassica napus*: ploidy does matter! Plant Cell 26:1448–1463. https://doi.org/10.1105/tpc.114.122788

Chapter 14

Quantification of Synapsis Using Immunolocalization in Embedded Nuclei of *Lolium*

Dylan W. Phillips

Abstract

The establishment, formation and disassembly of the synaptonemal complex (SC) is intimately associated with other essential processes that occur during prophase I of meiosis, including recombination. Labeling the SC using primary antibodies raised against key proteins, detected using secondary antibodies conjugated to fluorescent dyes, differentiate between synapsed and unsynapsed regions, revealing the dynamics of the process. Embedding meiotic nuclei in acrylamide pads preserves the three-dimensional (3D) organization of the chromosomes, which can be optically sectioned using confocal laser scanning microscopy to produce a faithful representation of the SC at the point of fixation. Deconvolution, and processing using Imaris allows the axes to be isolated from the nucleus and their features measured. Here, I describe a robust protocol to quantify the SC using immunofluorescence in *Lolium perenne* and *L. temulentum*.

Key words Meiosis, Lolium, Synapsis, Synaptonemal complex, Immunolocalization

1 Introduction

The intimate association of homologs via the establishment of the synaptonemal complex (SC) during prophase I is integral to the success of meiosis. The formation of this distinctive tripartite proteinaceous structure is remarkably conserved in sexually reproducing members of the animal and plant kingdom and provides the platform for recombination to occur [1]. The formation of the SC can provide details of the dynamics of pairing and alignment of homologous chromosomes, as well as bouquet formation and recombination. Historically the quantification of axis formation and synapsis was performed by serially sectioning meiotic nuclei that were imaged using electron microscopy (EM) [2]. Such methods involved the use of toxic reagents, required high levels of skill, specialist equipment and unbound resolve. Such methods have been superseded by techniques utilizing fluorescent microscopy.

Mónica Pradillo and Stefan Heckmann (eds.), *Plant Meiosis: Methods and Protocols*, Methods in Molecular Biology, vol. 2061, https://doi.org/10.1007/978-1-4939-9818-0_14,

Detection of key SC proteins using immunolocalization, coupled with the preservation of the 3D structure of meiotic nuclei and incremental developments in microscopy, has been packaged into a method that has opened up new possibilities in the study of the dynamics of the SC. The technique was adapted from a method developed to preserve *Drosophila* polytene chromosomes using an optically clear and permeable acrylamide gel matrix [3]. The technique was adapted to study the dynamics of telomeres in maize during meiosis [4] and used to examine proteins forming the SC of maize [5], rye [6], barley [7], and *Arabidopsis* [8]. The overarching benefit of using such an immunological approach is the simplicity of the method to produce the embedded meiotic nuclei, the preservation of the spatial organization of the chromosomes and associated proteins, ease of the imaging strategy and relative simplicity of the image analysis. This method also delivers quantitative data relating to SC formation and SC disassembly.

Lolium perenne is an outbreeding perennial forage crop ($2n = 2x = 14$) and is closely related to the annual self-fertile *Lolium temulentum* ($2n = 2x = 14$). Both species have large chromosomes which are very amenable to cytological analysis. This chapter outlines the steps involved in the fixation and successive preservation of meiotic nuclei in acrylamide pads, immunolocalization, imaging, and subsequent analysis. Using this method it is possible to obtain quantitative data relating to trace unpaired axial elements during leptotene and the progression of synapsis during zygotene (Fig. 1).

2 Materials

2.1 Plants

1. *L. perenne* (cultivar AberDart) and *L. temulentum* (line BA3081) are grown in 12 cm square pots containing four-parts fine grade compost and one part horticultural grit. All plants are grown in a controlled environment room (*see* **Note 1**).
2. *L. perenne* requires vernalization to induce flowering. Seeds are germinated and grown in noninductive conditions, 20 °C constant temperature and 24 h of light, for 5 weeks. Plants are moved to the primary induction conditions, 6 °C constant temperature and 10 h of light, for 8 weeks. Finally plants are moved to secondary induction conditions, 18 °C constant temperature and 20 h light. From sowing, plants take 17–20 weeks to reach meiosis.
3. *L. temulentum* is an annual species that has no vernalization requirement. Plants are grown at a constant temperature of 20 °C and 16 h of light. From sowing, plants take 6–8 weeks to reach meiosis.

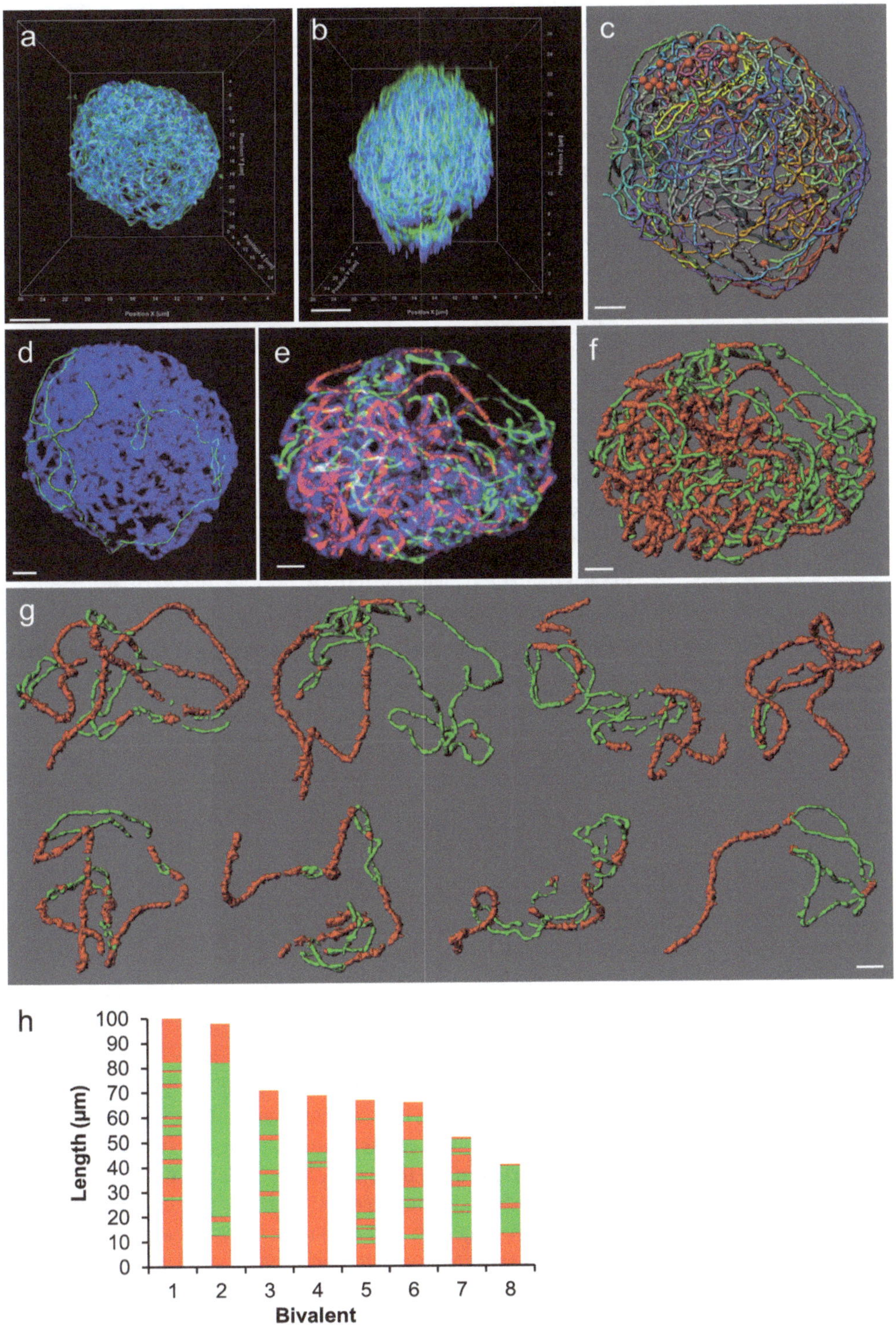

Fig. 1 Tracing synaptonemal complex formation in *L. temulentum* using immunolocalization. (**a**) Deconvolved projection of a leptotene nucleus embedded in polyacrylamide and captured using CLSM, showing unsynapsed axis associated ASY1 cores (green). (**b**) Leptotene nucleus shown in (**a**) with view rotated 90°

2.2 Fixation

1. 10× Buffer A salts: 150 mM PIPES, 800 mM KCl, 200 mM NaCl, 20 mM EDTA, 5 mM EGTA. Combine all the ingredients and adjust pH to 6.8 with KOH. Filter-sterilize and store at −20 °C for short term (few months) and − 80 °C for longer-term storage.
2. 0.4 M spermine: 0.4 M spermine in 50 mM PIPES. Adjust pH to 6.8 with KOH, filter-sterilize, aliquot and store at −20 °C.
3. 1 M spermidine: 1 M spermidine in 50 mM PIPES. Adjust pH to 6.8 with KOH, filter-sterilize, aliquot and store at −20 °C.
4. 1 M dithiothreitol (DTT): 1 M DTT in 10 mM sodium acetate. Adjust pH to 5.5, filter-sterilize, aliquot and store at −20 °C.
5. 2 M sorbitol: 2 M sorbitol in sterile water. Filter-sterilize, aliquot and store at −20 °C.
6. 1× Buffer A: Mix 5 ml of 10× Buffer A salts, 25 μl of 0.4 M spermine, 25 μl of 1 M spermidine, 50 μl of 1 M DTT, 8 ml of 2 M sorbitol and 37 ml of sterile water. Store at 4 °C for up to 1 month.
7. Fixative: 2% Paraformaldehyde (w/v) in 1× Buffer A (*see* **Note 2**). Aliquot and store at −20 °C.
8. Cavity slide, polished round depressions 15 mm in diameter and 1 mm deep.
9. Fine forceps (Fine Scientific Tools).
10. Stainless steel insect pins, size 1, with holder (Fine Scientific Tools).
11. Drawn-out glass pasture pipette.
12. Small humid chamber (*see* **Note 3**).
13. Dissecting microscope.

Fig. 1 (continued) demonstrating preservation of the 3D nuclear structure. (**c**) All axial elements isolated using Imaris from leptotene nucleus shown in (**a**). Each element has been assigned a false color and ends of each element marked with a red sphere. (**d**) Maximum intensity projection showing single axial element measuring 117 μm in length isolated from (**a**). (**e**) Deconvolved maximum intensity projection of a zygotene nucleus embedded in polyacrylamide and captured using CLSM, showing ASY1 cores (green) and central element protein ZYP1 (red) denoting synapsed regions. (**f**) Zygotene nucleus shown in (**e**) processed using Imaris with each of the eight synapsing bivalents isolated, showing generated surfaces for ASY1 (green) and ZYP1 (red). (**g**) Each bivalent isolated using Imaris from nucleus shown in (**e**) displayed individually and ranked in descending length. Generated surfaces represent ASY1 (green) and ZYP1 (red). (**h**) Stacked bar graph showing the cumulative lengths of ZYP1 (red) and ASY1 (green) fragments constituting each bivalent isolated from nucleus shown in (**e**). Bivalents ordered by descending total length. Bivalents 7 and 8 terminate in the vicinity of the nucleolus and are likely to be the chromosome containing the major rDNA locus. (**a**, **b**, **d**, and **e**) Chromosomes counterstained with DAPI (blue) and scale bar is equal to 5 μm (**a**, **b**) and 2 μm (**c**–**g**)

2.3 Acrylamide Pads

1. 2× Buffer A: Mix 200 μl of 10× Buffer A salts, 1 μl of 0.4 M spermine, 1 μl of 1 M spermidine, 2 μl of 1 M DTT, 320 μl of 2 M sorbitol, and 480 μl of sterile water. Store at 4 °C.
2. Acrylamide solution: 1:1 mix of 30% acrylamide and bisacrylamide (29:1) (Bio-Rad) and 2× Buffer A. Prepare fresh and store on ice.
3. 20% ammonium persulfate: 250 μl of 20% (w/v) ammonium persulfate in water. Prepare just before use, store on ice.
4. 20% sodium sulfite: 250 μl of 20% (w/v) sodium sulfite in water. Prepare just before use, store on ice.
5. Brass rod (dimensions ≈ 3 mm × 100 mm).
6. Small 35 × 10 mm petri dishes.
7. Coverslip (Thermo Scientific Menzel coverslip 22 × 22 mm #1).

2.4 Immunolocalization and Mounting

1. Phosphate buffered saline (PBS; 10×): 1.37 M NaCl, 27 mM KCl, 100 mM Na_2HPO_4, 18 mM KH_2PO_4. Adjust pH to 7.4 and sterilize by autoclaving.
2. Permeabilization buffer: 1× PBS, 1% Triton X-100, 1 mM EDTA.
3. Blocking buffer: 3% bovine serum albumin (BSA) in 1× PBS, 0.1% Tween 20, 1 mM EDTA.
4. Primary antibodies: diluted to the relevant level (e.g., 1:100 or 1:200) in blocking buffer.
5. Wash buffer: 1× PBS, 0.1% Tween 20, 1 mM EDTA.
6. Secondary antibodies: anti-rabbit, anti-rat, or anti-guinea pig antibodies conjugated to Alexa Fluor (ThermoFisher Scientific) diluted to 1:250 in blocking buffer.
7. Mounting media: 2.5% DABCO (1,4-diazobicyclo(2,2,2) octane), 200 mM Tris–HCl (pH 8), 80% glycerol, 1 μg/ml 4, 6-diamidino-2-phenylindole (DAPI) (*see* **Note 4**).
8. Disposable coverslips (*see* **Note 5**).
9. Large humid chamber (*see* **Note 6**).
10. Standard microscope slides.
11. Clear nail polish.

3 Methods

3.1 Sample Collection and Fixation

1. With aid of a dissecting microscope, use fine forceps and needle to harvest meiotic anthers from both the primary and secondary florets of either *L. perenne* or *L. temulentum* into a well of a cavity slide containing 20 μl of 1× Buffer A (*see* **Notes 7** and **8**).

2. Remove 1× Buffer A using the drawn out pasture pipette, replace with 50 μl of fixative, and leave in a small humid chamber for 15 min at room temperature.
3. Remove fixative and wash twice with 50 μl of 1× Buffer A at room temperature.
4. Following the final wash add 20 μl of 1× Buffer A and thoroughly disrupt the anthers by tapping for approximately 1 min with a brass rod. The mixture should turn a milky white color (*see* **Note 9**).
5. Add a further 10 μl of 1× Buffer A to each well and mix by gently pipetting up and down (*see* **Note 10**).

3.2 Embedding Meiotic Nuclei

1. For each well place two small petri dish lids on a dust free and level surface. On each lid place a single 22 × 22 coverslip.
2. Place 10 μl of the meiocyte suspension on the center of each coverslip.
3. Working rapidly, take a prepared 0.5 ml disposable tube containing 10 μl of acrylamide solution, add 0.5 μl of 20% ammonium persulfate and add 0.5 μl of 20% sodium sulfite. Mix briefly by pipetting.
4. Immediately add 5 μl of the solution prepared in **step 3** to the meiocyte suspension and mix by pipetting up and down three times.
5. Using forceps place a second 22 × 22 coverslip on top of the first at a rotated 45° angle, trapping the meiocyte solution between, producing an arrangement resembling an octagram (*see* **Note 11**).
6. The solution will take approximately 30 min to polymerize at room temperature (*see* **Note 12**).
7. Carefully pry apart the two coverslips using a razor blade. Normally the pad sticks fully to one of the coverslips. If it tears retain the coverslip containing the largest fragment of pad.
8. Place the coverslip containing the pad into the base of the small petri dish, with the pad facing upward. Proceed immediately to the immunolocalization step.

3.3 Immunolocalization

1. Add 2 ml of permeabilization buffer to the petri dish and incubate for 1 h at room temperature.
2. Briefly rinse with an excess of wash buffer, before adding 2 ml of blocking buffer. Incubate for 2 h at room temperature.
3. Apply 50 μl of the diluted primary antibodies to the center of the acrylamide pad and cover with a disposable coverslip. Apply the petri dish lid and place into a large humid chamber.

Incubate for 1 h and move to 4 °C for 48 h or longer. Remove from fridge and incubate for a further 1 h.

4. Carefully remove the disposable coverslip, collect diluted primary antibodies (*see* **Note 13**) and wash with wash buffer four times for 30 min at room temperature, with gentle agitation.
5. Apply 100 μl of diluted secondary antibodies buffer to the center of the acrylamide pad and cover with a disposable coverslip. Return petri dish to a large humid chamber, incubate in the dark for 1 h at room temperature, and move to 4 °C for 16 h or longer. Remove from fridge and incubate for a further 1 h at room temperature (*see* **Note 14**).
6. Carefully remove the disposable coverslip and wash with wash buffer four times for 30 min at room temperature, with gentle agitation. Proceed directly to mounting.

3.4 Mounting

1. Carefully remove the coverslip and place with the pad facing up onto a clean, labeled, microscope slide.
2. Place five drops of mounting media evenly over the surface of the pad and incubate in the dark for 30 min.
3. Apply a second 22 × 22 mm coverslip directly on top of the gel pad. Remove excess mounting media using small pieces of Whatman No.1 filter paper placed at each edge of the coverslip sandwich.
4. Realign the two coverslips so that they are perfectly registered and seal with ample clear nail varnish. Clean any residual mounting media using lens tissue and store at 4 °C.

3.5 Imaging

3D imaging of the meiotic nuclei preserved in the acrylamide pads can be accomplished using any fluorescent microscope system equipped with a motorized focus drive. For tracing SC formation a confocal laser scanning microscope provides sufficient resolution. The nuclei shown in Fig. 1 are imaged using a Leica TCS SP5 II confocal laser scanning microscope controlled by Leica LAS-AF software. Each channel is collected sequentially via a conventional photomultiplier tube or a Leica HyD hybrid detector. Nuclei are optically sectioned with between 80 and 100 sections (step size 0.15 μm), with a voxel dimension of 0.0677 × 0.0677 × 0.147 μm.

3.6 Image Analysis

Z-stacks are deconvolved using AutoQuant X3 "Adaptive Point Spread Function" (blind) method applied with 25 iterations (Media Cybernetics). SC components are isolated using Imaris 8.2 (Bitplane).

1. Uniformly adjust each channel to remove any background, and increase brightness so that the signals are clearly visible.

2. For each bivalent create a new group folder, to this create a two new "Surfaces" and skip automatic surface creation, allowing Surfaces to be created manually.
3. Identify an end of a bivalent and begin to start to manually add Surfaces, tracing the bivalent. Use separate surfaces to differentiate synapsed and unsynapsed regions (*see* **Note 15**).
4. Continue until all axes are isolated and use the "Measure Points" tool to quantify the length of the isolated axis.

4 Notes

1. Meiosis is sensitive to environmental conditions, particular temperature. Ensure growth conditions are tightly controlled and consistently applied.
2. The simplest method to dissolve the paraformaldehyde is to place the solution in an air tight bottle and leave overnight in a 65 °C water bath. The fixative can be stored at −20 °C for over 1 year with no obvious deterioration.
3. Small humid chamber comprised of a 90 × 15 mm petri dish with Whatman Grade 1 filter paper in the base and lid, dampened with water. Lay two toothpicks in the inverted lid, this raises the cavity slide. In this instance the base is used as a lid.
4. In my experience, DABCO mounting media provides an equal level of photostability when compared to the common commercially bought equivalents.
5. Disposable coverslips are made of standard autoclave bags cut to ≈ 22 × 22 mm.
6. Large humid chamber comprises of a clip-lid storage box containing a piece of suitably size blue paper roll that is dampened with water.
7. Spikes collected should have the inflorescent enclosed by the flag leaf. It is possible to estimate the length of the inflorescence by placing gentile pressure using thumb and forefinger on the spike. Those approximately 7 cm – 15 cm in length are meiotic. A single anther from the primary and secondary floret of the central spikelet is collected and staged using acetocarmine staining. If there is a short hiatus between staging and harvesting of anthers then inflorescences are stored on ice in a large 150 × 25 mm petri dish containing dampened filter paper.
8. All the anthers from a single spike are harvested into a single well. It is usual to harvest multiple spikes and cavity slides already containing anthers can be stored briefly on ice in a small humid chamber (*see* **Note 3**) till the next step.

9. The anthers of *L. perenne* and *L. temulentum* are large and meiocytes can be extruded using fine forceps and a needle at this point. The tapping method requires little skill, releases a large number of meiocytes and is quick.
10. Gently pipetting up and down aids the release of the meiocytes from the columns. During this process avoid adding any air bubbles to the meiocytes suspension.
11. If using a super resolution system with limited working distance then gentle pressure is applied to the sandwich to reduce the thickness of the final acrylamide pad. The weight of a few low denomination coins is sufficient.
12. Under different climatic conditions the rate of polymerization can vary. Once set the pad pulls slightly away from the edge of the coverslip sandwich, once this occurs the pad can be opened.
13. The diluted primary antibody can be reused two or three times more without any noticeable impact on the end result.
14. Once the secondary antibodies have been added the pads are kept in the dark as much as practically possible during all incubation, wash and mounting steps.
15. It is fairly simple to isolate bivalents from nuclei at stages beyond mid-zygotene. Those at early leptotene to early zygotene are challenging and should only be attempted once the skill has been developed. When tracing the axis the nucleus must be rotated often to achieve the best possible view and use of the "Ortho slicer" is recommended. Isolating the bivalents and measuring a mid-zygotene nucleus takes approximately 1 h.

Acknowledgments

Financial support is provided through a UK Biotechnology and Biological Sciences Research Council (BBSRC) Strategic Programme Grant (BB/CSP1730/1).

References

1. Loidl J, Bonini N (2016) Conservation and variability of meiosis across the eukaryotes. Annu Rev Genet 50:293–316
2. Jenkins G (1985) Synaptonemal complex formation in hybrids of *Lolium temulentum* x *Lolium perenne* (L) – 1. High chiasma frequency diploid. Chromosoma 92:81–88
3. Urata Y, Parmelee SJ, Agard DA, Sedat JW (1995) A three-dimensional structural dissection of *Drosophila* polytene chromosomes. J Cell Biol 131:2279–2295
4. Bass H, Marshall W, Sedat J, Agard D, Cande W (1997) Telomeres cluster *de novo* before the initiation of synapsis: a three-dimensional spatial analysis of telomere positions before and during meiotic prophase. J Cell Biol 137:5–18
5. Wang C, Carlton P, Golubovskaya I, Cande W (2009) Interlock formation and coiling of meiotic chromosome axes during synapsis. Genetics 183:905–915
6. Phillips D, Mikhailova E, Timofejeva L, Mitchell J, Osina O, Sosnikhina S, Jones RN,

Jenkins G (2008) Dissecting meiosis of rye using translational proteomics. Ann Bot 101:873–880

7. Phillips D, Nibau C, Wnetrzak J, Jenkins G (2012) High resolution analysis of meiotic chromosome structure and behaviour in barley (*Hordeum vulgare* L.). PLoS One 7(6):e39539

8. Hurel A, Phillips D, Vrielynck N, Mezard C, Grelon M, Christophorou N (2018) A cytological approach to studying meiotic recombination and chromosome dynamics in *Arabidopsis thaliana* male meiocytes in three dimensions. Plant J 95:385–396

Chapter 15

Following the Formation of Synaptonemal Complex Formation in Wheat and Barley by High-Resolution Microscopy

Benoit Darrier, Mikel Arrieta, Sybille U. Mittmann, Pierre Sourdille, Luke Ramsay, Robbie Waugh, and Isabelle Colas

Abstract

Wheat and barley have large genomes of 15 Gb and 5.1 Gb, respectively, which is much larger than the human genome (3.3 Gb). The release of their respective genomes has been a tremendous advance the understanding of the genome organization and the ability for deeper functional analysis in particular meiosis. Meiosis is the cell division required during sexual reproduction. One major event of meiosis is called recombination, or the formation of crossing over, a tight link between homologous chromosomes, ensuring gene exchange and faithful chromosome segregation. Recombination is a major driver of genetic diversity but in these large genome crops, the vast majority of these events is constrained at the end of their chromosomes. It is estimated that in barley, about 30% of the genes are located within the poor recombining centromeric regions, making important traits, such as resistance to pest and disease for example, difficult to access. Increasing recombination in these crops has the potential to speed up breeding program and requires a good understand of the meiotic mechanism. However, most research on recombination in plant has been carried in *Arabidopsis thaliana* which despite many of the advantages it brings for plant research, has a small genome and more spread out of recombination compare to barley or wheat. Advance in microscopy and cytological procedures have emerged in the last few years, allowing to follow meiotic events in these crops. This protocol provides the steps required for cytological preparation of barley and wheat pollen mother cells for light microscopy, highlighting some of the differences between the two cereals.

Key words 3D microscopy, Barley, Meiosis, Synapsis, Wheat

1 Introduction

There is a vast number of cytological and immunochemical methods to study meiosis in model organisms but if all are based on the same principles, some work better in one organism than another. In plant, cell wall is also complicating the penetration of probes or antibodies which requires extra steps of permeabilization or digestion [1, 2]. Meiosis is the specific cell division that produce male

Mónica Pradillo and Stefan Heckmann (eds.), *Plant Meiosis: Methods and Protocols*, Methods in Molecular Biology, vol. 2061, https://doi.org/10.1007/978-1-4939-9818-0_15, © Springer Science+Business Media, LLC, part of Springer Nature 2020

and female gamete and is highly controlled [3, 4] and cytology is a powerful method for functional analysis of this process. The duration of meiosis in barley is about 43 h [5, 6], while in bread wheat, it lasts for 24 h despite having to process three times more genetic material [7]. An important event in meiosis is chromosome synapsis, that aligns the homologues together which allows for homologous recombination [8–10]. Any defect in chromosome synapsis leads to reduced recombination and disastrous chromosome segregation [11–14]. This protocol takes advantage of 3D structured illumination microscopy (3D-SIM) to follow the events of synapsis in both barley and wheat. We use TaASY1 (Asynaptic 1) [15], an antibody raised in wheat and developed by Boden et al. [16] to monitor the formation of chromosome axes during the first step of the synaptonemal complex formation. To monitor synapsis, we use a custom antibody made in our lab against the barley protein HvZYP1 [10, 12]. The protocol details the steps to prepare pollen mother cells isolation in both barley and wheat, for investigating the progression of synapsis, highlighting some of the difference between the two organisms and associated difficulties.

2 Materials

2.1 Plant

1. Barley *cv. Bowman* seeds.
2. Wheat cv *Chinese spring* seeds.

2.2 Meiosis Staging and Plant Fixation

1. Stereomicroscope Microtec HM-4TR; equipped with 10× focusing eyepiece with graticule 10:100 (HM-415FG STG) and HM-3L4N LED stand (*see* **Note 1**).
2. Microscope Microtec LM-2TR equipped with wide field eyepiece 10×/FN 20 with graticule 10:100 (LM-E10FG) and calibrating stage micrometer 1 mm in 100 parts (PM-STG) and objective 10× and 40× (*see* **Note 1**).
3. Dissecting tools: Very Fine forceps (or insulin needles), single edge razor blade.
4. Glass microscope slides and 18 × 18 glass coverslip.
5. Acetocarmine: Dissolve 0.5 g of Carmine in 45% acetic acid, bring to boil and stir for 15–20 min. Removed from heat, let it cool and keep in the fridge overnight. Filter to remove any precipitates and store in a dark bottle (*see* **Note 2**).
6. Fixative: 4% formaldehyde fixative was made by dissolving 1 g of paraformaldehyde in 12.5 ml of water, heat up to 60 °C for few minutes and add 2–3 drops of 5 M NaOH until the solution clears up. Let it cool on ice and add 12.5 ml of 2× PBS (*see* **Note 3**).

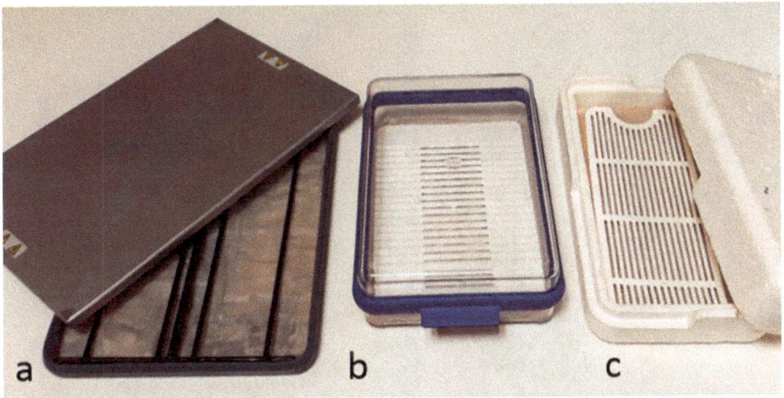

Fig. 1 Example of humid chambers. (**a**) Humid chamber from an old Boekel Scientific InSlide Out™ Slide Hybridizer (model 241000); (**b**) converted lunch box from supermarket; (**c**) polystyrene humid chamber box for hybridization

2.3 Slide Preparation and Immunostaining

1. Working buffer: 1× PBS containing 0.5% Triton™ X100, 0.5% Tween® 20. Also add 0.5% Lipsol if working with wheat.
2. Embryo Dish (or cavity slide) and Glass rod.
3. Microscope Polysine® slide.
4. Flat slide holder (Brunel Microscopes Ltd.).
5. PAP pen liquid blocker.
6. Blocking solution: Working solution containing 5% goat serum albumin and 5% donkey serum albumin. Make it fresh (*see* **Note 4**).
7. Primary antibodies: TaASY1 (Agrisera) and HzZYP1 [10] are used at 1:1000 and 1:500 respectively in blocking buffer.
8. Secondary antibodies: Secondary Antibody, Alexa Fluor® 488 or 568 conjugate (Thermo Fisher Scientific) against rat and rabbit are used at 1:300 in blocking buffer [10].
9. Antifading agent: Vectashield antifade mounting medium containing DAPI (Vector Laboratories, catalogue number: H-1200).
10. Humidity chamber (*see* Fig. 1).
11. Glass Coplin staining jar (any brand or size).
12. Glass coverslip, 22 × 44 mm, thickness No 1.5 (VWR or similar) and coverslip made of Parafilm (or similar).

3 Methods

3.1 Plant Material

Both barley and wheat are grown from seeds (three per pot) in compost-based pots (7 in. diameter) in temperature-controlled glasshouse under the following conditions: 16 h of light at

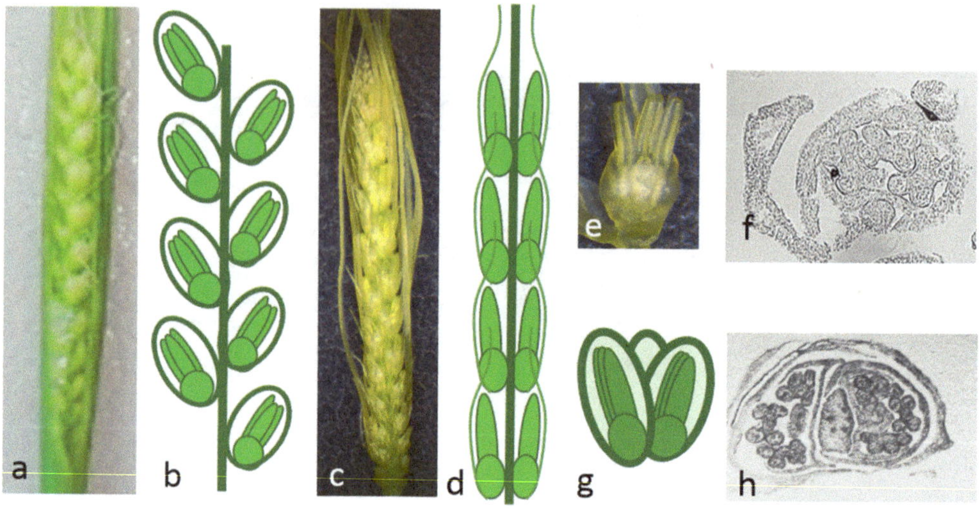

Fig. 2 Wheat and Barley plants and staging. (**a**) Wheat spike and (**b**) corresponding diagram of the florets organization; (**c**) Barley spike and (**d**) corresponding diagram of the florets organization; (**e**) barley spikelet; (**f**) Cross section of barley flower; (**g**) diagram of a wheat spikelet; (**h**) cross section of wheat spikelet

18–20 °C and 8 h of dark at 14–16 °C (*see* **Note 5**). The humidity is kept between 65% and 70%. Barley staging starts from 6 weeks of growing by looking at spikes of 1.2–2 cm (*see* **Note 6**). Wheat staging starts at around 8 weeks by looking at spikes of about 5–6 cm (*see* **Note 7**). In both species, the later stage of meiosis is in the middle of the spike with the earlier stages toward the top and the bottom, and this arrangement of florets is used to define the spike meiotic stage.

3.2 Meiosis Staging and Plant Fixation

1. Open the stem to release the spike (or inflorescence) (Fig. 2a, c).
2. Even though the morphology of barley and wheat is a bit different (Fig. 2b, d), each developing spikelet has three anthers of identical size (Fig. 2e–h). Wheat will require a bit more dissection than barley where anthers are not embedded within plant tissue.
3. Under the stereomicroscope, remove one of the three developing anthers and place it onto a microscope slide.
4. Add a drop of acetocarmine and squash it slightly under a glass coverslip.
5. Under the microscope, locate the anther with 10× objective and identify the meiotic stage by using the 40× objectives.
6. Transfer the remaining two anthers in an embryo dish containing 200 μl of working buffer at room temperature or on ice in case of warm weather. Pool the anthers of at least half of the spike.

7. Replace the working buffer by fixative and incubate for 20 and 30 min for barley and wheat respectively (*see* **Note 7**).
8. Wash the sample twice for 5 min with 1× PBS and proceed to the slide preparation (*see* **Note 8**).

3.3 Slide Preparation and Immunostaining

1. On a Polysine slide, make a well in the center with the PAP pen and let it dry fully.
2. Remove most of the working buffer and tap out the fixed anthers repeatedly with a glass rod to release the meiocytes. Add 30 μl of working buffer, mix gently and transfer 25 μl of the solution onto the prepared slide.
3. Let it dry slowly and add 30 μl of the primary antibody freshly prepared in blocking solution.
4. Add a soft plastic coverslip and leave for 30 min (up to 1 h) at RT in the dark wet chamber, then move the box to a cold room (or fridge) for at least 24 h (up to 48 h).
5. Warm up the slides at RT for at least 30 min (up to 1 h), then wash them twice 5 min in 1× PBS on orbital shaker.
6. Apply 40 μl of secondary antibody solution, cover with a soft plastic coverslip and leave to incubate at RT for 90 min to 2 h.
7. Wash the slides as previously in **step 6**, remove the excess PBS and add one drop of Vectashield containing DAPI.
8. Add a full-length glass cover slip, press gently upside down on paper towel to remove the Vectashield excess, and seal the four slides with nail varnish.
9. Let the varnish to dry and store in the dark at 4 °C.

3.4 Imaging

1. 3D-SIM images (Fig. 3) are acquired on a DeltaVision OMX Blaze (GE Healthcare) fitted with an Olympus PlanApo N 60× 1.42 NA oil objective. Settings for image acquisition are used as in [12]. At least 32-bit 1024 × 1024 image ($z = 0.125$) is preferred.
2. The volume of the wheat meiocytes is 2.5× larger than barley meiocytes. It takes about 5–7 min per cell to acquire the whole barley cell, at least twice as much for wheat nuclei. For example, a total of 93 total images (31 focal planes per channel; 3 wavelengths) were taken for barley meiocyte at leptotene/zygotene stage images could be taken with but 399 images (133 focal planes per channel; 3 wavelength) are needed for wheat meiocytes at the same stage.
3. Imaging Software used are Fiji (ImageJ) [17] and Imaris 8.0.2.
   (a) In Fiji, images are opened using the BioFormat importer, which is opening any type of microscope images (Leica, Zeiss, Nikkon, Deltavision, etc.).

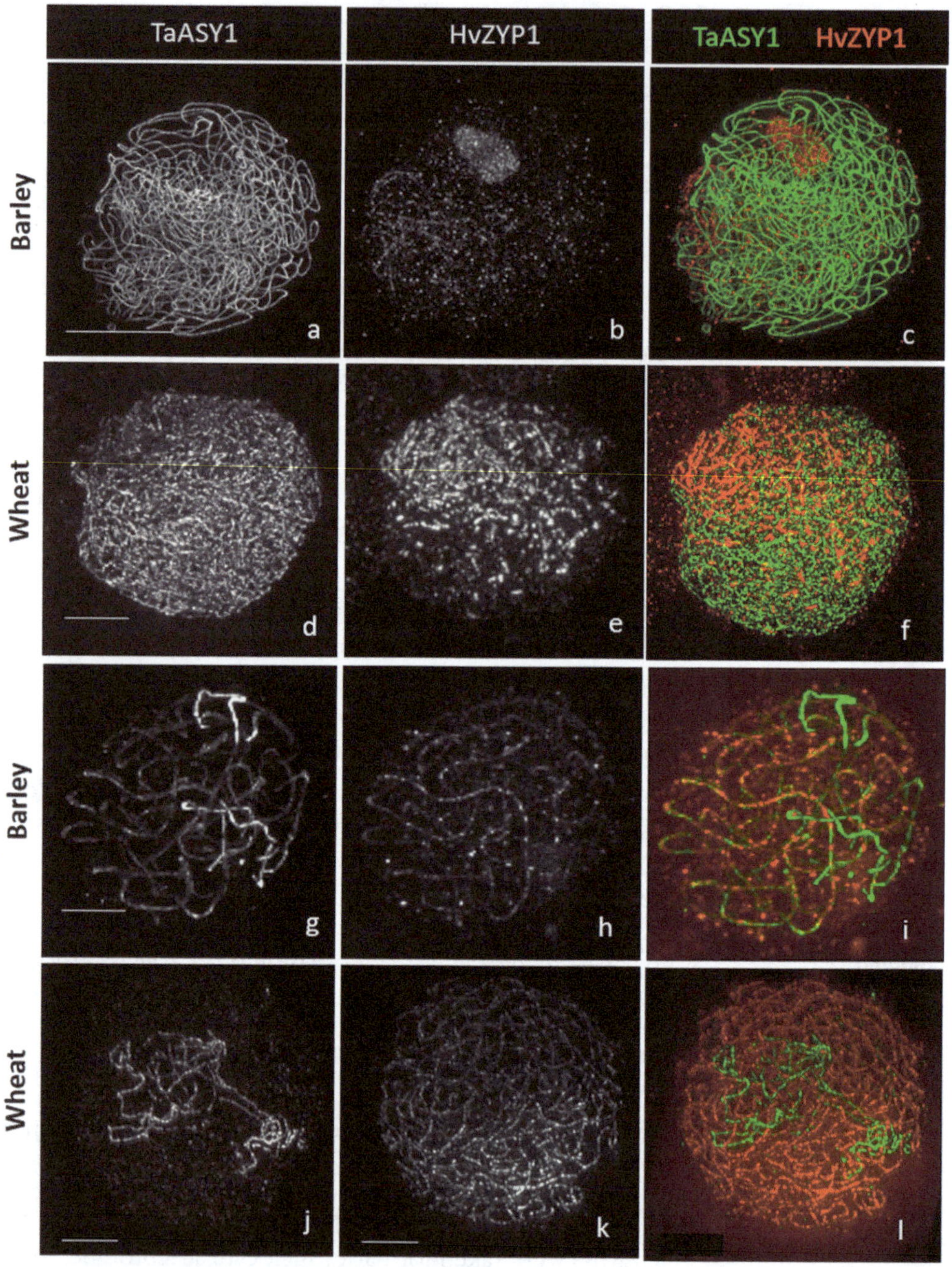

Fig. 3 Synapsis in barley and wheat by 3D-SIM. (**a–c**) Barley meiocyte at the beginning of synapsis, leptotene/zygotene stage (**d–f**) Wheat meiocyte at the beginning of synapsis, leptotene/zygotene stage. (**g–i**) Barley meiocyte almost fully synapsed, late zygotene (**j–l**) Wheat meiocyte almost fully synapsed, late zygotene. TaASY1 (green), HvZYP1 (Red). Scale bar 5 μm

(b) Select the image to open; in the new Window you could choose the option "split channel" to open the three channels in separate windows.

(c) To make a projection of each channel, go to Image/Stacks/Z project.

(d) To change the channel color, go to Image/color/channels tool and choose your color.

(e) To merge multiple channel together, go to Image/color/merge channels.

(f) To save your image, go to File/save as and choose your extension.

(g) If images needs deconvolution, we used the plugin DeconvolutionLab. OMX images do not need deconvolution.

4. In Imaris, images are opened in view mode that allows 3D visualization. This imaging software is very intuitive, and *z*-projection were done using "Easy 3D" for each channel or merged channels. Images are saved using the "snapshot" option choosing for the correct dimension (e.g., 512 × 512) and image printing resolution (600 dpi at least).

4 Notes

1. Over brands can be use if having equivalent eyepieces and objectives.
2. The preparation of acetocarmine must be done under fume hood. The use of acetocarmine does not require a fume hood if used in a well-ventilated area for a short period of time (no longer than 2 h per day).
3. Formaldehyde is a carcinogen and should be worked under fume hood at all time. Paraformaldehyde will not dissolve without NaOH and will lose efficiency if boiled. Check the local rules for disposal by professional chemical waste disposal compagnies.
4. Goat and donkey serums are stored in the freezer but avoid freeze–thaw cycles.
5. Different growing conditions (even a small difference of temperature) might affect growing time and therefore meiosis start.
6. By prophase there is no emergence of flag leaf, but by metaphase stages, the flag leaf starts emerging in wheat.
7. The whole spike can also be fixed by increasing each time by 10 min and sample can be stored for a couple of days before

slide preparation although we recommend to use the sample immediately.

8. Washing steps can be conducted using a combination of buffers with different stringencies depending on the species.

Acknowledgments

The research leading to these results has received funding from the European Community's Seventh Framework Programme *FP7/2007-2013* under grant agreement no. 222883 MeioSys and ERC advanced grant "Shuffle" (Project ID: 669182). Use of the OMX microscope was supported by the Euro-BioImaging PCS and through the MRC Next Generation Optical Microscopy Award (Ref: MR/K015869/1) and part of this work was performed in the frame of the Proof of Concept Studies (PCS) for the ESFRI research infrastructure project Euro-BioImaging at the PCS facility OMX (WTB Dundee). L.R. and R.W. were funded from the Scottish Government's Rural and Environment Science and Analytical Services Division Work Program 5.2. B.D. was funded by INRA and Auvergne and his training at the James Hutton Institute was supported by a funding from INRA-DARESE (Direction de l'Action Régionale, de l'Enseignement Supérieur et de l'Europe) in the course of EIR-A (Ecole Internationale de Recherche d'Agreenium). S.M. was funded by Biotechnology and Biological Science Research Council EASTBIO PhD studentship program and M.A. was supported by the European Community's Seventh Framework Programme FP7-PEOPLE-2013-ITN COMREC-606956.

References

1. Bass HW, Birchler J (2011) Plant cytogenetics: genome structure and chromosome function. Springer, New York, p 350
2. Pawlowski W, Grelon M, Armstrong S (2013) Plant meiosis: methods and protocols, methods in molecular biology, 990. Springer, New York, p 237
3. Baudat F, Imai Y, de Massy B (2013) Meiotic recombination in mammals: localization and regulation. Nat Rev Genet 14(11):794–806
4. Mercier R, Mézard C, Eric Jenczewski E et al (2014) The molecular biology of meiosis in plants. Annu Rev Plant Biol 66:297–327
5. Finch RA, Bennett MD (1972) The duration of meiosis in diploid and autotetraploid barley. Can J Genet Cytol 14(3):507–515
6. Higgins JD, Perry RM, Barakate A, Ramsay L, Waugh R, Halpin C, Susan J, Armstrong SJ, Franklin FCH (2012) Spatiotemporal Asymmetry of the Meiotic Program Underlies the Predominantly Distal Distribution of Meiotic Crossovers in Barley. The Plant Cell 24 (10):4096–4109
7. Bennett MD, Chapman Vand Riley R (1971) The duration of meiosis in pollen mother cells of wheat, rye and triticale. Proc R Soc London. Series B, Biol Sci 178(1052):259–275
8. Zickler D, Kleckner N (1999) Meiotic chromosomes: integrating structure and function. Annu Rev Genet 33:603–754
9. Phillips D, Nibau C, Wnetrzak J et al (2012) High resolution analysis of meiotic chromosome structure and behaviour in barley (Hordeum vulgare L.). PLoS One 7(6):e39539
10. Colas I, Darrier B, Arrieta M, Mittmann SU, Ramsay L, Sourdille P, Waugh R (2017)

Observation of extensive chromosome Axis Remodeling during the "diffuse-phase" of meiosis in large genome cereals. Observation of extensive chromosome axis remodelling during the 'diffuse-phase' of meiosis in large genome cereals. Frontiers in plant. Sciences **8**:1235. https://doi.org/10.3389/fpls.2017.01235

11. Barakate A, Higgins JD, Vivera S et al (2014) The synaptonemal complex protein ZYP1 is required for imposition of meiotic crossovers in barley. Plant Cell 26(2):729–740
12. Colas I, Macaulay M, Higgins JD et al (2016) A spontaneous mutation in MutL-homolog 3 (HvMLH3) affects synapsis and crossover resolution in the barley desynaptic mutant *des10*. New Phytol 212(3):693–707
13. Golubovskaya IN, Wang CJ, Timofejeva L et al (2011) Maize meiotic mutants with improper or non-homologous synapsis due to problems in pairing or synaptonemal complex formation. J Exp Bot 62(5):1533–1544
14. Pawlowski WP, Wang CJ, Golubovskaya IN et al (2009) Maize AMEIOTIC1 is essential for multiple early meiotic processes and likely required for the initiation of meiosis. Proc Natl Acad Sci U S A 106(9):3603–3608
15. Armstrong SJ, Caryl AP, Jones GH, Franklin FC (2002) Asy1, a protein required for meiotic chromosome synapsis, localizes to axis-associated chromatin in Arabidopsis and Brassica. J Cell Sci 115:3645–3655
16. Boden SA, Langridge P, Spangenberg G, Able JA (2009) TaASY1 promotes homologous chromosome interactions and is affected by deletion of Ph1. Plant J 57:487–497
17. Schindelin J, Arganda-Carreras I, Frise E, Kaynig V, Longair M, Pietzsch T, Preibisch S, Rueden C, Saalfeld S, Schmid B, Tinevez JY, White DJ, Hartenstein V, Eliceiri K, Tomancak P, Cardona A (2012) Fiji: an open-source platform for biological-image analysis. Nat Methods 9:676–682

Part III

Molecular and Biochemical Approaches to Study Meiosis

Chapter 16

Chromatin Immunoprecipitation of Meiotically Expressed Proteins from *Arabidopsis thaliana* Flowers

Christophe Lambing, Kyuha Choi, Alexander R. Blackwell, and Ian R. Henderson

Abstract

During meiosis recombination occurs between homologous chromosomes which can result in reciprocal exchanges of genetic information, called crossovers. Crossover rate is heterogeneous within the genome, with local regions having a significantly higher recombination rate relative to the genome average. These regions are termed hotspots and typically occur with widths of kilobases. Therefore, there is a need to profile recombination factors at a similar resolution during meiosis via techniques such as chromatin immunoprecipitation (ChIP). Here we describe a ChIP protocol, combined with high throughput sequencing (ChIP-seq) optimised for analysis of meiotically expressed proteins in *Arabidopsis thaliana* flowers. We provide methods to (1) isolate nuclei and prepare the chromatin for shearing, (2) immunoprecipitate DNA molecules cross-linked to a protein of interest, (3) to size-select and purify immunoprecipitated DNA molecules, and (4) to prepare DNA sequencing libraries suitable for high-throughput sequencing. Together, these methods allow the detection of binding sites for meiotic proteins in the Arabidopsis genome at high resolution, which will provide insights into relationships between meiotic chromosome organization, chromatin and recombination.

Key words Meiosis, Chromatin immunoprecipitation, Chromosome axis, Recombination, Arabidopsis

1 Introduction

Meiotic recombination initiates with the formation of DNA double strand breaks (DSBs) generated by SPO11 and accessory proteins [1, 2]. DSBs occur in the context of chromatin loops tethered to a meiotic chromosome axis, which promotes inter-homolog recombination and crossover formation [3, 4]. Meiotic recombination can result in allele reshuffling via crossovers and this contributes to genetic diversity [5]. Analysis of crossover frequency in plant genomes has revealed that recombination rate is heterogeneous, with narrow 1–2 kilobase (kb) hot spots exhibiting higher crossover frequency relative to the genome average [6–9]. However, the

Mónica Pradillo and Stefan Heckmann (eds.), *Plant Meiosis: Methods and Protocols*, Methods in Molecular Biology, vol. 2061, https://doi.org/10.1007/978-1-4939-9818-0_16, © Springer Science+Business Media, LLC, part of Springer Nature 2020

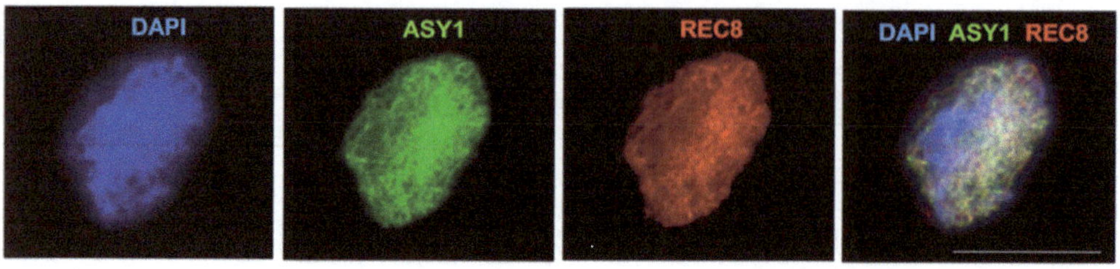

Fig. 1 Immunostaining of ASY1 and REC8 components of the chromosome axis in Arabidopsis. Immunostaining of ASY1 (green) and REC8 (red) on a pollen mother cell at leptotene stage. The chromatin is stained with DAPI (blue). Scale bar = 10 μM

genetic and epigenetic features that determine hotspot locations and activity remain to be fully understood.

Meiotic chromosomes are associated with a proteinaceous structure called the chromosome axis [10]. Several chromosome axis proteins have been identified in plants, including HORMA-domain containing proteins ASYNAPTIC 1 (ASY1)/HOMOLOGOUS PAIRING ABERRATION 2 (PAIR2) [11, 12], coiled-coil proteins ASYNAPTIC3 (ASY3)/PAIR3 [13, 14] and ASYNATPTIC4 (ASY4) [15, 16] and the cohesin subunits REC8/Absence of First Division 1 (AFD1) [17, 18] and STRUCTURAL MAINTENANCE OF CHROMOSOME 3 (SMC3) [19] (Fig. 1). Meiotic recombination occurs in association with the chromosome axis, and axis proteins are required to stabilize the interhomolog strand invasion protein DNA MEIOTIC RECOMBINASE 1 (DMC1) on chromatin and promote joint molecule formation between chromosomes [11–20]. In budding yeast [21] and *Caenorhabditis elegans* [22], meiotic axis proteins also play an important role in mediating crossover interference, which causes crossovers to be more widely spaced than expected at random, and their positioning along chromosomes.

The use of cytogenetic, molecular genetic, and biochemical approaches to study meiosis has helped to identify numerous proteins involved in the recombination pathways [23]. In addition, cytogenetic methods are useful to study meiotic progression, for example a recent study revealed that the chromosome axis remodels concomitantly with progression of meiosis [24]. However, these techniques have limited spatial resolution and are not suitable to determine the DNA binding sites of recombination factors and of chromosome axis components at the fine scale. One approach to study protein localization at the fine scale is chromatin immunoprecipitation sequencing (ChIP-seq) (Fig. 2) [25]. This technique detects the binding site of proteins with a resolution of ~200–400 bp [26]. The success of this experimental approach depends on the availability of an antibody with high affinity and specificity that can be used to "pull down" formaldehyde cross-linked protein–DNA complexes [26]. It is also possible to generate

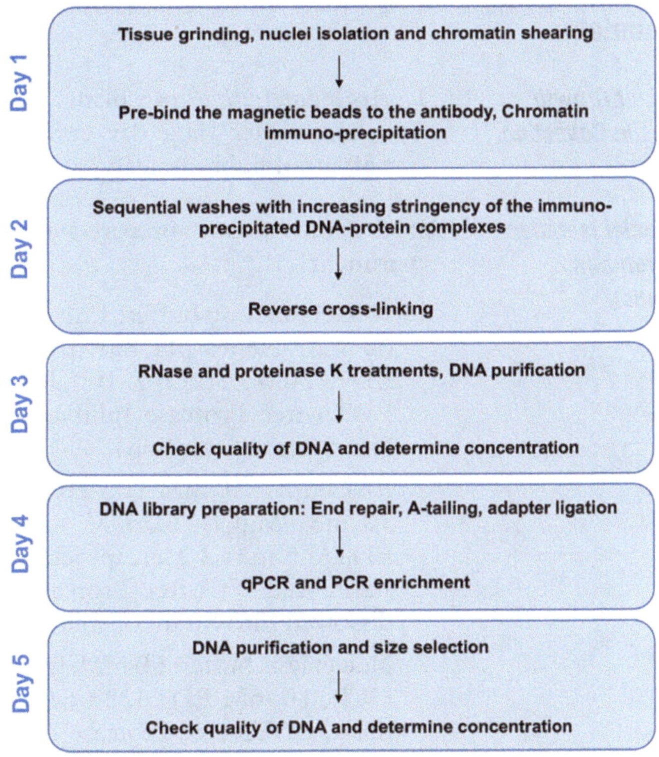

Fig. 2 Schematic representation of ChIP-seq procedure

tagged transgenic lines expressing a protein of interest translationally fused with an epitope. Several commercial antibodies are available to detect these epitopes with high specificity and which have been proven to work for ChIP-seq [25, 26]. However, it can be challenging to generate stable transgenic lines with a functional epitope-tagged protein that complements the mutant phenotypes. As an alternative, it is possible to raise an antibody against a recombinant protein of interest, or against synthesized peptides, in order to immunoprecipitate factors of interest.

Here we report detailed methodologies for ChIP-seq of meiotic proteins in Arabidopsis, which can be used to study the genome-wide profile of chromosome axis and recombination proteins. Arabidopsis flowers contain a larger number of male germ cells and comparatively few female germ cells [27]. We optimised the method to precipitate meiotic proteins from floral buds at stage 9 of flower development, which corresponds predominantly to male meiosis [28]. We describe a procedure for nuclei isolation, chromatin separation and shearing, chromatin immunoprecipitation, and DNA recovery. We also describe a method for DNA library preparation suitable for next-generation sequencing. Finally, we provide information on critical steps including chromatin preparation, PCR enrichment, and DNA size selection during the DNA library preparation.

2 Materials

2.1 Plant Growth and Tissue Collection

1. *Arabidopsis thaliana* plants are grown in controlled growth rooms under long day conditions (16 h light–8 h dark at 150 μmol light intensity) at 20 °C.

2.2 Nuclei Isolation and Chromatin Preparation

All the buffers are prepared fresh on the day of the ChIP experiment.

1. Nuclei isolation buffer: 1 M sucrose, 5 mM KCl, 5 mM $MgCl_2$, 60 mM Hepes pH 8.0 (pH adjusted with NaOH), 5 mM EDTA, 0.6% Triton X-100, 0.4 mM PMSF, 1 μM pepstatin A, EDTA-free Protease Inhibitor Cocktail (one tablet dissolved into 50 ml of solution).
2. EB2 buffer: 10 mM Tris–HCl pH 8.0 (pH adjusted with HCl), 10 mM $MgCl_2$, 0.25 M sucrose, 1% Triton X-100, 1 mM EDTA, 5 mM 2-Mercaptoethanol, 0.1 mM PMSF, 1 μM pepstatin A, EDTA-free Protease Inhibitor Cocktail (one tablet dissolved into 50 ml of solution).
3. Nuclei lysis buffer: 50 mM Tris–HCl pH 8.0 (pH adjusted with HCl), 10 mM EDTA, 1% SDS, 0.1 mM PMSF, 1 μM pepstatin A, EDTA-free Protease Inhibitor Cocktail (one tablet dissolved into 50 ml of solution).
4. ChIP dilution buffer: 20 mM Tris–HCl pH 8.0 (pH adjusted with HCl), 167 mM NaCl, 1.1 mM EDTA, 1.1% Triton X-100, 1 μM pepstatin A, EDTA-free Protease Inhibitor Cocktail (one tablet dissolved into 50 ml of solution).

2.3 Chromatin Immunoprecipitation

All the buffers are prepared fresh on the day of the ChIP experiment.

1. ChIP dilution buffer: 20 mM Tris–HCl pH 8.0 (pH adjusted with HCl), 167 mM NaCl, 1.1 mM EDTA, 1.1% Triton X-100, 1 μM pepstatin A, EDTA-free Protease Inhibitor Cocktail (one tablet dissolved into 50 ml of solution).
2. Low-salt buffer: 150 mM NaCl, 20 mM Tris–HCl pH 8.0 (pH adjusted with HCl), 2 mM EDTA, 1% Triton X-100, 0.1% SDS, 1 μM pepstatin A, EDTA-free Protease Inhibitor Cocktail (one tablet dissolved into 50 ml of solution).
3. High-salt buffer: 500 mM NaCl, 20 mM Tris–HCl pH 8.0 (pH adjusted with HCl), 2 mM EDTA, 1% Triton X-100, 0.1% SDS, 1 μM pepstatin A, EDTA-free Protease Inhibitor Cocktail (one tablet dissolved into 50 ml of solution).
4. TE: 10 mM Tris–HCl pH 8.0 (pH adjusted with HCl), 1 mM EDTA.
5. Elution buffer: 100 mM $NaHCO_3$, 1% SDS.

3 Methods

3.1 Tissue Collection and Cross-Linking

1. Collect unopened Arabidopsis floral buds at stage 9 of flower development [26] from 5–6 week old plants in 2 ml SuperLock microcentrifuge tubes. It is recommended to keep the tube on ice during tissue collection. Once full, the tubes (a 2 ml Eppendorf tube full of floral buds weighs ~500 mg) are immediately frozen in liquid nitrogen and stored at −80 °C.
2. On the day of performing ChIP-seq: transfer 1 g of buds (corresponding to two 2 ml Eppendorf tubes of floral buds from **step 1**) from −80 °C to a mortar containing liquid nitrogen. Grind the buds to a fine powder in liquid nitrogen using a pestle. When all liquid nitrogen is evaporated, add more liquid nitrogen in the mortar and repeat the grinding step 2–3 more times until the buds are ground to a very fine powder.
3. Transfer the ground floral powder to an ice-cold mortar and add 25 ml of nuclei isolation buffer. Resuspend the powder in solution using a pestle.
4. Repeat **step 2** to grind an additional gram of buds in a mortar with liquid nitrogen. Add the ground power to the previous homogenate and resuspend the solution with a pestle.
5. Transfer the homogenate to a 50 ml conical tube and add 680 μl of 37% formaldehyde solution. The final ratio of formaldehyde should be 1% by volume.
6. Incubate the conical tube on a roller mixer at 12 rpm at room temperature for 25 min (*see* **Note 1**).
7. Stop the cross-linking reaction by adding 1.7 ml of 2 M glycine to a final concentration of 130 mM and incubate the solution on a tube roller mixer at 12 rpm and room temperature for 25 min.
8. Keep the solution on ice until all the samples are processed. Between 5 and 10 g of buds is usually needed to perform one ChIP experiment and **steps 2–7** are repeated until all the tissue material has been processed. The amount of starting material depends on the abundance of the protein on the chromatin. For example, axis proteins are abundant and 6 g of floral buds is usually enough for one experiment. In contrast, recombination proteins may be less abundant and 10 g of floral buds is recommended for one experiment.

3.2 Nuclei Isolation and Chromatin Preparation

During the procedure keep the samples on ice.

1. Filter the homogenate through one layer of miracloth into a 50 ml conical tube. Repeat the filtration once more with a fresh miracloth.

2. Centrifuge at 3000 × *g* for 20 min at 4 °C.
3. Discard the supernatant and resuspend the pellet in 1 ml of EB2 solution (*see* **Note 2**). Transfer the homogenate to a 1.5 ml Clear-view Snap-Cap low-retention microtube.
4. Centrifuge at 12,000 × *g* for 10 min at 4 °C.
5. Discard the supernatant (*see* **Note 3**). Resuspend the pellet in 600 μl of Nuclei Lysis Buffer and divide the sample into two new 1.5 ml Clear-view Snap-Cap low-retention microtube (300 μl into each).

3.3 Chromatin Shearing

During these procedures keep the samples on ice.

1. Sonicate the chromatin using a Bioruptor (High power, 30 s ON—30 s OFF) for 15 min (*see* **Note 4**).
2. Add 200 μl of ChIP dilution buffer to each tube and mix the solution by gently pipetting.
3. Sonicate the chromatin using a Bioruptor (High power, 30 s ON—1 min OFF) for 25 min.
4. Centrifuge the microtubes at 2400 × *g* for 10 min at 4 °C.
5. Pool the supernatants (~450 μl) from the microtubes for each sample, in order to homogenize the sonicated chromatin between tubes. For example, if 10 g of buds is ground for a wild-type sample, ten microtubes of chromatin are used for sonication and 10× 450 μl is recovered and pooled after sonication. 5% of the volume of sonicated chromatin is removed and stored at −20 °C. This sample corresponds to the input sample and will be used again from **step 2** in Subheading 3.5 onward.
6. Proceed immediately with the remaining sonicated chromatin to Chromatin immunoprecipitation (*see* Subheading 3.4.2), or store the chromatin at −20 °C. Chromatin can be stored for a few months at −20 °C without any loss in integrity.

3.4 Chromatin Immunoprecipitation

During all the procedures keep the samples on ice when possible.

3.4.1 Prebinding of the Antibody to Dynabeads Protein G

1. Transfer 50 μl of Dynabeads Protein G (*see* **Note 5**) to a 2 ml microcentrifuge tube and add 1.5 ml of ChIP dilution buffer. One tube is used per gram of buds. For example, prepare ten tubes of 50 μl of Dynabeads if 10 g of buds are used.
2. Rotate for 5 min at 12 rpm at 4 °C.
3. Place the microtubes in a magnetic separation rack and incubate for 2 min. All the beads should be concentrated to the side of the tube adjacent to the magnet.

4. Carefully remove the supernatant without disturbing the beads and discard.
5. Add 1.5 ml of ChIP dilution buffer.
6. Remove the microtubes from the magnetic separation rack and gently flick the tubes to resuspend the beads in solution. Repeat **steps 2–6** twice.
7. Add 5–10 μg of antibody depending on the antibody specification (*see* **Note 5**).
8. Incubate the antibody solution on a rotator at 12 rpm between 6 h and overnight at 4 °C. The prebinding of the antibody to the beads can be prepared on the day before chromatin precipitation. In that case the incubation time corresponds to the time necessary to prepare the chromatin (~6–8 h). Alternatively, the prebinding of the antibody to the beads can be prepared the day before the preparation of chromatin and in that case the incubation can be carried out overnight. Overnight incubation has no negative impact on the downstream steps.
9. After incubation, place the microtubes in a magnetic separation rack and incubate for 2 min.
10. Carefully remove the supernatant without disturbing the beads.
11. Add 1 ml of ChIP dilution buffer.
12. Remove the microtubes from the magnets and gently flick the tubes to resuspend the beads in solution.
13. Rotate for 5 min at 12 rpm at 4 °C.
14. Place the microtubes in a magnetic separation rack and incubate for 2 min. Carefully remove the supernatant without disturbing the beads.
15. Add 1 ml of ChIP dilution buffer.
16. Remove the microtubes from the magnetic separation rack and gently flick the tubes to resuspend the beads in solution.
17. Repeat **steps 13–17** once. The resuspended beads correspond to the prebound antibody.

3.4.2 Immunoprecipitation of Chromatin with Prebound Antibody

1. Transfer ~420 μl of sonicated chromatin to the microtube containing the beads with prebound antibody.
2. Incubate the solution on a rotator at 12 rpm for 14–16 h at 4 °C.
3. Place the microtubes in a magnetic separation rack and incubate for 2 min.
4. Carefully remove the supernatant without disturbing the beads.
5. Add 1.5 ml of low-salt buffer to each tube.

6. Remove the microtubes from the magnetic separation rack and gently flick the tubes to resuspend the beads in solution.
7. Rotate for 5 min at 12 rpm at 4 °C.
8. Place the microtubes in a magnetic separation rack and incubate for 2 min.
9. Carefully remove the supernatant without disturbing the beads.
10. Repeat **steps 5–9** one more time.
11. Add 1.5 ml of high-salt buffer to each tube.
12. Remove the microtubes from the magnetic separation rack and gently flick the tubes to resuspend the beads in solution.
13. Rotate for 5 min at 12 rpm at 4 °C.
14. Place the microtubes in a magnetic separation rack and incubate for 2 min.
15. Carefully remove the supernatant without disturbing the beads.
16. Repeat **steps 11–15** once more (*see* **Note 6**).
17. Add 1 ml of TE buffer.
18. Remove the microtubes from the magnetic separation rack and gently resuspend the beads in solution by pipetting.
19. Transfer the homogenate to a fresh 1.5 ml Clear-view Snap-Cap low-retention microtube.
20. Place the microtubes in a magnetic separation rack and incubate for 2 min.
21. Carefully remove the supernatant without disturbing the beads.
22. Elute the chromatin from the beads by adding 200 μl of elution buffer.
23. Briefly vortex the tubes and then incubate at 65 °C for 10 min.
24. Place the microtubes in a magnetic separation rack and incubate for 2 min.
25. Transfer the eluted solution to a new 1.5 ml Clear-view Snap-Cap low-retention microtube without disturbing the beads. The solution contains the immunoprecipitated chromatin.
26. Add another 200 μl of elution buffer to the tubes with the beads.
27. Briefly vortex the tubes then incubate at 65 °C for 10 min.
28. Place the microtubes in a magnetic separation rack and incubate for 2 min.
29. Take the eluted solution and combine it with the previous eluted solution. In total ~400 μl of eluted solution is recovered for each tube.

3.5 Reverse Cross-Linking and DNA Purification

1. Add 20 μl of 5 M NaCl to the 400 μl of eluted immunoprecipitated chromatin solution. Mix the solution by gently pipetting and incubate the tubes at 65 °C overnight to reverse the cross-linking.
2. Thaw the input sample (from **step 5** in Subheading 3.3) and add elution buffer to a final volume of 400 μl, then add 20 μl of 5 M NaCl to a final concentration of 0.25 M. Mix the solution by gently pipetting and incubate the tubes at 65 °C overnight to reverse the cross-linking.

   For all the subsequent steps both input and ChIP samples are processed the same way.
3. Add 10 μl of 0.5 M EDTA, 20 μl of 1 M Tris–HCl pH 7.0 and 3 μl of 20 mg/ml RNase A. Gently mix the solution by pipetting. Incubate for 30 min at 37 °C.
4. Add 2 μl of 20 mg/ml proteinase K. Gently mix the solution by pipetting. Incubate for 2 h at 45 °C.
5. Add 450 μl of phenol–chloroform–isoamyl alcohol and vortex for 15 s.
6. Centrifuge at 14,000 × *g* for 25 min at room temperature.
7. Transfer the supernatant to a fresh 1.5 ml Clear-view Snap-Cap low-retention microtube.
8. Repeat **steps 6** and 7.
9. Add 1/10 volume of 3 M sodium acetate pH 5.2 (typically 40 μl of 3 M sodium acetate is added to the solution), 2 μl of GlycoBlue and 2.5 volumes of 100% ethanol (typically 1 ml of ethanol is added to the solution).
10. Mix the tubes by inverting ten times and then incubate at −20 °C overnight.
11. Centrifuge the tubes at 14,000 × *g* for 20 min at 4 °C.
12. Discard the supernatant and add 500 μl of ice-cold 70% ethanol.
13. Centrifuge the tubes at 14,000 × *g* for 10 min at 4 °C.
14. Discard the supernatant and leave the DNA pellet to air dry for 15 min at room temperature.
15. Resuspend the DNA pellet from one tube in 20 μl of nuclease-free water by gently pipetting. Use this resuspended solution to resuspend the DNA pellet from a second tube corresponding to the same ChIP sample and repeat this for all the tubes of the same sample. For example, a total of ten Eppendorf tubes from 10 g of starting material.
16. Store the ChIP and input DNA samples at −20 °C.

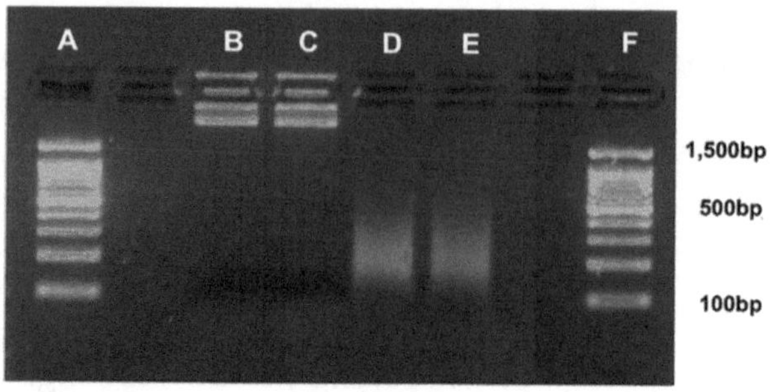

Fig. 3 Agarose gel electrophoresis of ChIP DNA before and after sonication. A 2% agarose gel with SYBR Gold nucleic acid stain showing, (**a**) 100 bp DNA ladder, (**b**) 15 ng of sample 1 DNA before sonication, (**c**) 15 ng of sample 2 DNA before sonication, (**d**) 15 ng of sample 1 DNA after sonication, (**e**) 15 ng of sample 2 DNA after sonication, and (**f**) 100 bp DNA ladder

3.6 DNA Quantification and Quality Control

Between 10 and 40 ng of DNA are routinely recovered from ChIP of meiocytes using this protocol. The amount of DNA recovered depends on the quality of the antibody, the amount of starting material and the abundance of protein on chromatin.

1. ChIP DNA is quantified using Qubit dsDNA HS assay kit.
2. Shearing of DNA is checked by loading 10–15 ng of input DNA on a 2% agarose gel and visualized after electrophoresis using SYBR gold nucleic acid gel stain (Fig. 3). DNA fragments ranging between 100 and 800 bp with enrichment at ~200 bp are suitable for DNA library construction.
3. The size of the immunoprecipitated DNA is checked using a High Sensitivity DNA kit with Agilent 2100 Bioanalyzer. This is particularly recommended when the amount of ChIP DNA recovered is relatively low (e.g., <20 ng).

3.7 DNA Library Preparation for Next-Generation Sequencing

3.7.1 End Repair

1. Add the following components in a 0.2 ml tube: 10 μl of 10× T4 DNA ligase reaction buffer, 4 μl of 10 mM dNTP mix, 5 μl of T4 DNA polymerase, 1 μl of DNA Polymerase I Large (Klenow) Fragment, 5 μl of T4 polynucleotide kinase, 10 ng of ChIP DNA or input DNA, and nuclease-free water to a final volume of 100 μl.
2. Gently mix the solution by pipetting, spin down (15,000 × *g*, 5 s) and incubate the solution at 20 °C for 30 min.
3. Spin down (15,000 × *g*, 5 s) the solution. Transfer the solution to a 1.5 ml Clear-view Snap-Cap low-retention microtube and add 1.8 volumes of Agencourt AMPure XP (180 μl). Gently mix the solution by pipetting ten times.
4. Incubate the tube for 10 min at room temperature.

5. Place the microtubes in a magnetic separation rack and incubate for 5 min. All the beads should be concentrated to the side of the tube adjacent to the magnet.
6. Carefully remove the supernatant without disturbing the beads.
7. Add 400 μl of 75% ethanol and incubate the tube for 1 min at room temperature.
8. Carefully remove the supernatant without disturbing the beads and repeat **steps** 7 and **8**.
9. Air-dry the beads for 10 min at room temperature.
10. Remove the microtube from the magnetic separation rack. Add 32 μl of nuclease-free water and resuspend the beads by gently pipetting ten times. Incubate the tube for 2 min at room temperature.
11. Place the microtubes in a magnetic separation rack and incubate for 5 min.
12. Carefully remove the supernatant (32 μl) without disturbing the beads and transfer it to a 0.2 ml Eppendorf tube.

3.7.2 dA-Tailing

1. Add the following components: 5 μl of NEBuffer 2, 10 μl of 1 mM dATP, and 3 μl of Klenow Fragment (3′ → 5′ exo-).
2. Gently mix by pipetting, spin down (15,000 × *g*, 5 s) the solution and incubate the solution at 37 °C for 30 min.
3. Spin down (15,000 × *g*, 5 s) the solution. Transfer the solution to a 1.5 ml Clear-view Snap-Cap low-retention microtube and add 1.8 volumes of Agencourt AMPure XP (90 μl). Gently mix the solution by pipetting ten times.
4. Incubate the tube for 10 min at room temperature.
5. Place the microtubes in a magnetic separation rack and incubate for 5 min. All the beads should be concentrated to the side of the tube adjacent to the magnet.
6. Carefully remove the supernatant without disturbing the beads.
7. Add 400 μl of 75% ethanol and incubate the tube for 1 min at room temperature.
8. Carefully remove the supernatant without disturbing the beads and repeat **steps** 7 and **8**.
9. Air-dry the beads for 10 min at room temperature.
10. Remove the microtube from the magnetic separation rack. Add 25 μl of nuclease-free water and resuspend the beads by gently pipetting ten times. Incubate the tube for 2 min at room temperature.

11. Place the microtubes in a magnetic separation rack and incubate for 5 min.
12. Carefully remove the supernatant (25 μl) without disturbing the beads and transfer it to a fresh 0.2 ml Eppendorf tube.

3.7.3 Ligation of Illumina Adapters

1. Add the following components: 4.5 μl of diluted Illumina adapters (1 μl of Illumina adapter diluted in 249 μl of nuclease-free water), 12.5 μl of 4× QS buffer, 5 μl of Quick-Stick ligase, and 3 μl of nuclease-free water.
2. Gently mix the solution by pipetting, spin down (15,000 × *g*, 5 s) and incubate the solution at 20 °C for 30 min.
3. Spin down (15,000 × *g*, 5 s) the solution. Transfer the solution to a 1.5 ml Clear-view Snap-Cap low-retention microtube and add 1 volume of Agencourt AMPure XP (50 μl). Gently mix the solution by pipetting ten times.
4. Incubate the tube for 10 min at room temperature.
5. Place the microtubes in a magnetic separation rack and incubate for 5 min. All the beads should be concentrated to the side of the tube adjacent to the magnet.
6. Carefully remove the supernatant without disturbing the beads.
7. Add 400 μl of 75% ethanol and incubate the tube for 1 min at room temperature.
8. Carefully remove the supernatant without disturbing the beads and repeat **steps** 7 and **8**.
9. Air-dry the beads for 10 min at room temperature.
10. Remove the microtubes from the magnetic separation rack. Add 20 μl of nuclease-free water and resuspend the beads by gently pipetting ten times. Incubate the tube for 2 min at room temperature.
11. Place the microtubes in a magnetic separation rack and incubate for 5 min.
12. Carefully remove the supernatant (20 μl) without disturbing the beads and transfer it to a fresh 0.2 ml tube.

3.7.4 PCR Enrichment

1. Prepare a real time quantitative PCR (qPCR) reaction as described in Table 1. Oligonucleotides with sequences matching the Illumina adapters are used to amplify DNA templates that will subsequently form clusters on a flowcell of the Next-Generation sequencing platform. Real-time qPCR allows the quantification of DNA molecules in the sample in order to determine the number of PCR cycles to use during the enrichment step (*see* **Note** 7, Fig. 4).

Table 1
Real time qPCR reaction composition

Components	Volume (μl)
DNA template	1
2× KAPA SYBR mix	5.5
Primer set (Illumina)	1
Nuclease-free water	2.5
Total	10

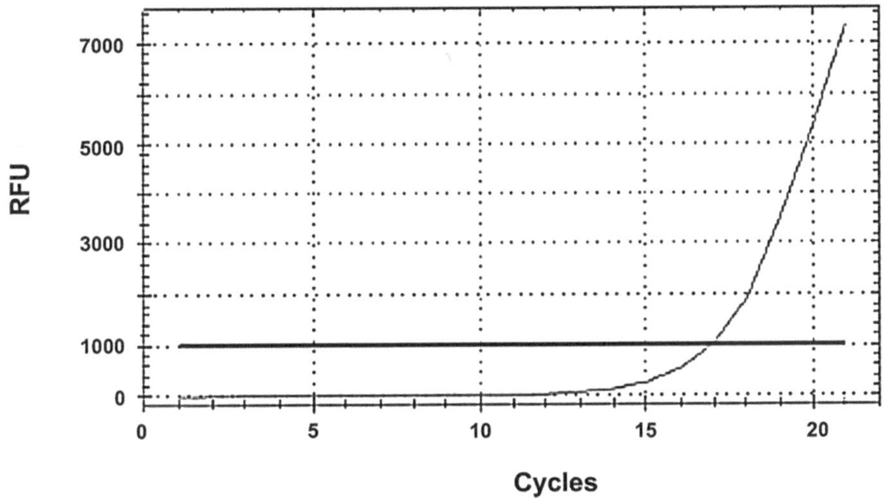

Fig. 4 Real-time qPCR amplification chart of a ChIP-seq DNA library. It is recommended to perform a real time qPCR of the ligated DNA library sample to determine the number of PCR cycles to use during the enrichment step. The number of cycles should be below 20 and should correspond to the number of cycles required to observe an increase of relative fluorescence units (RFU) minus 1 cycle. For example, on the chart the number of DNA molecules significantly increases after 17 cycles (arrow), as seen by the curve exponentially increasing and crossing the horizontal green line marking 1000 RFU. Hence, a number of 16 cycles should be used for the subsequent PCR enrichment step

2. Run real time qPCR in a thermal cycler with the parameters:
   1 cycle: 95 °C for 3 min;
   20 cycles: 95 °C for 30 s, 63 °C for 30 s, 72 °C for 30 s;
   Dissociation curve: from 65 to 95 °C with an increment of 0.5 °C after 5 s each time.
3. Prepare a PCR reaction as described in Table 2.
4. Run PCR for DNA enrichment in a thermal cycler with the parameters:
   1 cycle: 95 °C for 3 min;
   X cycles (number of cycles is determined by the real time qPCR): 95 °C for 30 s, 60 °C for 30 s, 72 °C for 30 s;
   1 cycle: 72 °C for 5 min.

Table 2
PCR reaction composition

Components	Volume (μl)
DNA template	19
2× KAPA SYBR mix	20
Primer set (Illumina)	1
Total	40

5. Spin down (15,000 × *g*, 5 s) the solution and add 10 μl of nuclease-free water to the PCR product.
6. Transfer the solution to a 1.5 ml Clear-view Snap-Cap low-retention microtube.

3.7.5 DNA Purification and Size Selection

1. Remove fragments that are >500 bp by adding 0.6 volume Agencourt AMPure XP (30 μl). Gently mix the solution by pipetting ten times.
2. Incubate the tube for 10 min at room temperature.
3. Place the microtubes in a magnetic separation rack and incubate for 5 min. All the beads should be concentrated to the side of the tube adjacent to the magnet.
4. Transfer the supernatant to a new 1.5 ml Clear-view Snap-Cap low-retention microtube.
5. Add 1.4 volumes of Agencourt AMPure XP (70 μl for a 50 μl PCR reaction). Gently mix the solution by pipetting ten times.
6. Incubate the tube for 10 min at room temperature.
7. Place the microtubes in a magnetic separation rack and incubate for 5 min. All the beads should be concentrated to the side of the tube adjacent to the magnet.
8. Carefully remove the supernatant without disturbing the beads.
9. Add 200 μl of 75% ethanol and incubate the tube for 1 min at room temperature.
10. Carefully remove all the solution without disturbing the beads and repeat **steps 9** and **10**.
11. Air-dry the beads for 10 min at room temperature.
12. Remove the microtubes from the magnetic separation rack. Add 35 μl of nuclease-free water and resuspend the beads by gently pipetting ten times. Incubate the tube for 2 min at room temperature.
13. Place the microtubes in a magnetic separation rack and incubate for 5 min.

14. Carefully remove the solution (35 μl) without disturbing the beads and transfer it to a new 1.5 ml Clear-view Snap-Cap low-retention microtube.
15. Add 1 volume of Agencourt AMPure XP (35 μl). This step will remove DNA fragments that are <250 bp, which frequently consist of adapter dimers (*see* **Note 8**). Gently mix the solution by pipetting ten times.
16. Incubate the tube for 10 min at room temperature.
17. Place the microtubes in a magnetic separation rack and incubate for 5 min. All the beads should be concentrated to the side of the tube adjacent to the magnet.
18. Carefully remove the supernatant without disturbing the beads.
19. Add 200 μl of 75% ethanol and incubate the tube for 1 min at room temperature.
20. Carefully remove all of the solution without disturbing the beads.
21. Repeat **steps 19** and **20**.
22. Air-dry the beads for 10 min at room temperature.
23. Remove the microtubes from the magnetic separation rack. Add 10 μl of nuclease-free water and resuspend the beads by gently pipetting ten times. Incubate the tube for 2 min at room temperature.
24. Place the microtubes in a magnetic separation rack and incubate for 5 min.
25. Transfer 10 μl of solution to a new tube. This corresponds to the eluted DNA library.
26. Check the quality and quantify the concentration of the DNA library sample as described in **step 3** in Subheading 3.6 (*see* **Note 8**, Fig. 5).

4 Notes

1. The cross-linking step should be optimized for local laboratory conditions. Both under– (causing poor protein–DNA cross-linking) and over–cross-linking (reducing the accessibility of epitope) may reduce the yield of immunoprecipitated DNA. Therefore, it is necessary to add a sufficient volume of formaldehyde solution per gram of ground tissue. Increasing the mass of ground buds in 25 ml of 1% formaldehyde solution may decrease the yield of cross-linking reaction. Small scale ChIP experiments with different cross-linking conditions can be performed to compare the amount of DNA being precipitated in order to optimize the cross-linking step.

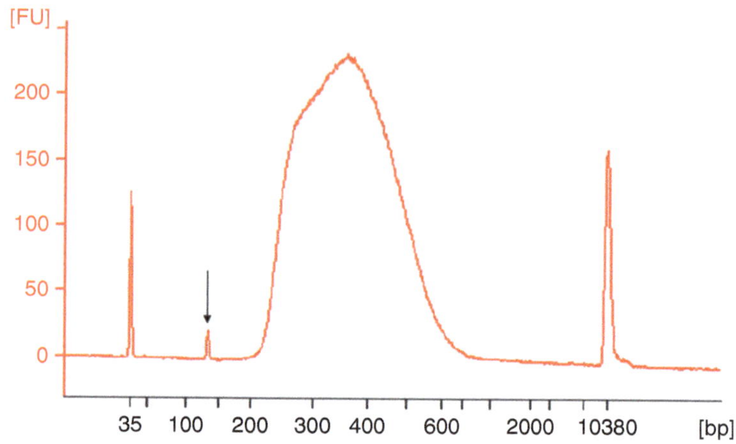

Fig. 5 Bioanalyzer electropherogram of a ChIP-seq DNA library from sonicated chromatin. The two peaks at 35 bp and 10.380 bp represents the lower and upper marker peaks from the Agilent High sensitivity DNA kits. These markers are important to calculate the size and concentration of the peak sample. The DNA library has a size ranging between 200 and 600 bp with enrichment for DNA molecules between 300 and 400 bp. Adapter dimers are marked with a black arrow and have a length between 120 and 140 bp. It is advisable to minimize the presence of adapter dimers before proceeding to next-generation sequencing

2. During nuclei isolation and chromatin preparation (Subheading 3.2) it is important to gently resuspend the pellet of nuclei in solution. For example, using a paintbrush and slowly resuspending the pellet by brushing the surface of the pellet to ensure good recovery of intact nuclei. Resuspension by pipetting is not recommended as it will create foam and risk breaking the nuclei.
3. After centrifugation, **step 4** in Subheading 3.2, make sure that the pellet of nuclei is mostly white and has little green residue, which indicates the presence of chloroplasts. Otherwise include a second wash with EB2 buffer. Alternatively, wipe off the green residue with a thin layer of Kimwipes tissue rolled around a spatula.
4. DNA sonication conditions should be optimised for local laboratory equipment. During sonication, ultrasonic waves generate heat and so the water bath should be kept at 4 °C to preserve the integrity of cross-linked protein-DNA complexes.
5. The choice of antibody depends on the epitope. It is recommended to test the antibody by western blot, immunoprecipitation, and immunofluorescence. If the antibody detects the epitope without significant background noise in these techniques it is a good indication that the antibody may also work for chromatin immunoprecipitation. Both Protein A and Protein G Dynabeads can be used for ChIP-seq. However, they have

different binding strengths to different immunoglobulins (IgG) and the choice of Dynabeads depends on the IgG class antibodies used for immunoprecipitating DNA.

6. More stringent wash conditions can be used to remove non-specific binding to the antibodies. For example, the second wash with High salt buffer (**step 16** in Subheading 3.4.2) can be replaced by a wash with Lithium chloride buffer (250 mM LiCl, 10 mM Tris–HCl pH 8.0, 1 mM EDTA, 1% Nonidet P-40, 1% sodium deoxycholate, 1 μM pepstatin A, EDTA-free Protease Inhibitor Cocktail (one tablet dissolved into 50 ml of solution)).
7. It is important to minimize the number of PCR cycles (<20) in **step 4** in Subheading 3.7.4. An increase in the number of PCR cycles cause an increased number of PCR-derived duplicate reads, which can cause bias when mapping reads and in the downstream bioinformatics analysis.
8. The final step of DNA size selection of the library (**step 15** in Subheading 3.7.5) is important to remove any adapter dimers. Adapter dimers cluster on the flow cell and may be sequenced more efficiently than the ligated ChIP DNA molecules, causing a reduction in the number of genome-mappable reads. Checking the final product of a DNA library on a bioanalyzer will indicate how much adapter dimer is present in the DNA library sample (Fig. 5). If the sample has a high molarity of adapter dimers, an additional step of size selection is recommended to remove them.

References

1. Robert T, Nore A, Brun C, Maffre C, Crimi B, Bourbon HM, de Massy B (2016) The TopoVIB-like protein family is required for meiotic DNA double-strand break formation. Science 351:943–949
2. Vrielynck N, Chambon A, Vezon D, Pereira L, Chelysheva L, De Muyt A, Mezard C, Mayer C, Grelon M (2016) A DNA topoisomerase VI-like complex initiates meiotic recombination. Science 351:939–943
3. Panizza S, Mendoza MA, Berlinger M, Huang L, Nicolas A, Shirahige K, Klein F (2011) Spo11-accessory proteins link double-strand break sites to the chromosome axis in early meiotic recombination. Cell 146:372–383
4. Stanzione M, Baumann M, Papanikos F, Dereli L, Lange J, Ramlal A, Trankner D, Shibuya H, de Massy B, Watanabe Y, Jasin M, Keeney S, Toth A (2016) Meiotic DNA break formation requires the unsynapsed chromosome axis-binding protein IHO1 (CCDC36) in mice. Nat Cell Biol 18:1208–1220
5. Villeneuve AM, Hillers KJ (2001) Whence meiosis? Cell 106:647–650
6. Yelina NE, Lambing C, Hardcastle TJ, Zhao X, Santos B, Henderson IR (2015) DNA methylation epigenetically silences crossover hot spots and controls chromosomal domains of meiotic recombination in Arabidopsis. Genes Dev 29:2188–2202
7. Si W, Yuan Y, Huang J, Zhang X, Zhang Y, Zhang Y, Tian D, Wang C, Yang Y, Yang S (2015) Widely distributed hot and cold spots in meiotic recombination as shown by the sequencing of rice F2 plants. New Phytol 206:1491–1502
8. Li X, Li L, Yan J (2015) Dissecting meiotic recombination based on tetrad analysis by single-microscope sequencing in maize. Nat Commun 24:6648

9. Demirci S, van Dijk AD, Sanchez Perez G, Aflitos SA, de Ridder D, Peters SA (2017) Distribution, position and genomic characteristics of crossovers in tomato recombinant inbred lines derived from an interspecific cross between Solanum lycopersicum and Solanum pimpinellifolium. Plant J 89:554–564
10. Kleckner N (2006) Chiasma formation: chromatin/axis interplay and the role(s) of the synaptonemal complex. Chromosoma 115:175–194
11. Armstrong SJ, Caryl AP, Jones GH, Franklin FC (2002) Asy1, a protein required for meiotic chromosome synapsis, localizes to axis-associated chromatin in Arabidopsis and Brassica. J Cell Sci 115:3645–3655
12. Nonomura K, Nakano M, Eiguchi M, Suzuki T, Kurata N (2006) PAIR2 is essential for homologous chromosome synapsis in rice meiosis I. J Cell Sci 119:217–225
13. Ferdous M, Higgins JD, Osman K, Lambing C, Roitinger E, Mechtler K, Armstrong SJ, Perry R, Pradillo M, Cunado N, Franklin FC (2012) Inter-homolog crossing-over and synapsis in Arabidopsis meiosis are dependent on the chromosome axis protein AtASY3. PLoS Genet 8:e1002507
14. Wang K, Wang M, Tang D, Shen Y, Qin B, Li M, Cheng Z (2011) PAIR3, an axis-associated protein, is essential for the recruitment of recombination elements onto meiotic chromosomes in rice. Mol Biol Cell 22:12–19
15. Osman K, Yang J, Roitinger E, Lambing C, Heckmann S, Howell E, Cuacos M, Imre R, Durnberger G, Mechtler K, Armstrong S, Franklin FCH (2018) Affinity proteomics reveals extensive phosphorylation of the Brassica chromosome axis protein ASY1 and a network of associated proteins at prophase I of meiosis. Plant J 1:17–33
16. Chambon A, West A, Vezon D, Horlow C, De Muyt A, Chelysheva L, Ronceret A, Darbyshire AR, Osman K, Heckmann S, Franklin FCH, Grelon M (2018) Identification of ASYNAPTIC4, a component of the meiotic chromosome axis. Plant Physiol 178:233–246
17. Cai X, Dong F, Edelmann RE, Makaroff CA (2003) The Arabidopsis SYN1 cohesin protein is required for sister chromatid arm cohesion and homologous chromosome pairing. J Cell Sci 116:2999–3007
18. Golubovskaya IN, Hamant O, Timofejeva L, Wang CJ, Braun D, Meeley R, Cande WZ (2006) Alleles of afd1 dissect REC8 functions during meiotic prophase I. J Cell Sci 119:3306–3015
19. Lam WS, Yang X, Makaroff CA (2005) Characterization of Arabidopsis thaliana SMC1 and SMC3: evidence that AtSMC3 may function beyond chromosome cohesion. J Cell Sci 118:3037–3048
20. Sanchez-Moran E, Santos JL, Jones GH, Franklin FC (2007) ASY1 mediates AtDMC1-dependent interhomolog recombination during meiosis in Arabidopsis. Genes Dev 21:2220–2233
21. Zhang L, Wang S, Yin S, Hong S, Kim KP, Kleckner N (2014) Topoisomerase II mediates meiotic crossover interference. Nature 511:551–556
22. Libuda DE, Uzawa S, Meyer BJ, Villeneuve AM (2013) Meiotic chromosome structures constrain and respond to designation of crossover sites. Nature 502:703–706
23. Lambing C, Heckmann S (2018) Tackling plant meiosis: from model research to crop improvement. Front Plant Sci 9:829
24. Lambing C, Osman K, Nuntasoontorn K, West A, Higgins JD, Copenhaver GP, Yang J, Armstrong SJ, Mechtler K, Roitinger E, Franklin FC (2015) Arabidopsis PCH2 mediates meiotic chromosome remodelling and maturation of crossovers. PLoS Genet 11:e1005372
25. Kaufmann K, Muino JM, Osteras M, Farinelli L, Krajewski P, Angenent GC (2010) Chromatin immunoprecipitation (ChIP) of plant transcription factors followed by sequencing (ChIP-SEQ) or hybridization to whole genome arrays (ChIP-ChIP). Nat Protoc 5:457–472
26. Kugou K, Ohta K (2009) Genome-wide high-resolution chromatin immunoprecipitation of meiotic chromosomal proteins in Saccharomyces cerevisiae. Methods Mol Biol 557:285–304
27. Ma H (2006) A molecular portrait of Arabidopsis meiosis. Arabidopsis Book 4:e0095
28. Smyth DR, Bowman JL, Meyerowitz EM (1990) Early flower development in Arabidopsis. Plant Cell 2:755–767

Chapter 17

Isolating Male Meiocytes from Maize and Wheat for "-Omics" Analyses

Stefanie Dukowic-Schulze, Nelson Garcia, Arun S. K. Shunmugam, Sateesh Kagale, and Changbin Chen

Abstract

Genome-wide gene expression studies have become a routine approach due to the advances in sequencing technologies, their ease of use, and increasing affordability. Simultaneous investigation of small RNA expression adds further valuable information but is not adopted as widely yet. Both RNA-seq and small RNA-seq benefit from the use of specific cell types. Here, we describe a protocol for the isolation of male meiotic cells from maize or wheat plants, along with the application of downstream RNA sequencing, extendable to other -omics approaches.

Key words Meiocytes, Meiosis, Maize, Wheat, Plants, Omics, Sequencing

1 Introduction

Most of our understanding of plant meiosis comes from the study of meiotic mutants [1]. These reverse genetics studies identified key genes for the meiotic process, and are now joined by broader forward genetics studies. Microarrays and RNA-seq can help identify genes important for meiosis due to their abundant or specific expression in reproductive tissues/cells. In order to pinpoint if genes are highly expressed in a specific cell at a specific stage, sample purity needs to be high. Transcriptome studies of whole anthers which contain multiple anther wall layers in addition to meiocytes can give an acceptable spatial and temporal approximation for meiocytes transcriptomes [2] but not for small RNA composition [3]. In general, using isolated meiocytes gives superior resolution and insight. Methods to isolate male meiocytes depend on the species and downstream application, and range from simple squeezing-out of meiocytes [4–7] over manual collection [6, 8–11] to flow cytometry [12]. Laser-assisted microdissection

Mónica Pradillo and Stefan Heckmann (eds.), *Plant Meiosis: Methods and Protocols*, Methods in Molecular Biology, vol. 2061,
https://doi.org/10.1007/978-1-4939-9818-0_17,

has also been used for plant meiocyte extraction but is not a top choice due to lower RNA quantity and quality [13].

Here we describe a protocol for isolating male meiocytes from maize or wheat for use in RNA sequencing approaches. Steps include initial identification of appropriate plants for sampling, dissection of anthers, release of meiocytes, and manual collection. The protocols for maize and wheat differ in their steps due to the differences in the biological material, and have been developed and analyzed previously [9, 10]. We also provide advice on meiocyte storage, RNA extraction, sequencing, and data processing. The protocol can be adapted to other plants, as we did for example initially in Arabidopsis [8] and also successfully performed for initial steps in rice (unpublished).

2 Materials

2.1 Plant Material

1. Maize plants (*Zea mays*) are grown in a well-controlled greenhouse or growth chamber (*see* **Note 1**). Optimal growing conditions are 16 h of light ($\geq$450 μM × m^{-2} × s^{-1}), with 24 °C during the day and 22 °C during the night setting. Seeds are planted directly into soil or pregerminated on a wet tissue until shoots are about 3–5 cm long. Seeds can be pregerminated in the lab, either in the dark or the light. This usually leads to a higher rate of viable seedlings. Pots for further growth are 25 cm in diameter, for 300–500 cm^3 of a 2:1 mix of field top soil and a professional all-purpose mix with vermiculite and perlite (Sunshine Mix #8 LC8, Sun Gro Horticulture). As regular care, we thoroughly water every other day, add a layer of sand to avert weed growth, apply ~20 g of 12-12-12 slow-release fertilizer granules at the three-leaf stage, and fertilize further biweekly with 1–2 g of Peters' 20-20-20 dissolved in water for watering.
2. Grow wheat plants in a controlled environmental chamber at 22 °C constant temperature with 16 h day length period (400 μM × m^{-2} × s^{-1}). Wheat seeds are grown in 4-in. plastic pots that are filled with Sunshine Mix #8 LC8. Only one seed is seeded per pot to facilitate proper tracking of meiosis stages. The plants are watered every day and fertilized with 20-20-20 fertilizer (3.0 g/l) added with micronutrient mix (0.3 g/l) (Plant Products Co. Ltd.).

2.2 Equipment

Items 1–6 are specifically for the maize protocol and **items 7–10** for the wheat protocol. All other items are common to both protocols.

1. Scalpel.
2. Tape.

3. Dissection needles (regular disposable 1 ml syringes work well).
4. Ethanol burner.
5. Glass Pasteur pipettes with cotton plugs (*see* **Note 2**).
6. Food-grade (nontoxic) rubber tubing (e.g., silicone) (*see* **Note 2**).
7. Bamboo skewer.
8. Insect pin.
9. Drummond micropipette with 5 μl capillary tube.
10. Polystyrene petri dish (60 mm × 15 mm, or square).
11. Forceps.
12. Permanent marker.
13. Plastic ware: 50 ml Falcon tubes, 1.5 ml microtubes, low-retention RNase-free 1.5 ml microtubes.
14. Microscope slides (glass for regular staining; plastic for meiocyte collection steps).
15. Plastic coverslips, 22 mm × 22 mm (*see* **Note 3**).
16. Precision pipettes (20 and 200 μl sizes).
17. Optional: Nail polish or professional slide sealant.
18. Microscopes (*see* **Note 4**): Dissection microscope (~10×), inverted microscope for meiocyte collection (40–100×) and a compound microscope for slide analysis (100–400×).
19. Petri dish (100 mm diameter, or old lids of pipette tip boxes) with wet tissue paper.

2.3 Self-Made Utensils

Items 1–3 are for maize meiocyte collection and **item 4** is for wheat meiocyte collection.

1. Plastic pestles for squashing maize anthers: Flatten regular 200 μl pipette tips at the pointed end by holding them shortly close to the flame of an ethanol burner, and then quickly pressing them down onto a clean microscope slide (*see* **Note 5**, Fig. 1a).
2. Oxidized iron pins: Create your own set of oxidized iron pins (alias rusty nails) by scraping thin long nails with sandpaper, mainly at the tip and the lower part. Next put them in, for example, glass jar lids, spray them with water and let them sit (preferentially outside) for several days or weeks (Fig. 1b).
3. Suction-enabling glass microcapillary system for meiocyte collection: The final device consists of a rubber tube connecting two modified Pasteur pipettes (Fig. 1c). Prepare 2–5 mouthpieces by breaking the thin part of the Pasteur pipette off, leaving part of the narrowing transition attached (*see* **Note 6**). Heat the cracked end over a flame to smooth the edges. Prepare

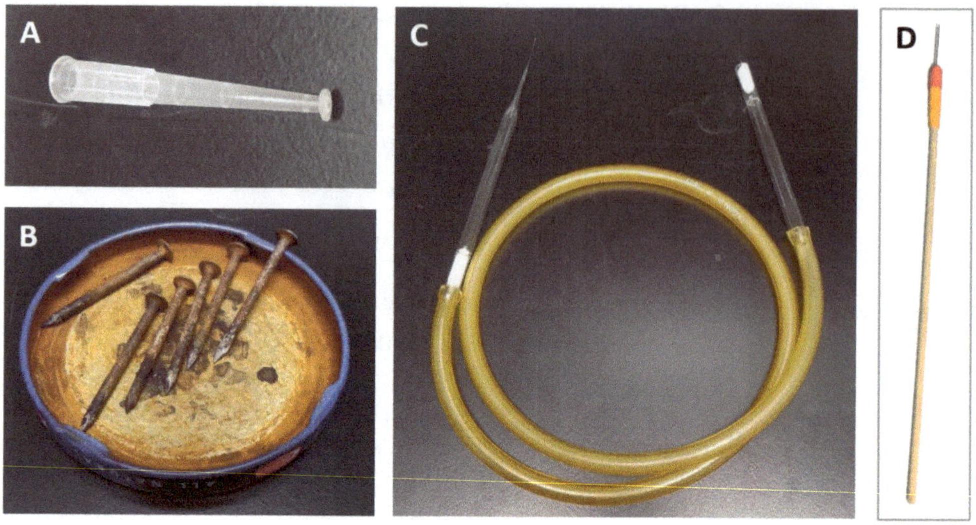

Fig. 1 Self-made utensils. (**a**) Modified pipette tip for squashing anthers. (**b**) Oxidized iron rods (rusty nails) for intensifying acetocarmine staining. (**c**) Maize meiocyte collection device. (**d**) Dissection needle out of insect pin and bamboo skewer

20–50 microcapillaries by heating the thin part of a Pasteur pipette over a flame and quickly extending it once the glass is melting (*see* **Note 7**). Clip off the tip that is now wider than the thin part of the outstretched pipette to create microcapillaries that are 3–10 cm long and ≤1 mm in diameter (*see* **Note 8**). Use ~50 cm of tubing to connect a mouthpiece with a microcapillary. For a slightly different way of producing the microcapillary system as well as an alternative usage, *see* [14].

4. Dissecting needle made from insect pin and bamboo skewer: Attach an insect pin onto a 3 mm thick bamboo skewer by gluing it to the tip. Additional tapes can be wrapped around the junction to fit the pin firmly to the skewer (Fig. 1d).

2.4 Buffers, Solutions, and Chemicals

Item 1 is needed for maize only; **items 2–9** for both.

1. Farmer's fixative: three parts of absolute ethanol to one part of acetic acid (v/v). For 500 ml, mix 375 ml of 100% ethanol and 125 ml of glacial acetic acid.
2. RNase decontamination spray (e.g., RNase Away or RNase Zap) (*see* **Note 9**).
3. Acetocarmine staining solution (self-made or purchased): 0.5–2% solution for maize, 2% solution for wheat. For 100 ml of a 0.5% solution dissolve 0.5 g of carmine in 55 ml of water (stir on heater, in fume cabinet), add 45 ml of glacial acetic acid while stirring and boil for about 30 min. Filter solution through Whatman paper and store in the dark (in cabinet; tinted glass or tin foil wrapped container).

4. Phosphate-buffered saline (PBS; 10×): 1.37 M NaCl, 0.027 M KCl, 0.1 M Na_2HPO_4, 0.018 M KH_2PO_4, pH 7.0. Dissolve 80 g of NaCl, 2 g of KCl, 14.4 g of Na_2HPO_4 and 2.4 g of KH_2PO_4 in 1 L of dH_2O. Adjust to pH 7.0 with HCl, autoclave and store at room temperature.
5. Optional: RNase inhibitors (e.g., Protector RNase Inhibitor, Sigma Aldrich; use as 100–200×) (*see* **Note 10**).
6. TRIzol Reagent (*see* **Note 11**).
7. Zymo DirectZol RNA Micro Kit.
8. Qubit fluorometer and appropriate kit reagents (or out-source).
9. Agilent Bioanalyzer (or out-source).

3 Methods

The protocol lists detailed steps for isolating pure populations of male meiocytes from maize or wheat, and processing them for RNA-seq and small RNA-seq. It has been extensively used with zygotene stage in maize and all of the prophase-I stages in wheat but can be adapted to other meiotic stages (*see* **Note 12**, Figs. 2 and 3). The method for maize has initially been called "Capillary collection of meiocytes (CCM)" when developed for Arabidopsis and the method for wheat has recently been published under the name "MeioCapture" [8–11]. Though some steps of the protocols differ, and hybrid protocols are possible, both approaches isolate prophase I meiocyte clusters, which are also called sporogenous archesporial columns (SACs).

Brief advice is given on downstream steps including RNA extraction, quality control of RNA, sequencing and bioinformatic analysis. Adaptations of the protocol aiming at other downstream experiments like bisulfite conversion or chromatin immunoprecipitation are possible and described elsewhere [9].

3.1 Maize

3.1.1 Determination of Meiotic Stage in Growing Maize Plants

1. Test appropriate maize plants (*see* **Note 13**) by cutting a vertical slit into the stalk with a scalpel. The incision should be ~10 cm long, starting 3–5 cm above the last node, and be ~1/3 of the stalk diameter deep. Cautiously extend the slit until you can see the tassel.
2. Gently free the tassel so you can see its major part. Take 2–3 spikelets from the mid part of the main tassel (*see* **Note 14**), place the spikelets in microtubes and add Farmer's fixative.
3. Gently push the intact tassel back into the center of the stalk, close the outer layers, and tightly and completely wrap the stalk with labeling tape. Label the name or number of the plant and the date on the tape bandage as well as on the tube.

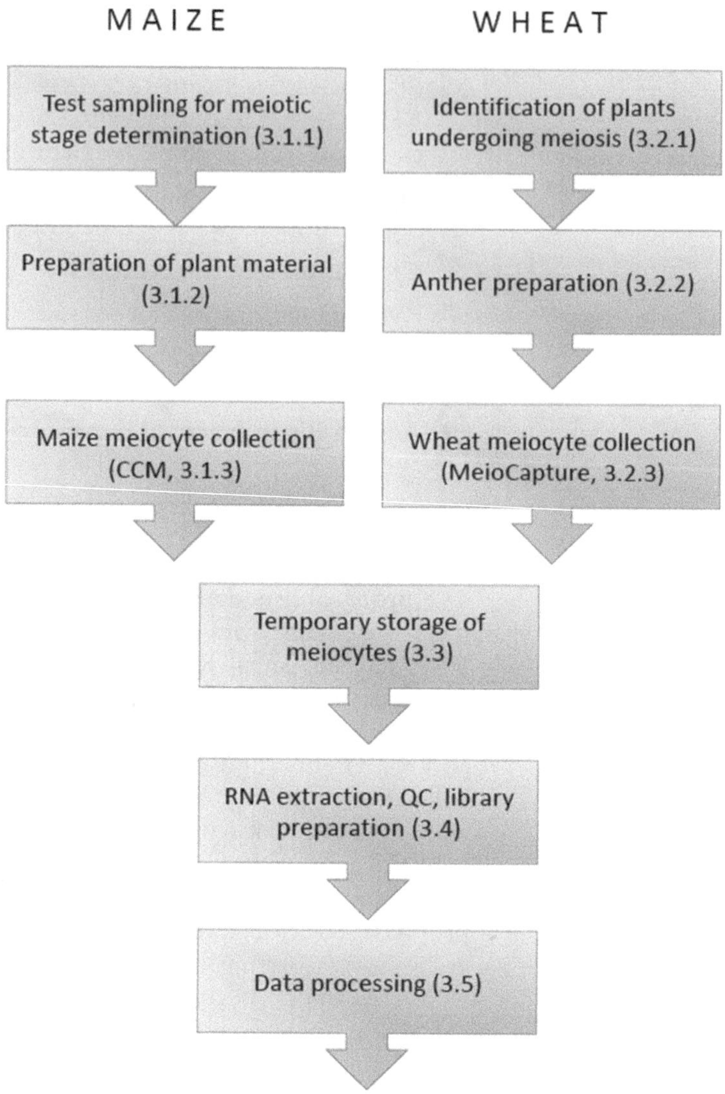

Fig. 2 Workflow for meiocyte collection and RNA-seq. Steps for capillary collection of meiocytes in maize and MeioCapture in wheat differ, adapted to the species. The workflow depicts separate and common steps, and refers to the respective protocol section

4. Dissect anthers out of maize spikelets with syringe needles (*see* **Note 15**).
5. Transfer the anthers on a microscope slide, add 20–30 μl of acetocarmine solution and swiftly progress with the following steps: With an oxidized nail, stir the drop of solution containing anthers, place anthers in center, then heat the drop over a flame for ~3 s, stir again with the nail and repeat until the drop looks purple (*see* **Note 16**).

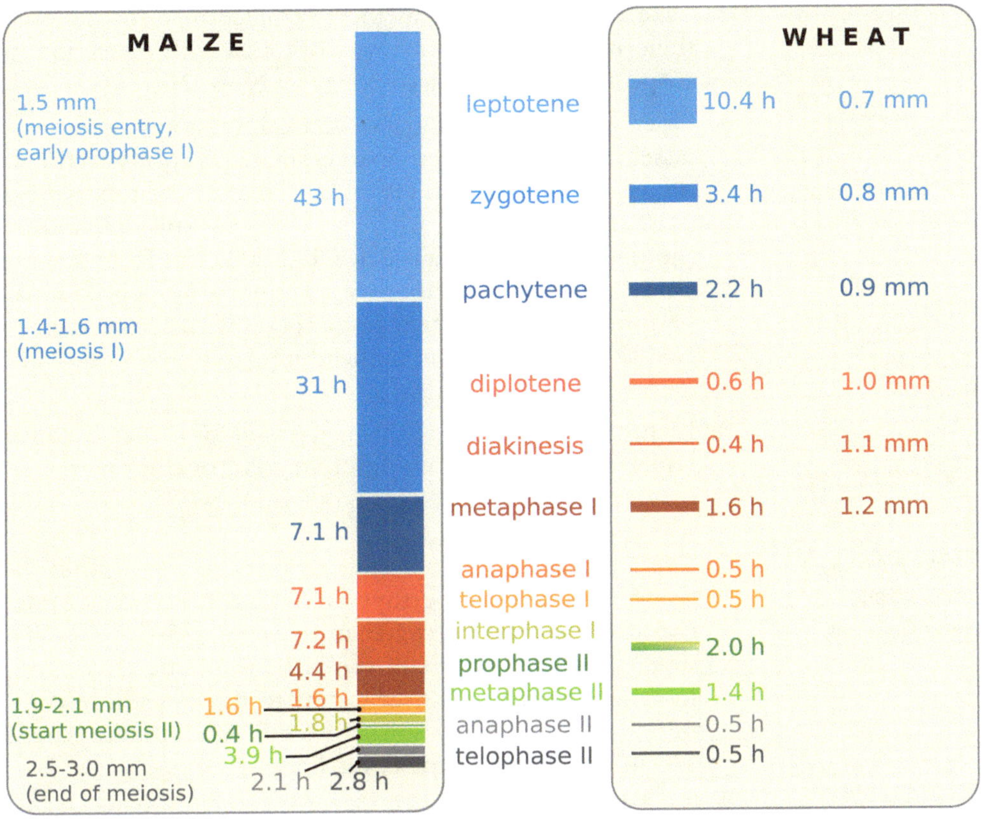

Fig. 3 Timing of male meiosis in maize and wheat. Adequate timing of meiocyte collection has to take the duration of meiotic stages into account. In general, the earlier the prophase I stage, the longer it lasts. Maize meiosis takes a total of ~120 h [15], wheat only ~24 h [16]. Information on corresponding anther lengths are taken from [10, 17]. Timing and anther length might vary due to specific inbred and environmental conditions

6. Squash the anthers with a disposable tip pestle, apply a coverslip, and press it down hard (*see* **Note 17**).
7. Examine slide under a bright-field microscope at 100–400× magnification (*see* **Note 18**). For examples of male meiotic stages in maize, *see* [9]. For duration of male meiotic stages in maize, *see* Fig. 3.

3.1.2 Preparation of Fresh Maize Material

Wear gloves and collect anthers in RNase-free tubes. Clean forceps and syringe needles in between with ethanol wipes. RNase Zap or similar solutions might destroy RNA and are only needed for precleaning equipment (and then wiping it thoroughly off, dry and/or wash with RNase-free water). At the steps below, RNA is still inside cells and not in high danger of degradation from exogenous sources. Swift processing is needed to minimize the time between taking spikelets out of the plant and immersing isolated meiocytes into TRIzol solution (*see* **Note 19**).

1. Take plants ready for meiocyte collection into the lab, place them in water, next to your work area with dissection microscope and inverted microscope (*see* **Note 20**).
2. Open the tape bandage and vertical incision to see the whole tassel. With a permanent marker, mark positions on the stalk at the center of the main tassel, as well as at the top and bottom (~10–20% from ends, where spikelets are still similar in size to the ones at the center). Place 1–2 spikelets from the marked positions in labeled tubes with Farmer's fixative. Incubate for ~5 min. Determine the meiotic stages with acetocarmine solution as in **steps 4–7** in Subheading 3.1.1 (*see* **Note 21**). Bandage the tassel up again.
3. Identify the range along the main tassel for meiocyte collection. Label it on the stalk with different color or thick marks. Always bandage up the stalk after taking samples.

3.1.3 Maize Meiocyte Collection (Capillary Collection of Meiocytes, CCM)

1. Use a plastic microscope slide or prepare a collection slide by fixing two plastic coverslips on a glass microscope slide with two drops of distilled water in between (~8–10 μl each). The small drops of water will prevent the plastic coverslips from moving out of place. Place it next to the dissection scope, add ~50 μl of 1× PBS (optionally with RNase inhibitors, *see* **Note 10**) in the center of one coverslip.
2. Take 10–20 spikelets directly out of the live plant, place them on a clean slide and close the tape over the incision. Under a dissection scope, dissect the anthers from the upper florets. Move dissected anthers right away into the drop of PBS on the collection slide.
3. Squash the dissected anthers carefully but thoroughly with a self-made flattened tip pestle. You will be able to see extruded meiocyte clusters with the dissection scope. Extend the drop to a neat circular area of ~10–15 mm in diameter.
4. Cautiously transfer the collection slide onto the inverted microscope.
5. Add more 1× PBS until a nice liquid dome forms (~4–7 mm in height). This needs ~700 μl in total.
6. Find a comfortable position and put the mouth piece of the collection device in your mouth. Hold the microcapillary part in a flat angle, tip at the bottom of the liquid dome (*see* **Notes 22** and **23**).
7. Move the microcapillary close to meiocytes and gently suck them in (*see* **Note 24**).
8. Blow meiocytes out onto the second coverslip and collect more. Do not fill the thicker part of the microcapillary more than ~5–10 mm. It is more difficult to release meiocytes from there.

9. Continue collection in the first drop (*see* **Note 25**).
10. Switch to the second coverslip with collected meiocytes, apply more PBS, using a pipette tip to facilitate a circular area.
11. Collect meiocytes from the second drop, avoiding remaining anther wall contaminations and diluting starch granules. While collecting, count the clusters of meiocytes or meiocytes themselves to note down the quantity per collection (*see* **Note 26**).

3.2 Wheat

Follow general guidelines for working with RNA (*see* **Notes 9** and **10**) and keep the duration of the procedure short, with a maximum of 1 h (*see* **Note 19**).

3.2.1 Identification of Appropriate Wheat Plant Material

Appropriate wheat plants are about 8 weeks old, and still contain the spike in the flag sheath leaf during the wheat booting stage. A slightly swollen area (boot) indicates the position of the spike that is ~3–4.5 cm during meiosis I. For better consistency between samples, use only primary tillers.

3.2.2 Anther Preparation

1. Prepare a round or square plastic petri dish by drawing or adding a printed grid of 6 × 6 squares on the bottom. Label the rows with your subset of interest (e.g., premeiotic G2/0.5–0.6 mm, leptotene/0.7 mm, zygotene/0.8 mm, pachytene/0.9 mm, diplotene/1.0 mm, diakinesis/1.1 mm, and metaphase I/1.2 mm) (*see* **Note 27**, Fig. 3). Put a drop of ~50 μl of 1× PBS in each square.
2. In another dish or on a microscope slide, dissect wheat heads (spikes) out of primary tillers from plants at appropriate stages.
3. Put the 4–5 florets per spikelet together into a drop of ~200 μl of 1× PBS (*see* **Note 28**).
4. Dissect the three anthers out of each floret and measure their length with an ocular micrometer or micrometer slide.
5. Transfer two of the three anthers to an individual drop of ~50 μl of 1× PBS at the appropriate position in the petri dish with the prepared grid, and one anther to a microscope slide for stage determination.
6. For each anther on a microscope slide, squash with a dissection needle, add 25 μl of 2% acetocarmine and add a coverslip.
7. Identify the meiotic stage at 40× magnification under a light microscope, as seen in Fig. 1 of [10]. If the size corresponds to the expected stage, leave their sibling anthers in the respective drop. Otherwise, label the drop with the sibling anthers with the identified stage or transfer them into the correct row in the grid.

3.2.3 Wheat Meiocyte Collection (MeioCapture)

1. As for the maize meiocyte collection (*see* **step 1** in Subheading 3.1.3), use a plastic microscope slide or prepare a collection slide by fixing two plastic coverslips on a glass microscope slide with two drops of distilled water in between (~8–10 μl each). The small drops of water will prevent the plastic coverslips from moving out of place. Place it next to the dissection scope, add ~50 μl of 1× PBS in the center of one coverslip.
2. Use a dissection microscope from now on, and add at least ten anthers of the same, confirmed stage to the drop on the slide, that is, the anther pairs from at least five drops on the petri dish with the grid.
3. For each anther, carefully cut the narrow end open with two self-made dissection needles (for assembly *see* **step 4** in Subheading 2.3; Fig. 1d). Hold each anther steady with one needle, gently rolling a second needle at a flat angle along the anther starting from the intact end to squeeze meiocyte columns (sporogenous archesporial columns—SACs) out of the anther (optimally all four meiocyte SACs). If needed, add more 1× PBS, and try to keep the drop as small and high as possible.
4. Make sure all SACs are floating free inside the liquid dome (*see* **Note 29**). Break the surface of the dome in case they are floating on the top.
5. Transfer the slide to an inverted microscope, add another drop of ~50 μl of 1× PBS next to the first drop containing free-floating SACs. With a 5 μl Drummond microdispenser, pipet individual SACs out of the first drop and transfer them to the second drop. Remove accidentally cotransferred debris from the second drop with the microdispenser at the end.

3.3 Temporary Storage of Meiocytes

1. Gently transfer purified meiocytes/SACs into a microtube with 300–500 μl of TRIzol (at room temperature or on ice): For the maize protocol, by blowing them out of the glass microcapillary; for the wheat protocol, by pipetting with a Drummond microdispenser. Do not exceed a ratio of 3:1 for TRIzol to added sample (*see* **Notes 30** and **31**).
2. Store meiocytes in TRIzol at 2–8 °C up to several weeks, adding more meiocytes from additional plants (*see* **Note 32**).

3.4 Preparation of Total RNA, Quality Control, and Library Preparation for Sequencing

1. Use a traditional TRIzol-based RNA extraction method or the TRIzol-based Zymo DirectZol™ RNA Micro Kit (*see* **Note 33**). The latter extracts both long and small RNAs efficiently and allows high concentration due to minimal elution volume of 6 μl (*see* **Note 34**).
2. Measure the amount and quality of the extracted RNA using for example a Qubit Fluorometer with reagents for high-sensitivity or broad-range RNA measurement, and an Agilent

Bioanalyzer for electronic output from a miniature gel electrophoresis, supplying the RIN (RNA integrity number, *see* **Note 35**) [18].

3. Choose library preparation protocols or kits according to your own preferences or recommendations from your local core facility for sequencing (*see* **Note 36**). Consider, if strand-specific sequencing would provide greater value to your project [19].

3.5 Tips on Data Processing Pipelines and Analysis

Supercomputers are not necessarily needed but speed up the process. What is however essential is flawless data management, pertaining to backups, diligent processing, and detailed and transparent documentation of all steps, parameters and files created. Bioinformatic experts might prefer data processing via a UNIX command line and in the statistical program "R", while there are also solutions with easy-to-use user interfaces, for example "Galaxy" (*see* **Note 37**). Processing of sequencing data depends heavily on the experience and preferences of yourself, your group, your bioinformatic collaborator and/or local sequencing center. We thus restrict this protocol to advice we ourselves found helpful. Key software and tools used are listed in Table 1.

1. In general:
   - There is by now a vast overload of -omics data available, published and constantly produced. Keep in mind that the production of this kind of data is the means and not the end goal. Avoid listing mere numbers of differentially regulated genes, specific upregulated or downregulated genes or miRNAs, and significant GO terms. Instead, dig through these data and previous literature to come up with new hypotheses and experimental approaches to test them.
   - For easy and fast visualization of aligned read counts, we recommend to create TDFs (Tiled Data Files) by using the "count" tool (*see* **Note 38**) from IGV (Integrative Genomics Viewer [24]) on BAM files. TDF files can quickly be loaded and browsed in IGV.
2. RNA-seq:
   - In case of adequate high-quality genomes, sequencing reads are aligned to the reference genome, and read counts of annotated genes extracted afterward. In case of less polished or even nonexistent reference genomes, de novo transcript assemblies can be generated to which sequencing reads can be aligned [29].
   - Essential tools for further data exploration are (1) a list of normalized gene counts, (2) lists of differentially expressed (DE) genes, and (3) files for easy visualization of reads aligned across a gene (e.g. TDF files, *see* above). R software

Table 1
Recommended computational tools for analysis

Tool	Description	Purpose	Citation and source
Unix command line	Shell; requires basic skills in programming; most versatile option to handle big data files, compare, sort and modify them	File formatting changes (e.g., "chr1" → "1") Sequence read analysis, statistics (with, e.g., the commands "grep," "awk," "sed") Implementation of many bioinformatic packages (for initial read trimming and aligning, and later steps, e.g., "samtools," "bedtools")	
Galaxy	GUI (graphic user interface), for easier and wider access to sequencing data processing, great workflow processing and integrated documentation; implemented tools can differ between public server/cloud/nonpublic server versions	As above	[20] https://usegalaxy.org/
R	Command line based statistics program; requires basic skill in programming; most R packages for specific tasks come with good documentation and "vignettes" that lead through an example	Handling of big data Statistical analysis Generating graphs	[21] https://www.r-project.org/
R Studio	Highly recommended GUI for R; with windows for scripts, command lines, image output, history and loaded data files	As above	[22] https://www.rstudio.com/
DEseq2	R package for RNA-seq normalization and differential gene expression analysis	Initial RNA-seq data mining	[23] https://bioconductor.org/packages/release/bioc/html/DESeq2.html
IGV (Integrative Genomics Viewer)	Genome browser for displaying sequencing reads, calculated peaks and known features	Visualization of RNA-seq (and other sequencing) data Great for verifying steps and outputs from	[24] http://software.broadinstitute.org/software/igv/

(continued)

Table 1 (continued)

Tool	Description	Purpose	Citation and source
		computational processing (all chromosomes covered? RNA-seq reads on annotated genes? replicate patterns similar?)	
AgriGO	Online tool for analysis of overrepresented GO (gene ontology) terms	Mining of RNA-seq data for functional clues	[25] http://bioinfo.cau.edu.cn/agriGO/
ShortStack	Tool for comprehensive analysis of plant small RNAs; needs to be installed and run on Unix	For aligning reads to a genome and creating counts for detected and supplied sRNA loci	[26] https://github.com/MikeAxtell/ShortStack
StringTie	Tool for complete and accurate reconstruction of a transcriptome from RNA-seq reads	For reconstruction of transcripts/genes using a reference genome and also an optional de novo assembly For analysis of spliced variants	[27] https://ccb.jhu.edu/software/stringtie/index.shtml
Kallisto	Fast RNA-seq quantification program	For quantifying abundances of transcripts from RNA-seq data Uncertainty in transcript abundances can be quantified using its bootstrap method	[28] https://pachterlab.github.io/kallisto/source

packages like DEseq, DEseq2, and EdgeR can generate both normalized gene counts and lists of DE genes [30, 31]. The TMM method (trimmed mean of M-values) is used by DEseq and EdgeR, while other typical alternatives or downstream possibilities are RPM (reads per million), TPM (transcript per million), and RKPM (reads per kilobase of transcript per million mapped reads) (*see* **Note 39**) [32, 33]. Alternatively, there are now competitive alignment-free transcript quantification tools like Sailfish, Salmon, and Kallisto that can be paired with alignment-free differential gene expression tools, as compared with more traditional approaches in [34].

- GO analysis: In the majority, computationally identified protein domains and homology to known genes in close and distant species underlie GO terms. GO terms should thus only be used as a means and not as conclusions. Investigate the genes underlying significant GO terms to get a better understanding of the reliability of the prediction (Have mutants been characterized?) and the implication for cellular processes in meiocytes. Also, GO slim categories which are broader might help in initial GO analysis. Many related full GO terms that might be significant in your samples stem from the same pool of differentially regulated genes, overemphasizing certain biological processes (e.g., transcription factor activity, transcription regulator activity, DNA binding, regulation of transcription, regulation of RNA metabolic process, regulation of gene expression).

3. Small RNA-seq:
   - When aligning sequencing reads to the genome, do not allow mismatches. MiRNAs are only 21–24 nt in size and consist of families with similar members, sometimes differing only in one base pair.
   - We suggest using the plant-specific ShortStack suite for genome alignment and identification of phased small RNA loci [26]. The built-in alignment algorithm also facilitates the placement of multiple mapping reads to loci that are likely, due to their read density [35] (*see* **Note 40**).
   - Good sources for annotation of small RNAs are miRBase (http://www.mirbase.org) [36], a dedicated database for miRNAs, and BioMart via Gramene (www.gramene.org/biomart/martview) [37]. Be aware that there is overlap between the databases but that they are not the same. MiRBase supplies annotations for the underlying miRNA genes and mature miRNAs, with one to two forms from opposite sides of the hairpin, then designated with an asterisk or a suffix of -5p or -3p. With the BioMart database on the other hand, you can also retrieve annotations for other small RNAs, including rRNAs, tRNAs, snRNA, and snoRNAs. Make sure to remove any duplicates in annotation files by sorting for chromosome location and letting Excel highlight duplicate entries.
   - We want to emphasize the importance and pitfalls regarding normalization for small RNAs here. So far, almost no literature exists to point out or showcase different results that arise from different routes of normalization. There might be no single way that solves the issue, so we encourage comparison of different normalization routes and connecting them with underlying biological meaning. The most critical

aspects to consider in the choice of normalization are: To which factor/reference should the data be normalized to? Is this factor biologically stable between conditions or will certain small RNAs be overrepresented or underrepresented? Several possibilities are normalizing against: (1) all aligned reads, (2) reads aligned to small RNAs (including rRNAs and tRNA), or (3) reads aligned to only miRNAs (*see* **Note 41**). The final counts in RPM will have different magnitudes depending on the reference point. Issues can for example result from general size difference of the miRNA pool between conditions or from small fluctuations in the huge pools of rRNAs and tRNAs that change the total contribution of the rather small miRNA pool.

4 Notes

1. Though maize plants grow best in the field, a continuous supply of plants from a greenhouse facilitates sufficient time for sampling and for effective identification of plants with appropriate meiotic stage. Important for growing maize in a greenhouse are optimal conditions, including high light ($\geq$450 μM $\times$ m^{-2} $\times$ s^{-1}) and big pots. We found it essential to mimic outside conditions with regular physical stress in the form of wind, and thus move pots and/or shake stalks manually about every other day. This results in stronger stalks and more consistent timing and positioning of meiotic tassels. Growth chambers can be used as well if they are high enough.
2. The outer diameter of the Pasteur pipettes (7 mm in our case) should be slightly bigger than the inner diameter of the rubber tubing (5 mm in our case).
3. A high liquid dome is needed for separating meiocytes which sink to the bottom from anther wall debris which floats on top. In case of wheat, the SACs float on the drops and anther debris sink to the bottom of the drops. Plastic coverslips or plastic microscope slides are the optimal solution and a cheaper alternative to specially coated glass microscope slides.
4. Use a light source for the dissection microscope that does not cause excessive heating (e.g., white LEDs). The microscope used for collection of meiocytes has to have the objectives underneath the slide. The control wheels for moving the slide on the stage should be positioned opposite the "good" hand (for right-handed persons on the left and vice versa). Also, the vision through the ocular pieces should reflect the actual slide directions and not be inverted or flipped in any way. It does take time to trick a brain into working with a mirrored or

flipped vision, and it is hard to untrain it again afterward (own experience).

5. We recommend preparing whole boxes of pipette tips with flattened tips ahead of time to speed up the actual meiotic stage tests and meiocyte collections.
6. Wear personal protective equipment consisting of lab coat, safety goggles and gloves (possibly cotton gloves underneath nitrile gloves). Use a heavy metal bar such as a file or other tool and hit the pipette forcefully, smacking down and holding the pipette steady in the air or putting the pipette on a stool or table so the thin part protrudes over it.
7. Wear personal protective equipment and use a broad forceps to hold the thin end in your good hand. It requires some practice to achieve the perfect timing and force for pipette extension. You notice when the glass melts by wiggling the forceps up and down. Then, in very quick succession, take the pipette out of the flame and stretch your arms apart to almost their full range while keeping them in front of your body.
8. Not all microcapillaries will work perfectly. Be prepared to try out several ones before finding an optimal one for collection. Do not settle on one that works in just an acceptable manner, aim for using the perfect ones. Factors to consider are the inner diameter of the microcapillaries (slightly bigger than the meiocyte clusters), the ease of mouth-pipetting (sucking in with high sensitivity, blowing out without needing too much force), and the form of the opening (best: straight ring perpendicular to pipette, usually suboptimal: slanted opening or ragged edge).
9. All plastic or glass equipment that might be contaminated with RNA degrading enzymes should be cleaned thoroughly before use. We recommend to first use 70% ethanol, followed by an RNase decontamination solution, which then needs to be fully wiped off since it can destroy RNA itself.
10. Since meiocytes are still intact during the collection procedures, RNase inhibitors do not have to be added, especially if the solutions themselves are RNase free.
11. Previously, the Chen lab used RNAlater to preserve both Arabidopsis and maize meiocytes for RNA applications. However, we encountered issues with precipitation, spinning down meiocytes and low RNA quality. We thus switched to TRIzol which gave us better results and easier handling, as in the case of wheat, where RNA integrity number (RIN) of meiocytes was 2.7 when stored in RNAlater, and 4.7 when stored in TRIzol.
12. Prophase I stages are easy to collect because of their long duration. Metaphase and anaphase are very short and are

therefore hard to separate temporally and spatially. Tetrads as the product of meiosis are again easy to collect due to their obvious shape and the extended period they persist.

13. Depending on the inbred line and meiotic stage of interest, phenotypic parameters indicating the meiotic stage of the hidden tassel differ. At the time of meiosis, maize tassels are still enclosed in the stalk. In the inbred lines B73, Mo17 and CML228, this corresponds to 6–8-week-old plants. The Leaf Collar Methods [38] is of great initial help in identifying plants undergoing meiosis. Early in meiosis, maize and wheat stalks feel softer above and on the last node, when pressing the stalk with a thumb against forefinger and middle finger placed at the back of the stalk. In order to improve timing and the success rate of determining appropriate plants, we recommend regular sow-out of plants at 2–3 fixed days of the week. Once a proper plant has been identified, the next plants will be ready with predictable timing. We also recommend keeping log entries of plants with line, seed lot, sow-out date, issues during growth, sampling and resampling date, meiotic stage, collection date, time and amount of cells collected. For a timeline that is helpful for timing resampling or maximal possible duration of meiocyte collection *see* Fig. 3.

14. Meiosis in male maize tassels progresses fairly synchronously and simultaneously between the anthers of the florets. The ~60–80% center part of the main tassel has the best conformity, resulting in the purest collections of certain stages. In wheat, the anthers from the same floret also follow synchronous meiotic progression which is exploited for meiocyte enrichment using MeioCapture.

15. Maize spikelets contain two florets with three anthers each. We recommend using only the upper florets which are usually bigger and farther progressed. Test the meiotic stage in upper and lower florets separately since this can help with stage identification. During prophase I, upper florets are 1–2 days further progressed than lower florets. The timetable for maize meiosis (Fig. 3) is from upper florets. Typically found correlations of stages between upper and lower florets in maize inbred lines B73, Mo17, and CML228 are depicted in Fig. 4a of [9].

16. Rusty nails can be reused. Blot liquid off with a tissue paper. Avoid using the nail with broken anthers or after squashing the anthers in order to minimize contamination of successive samples with previous meiocytes.

17. Wear gloves when working with acetocarmine solution. Place the slide on the upper half of a tissue paper, fold it on top of the slide, then press down by fixating the slide with one hand and applying strong pressure with the thumb or forefinger and

middle finger of your other hand. The tissue paper will blot away excessive staining solution.

18. Anther layer cells are smaller than meiocytes, especially when those are well spread. Tapetal cells have two nuclei. Since they are easy to spot, beginners often confuse them with meiocytes in telophase I or prophase II. Meiocytes in prophase I stages occur in worm-like clusters. The chromosome staining is often suboptimal in intact clusters. If sufficient pressure has been applied on the coverslip, separated and well-spread meiocytes are present which are optimal for identifying the meiotic stage. Never trust a single identified meiocyte, find others on the same slide to confirm the determined stage. Shorter later stages can occur in parallel in anthers from the same or close spikelets. Also, biotic and abiotic stress factors can mess with the synchrony and progression of meiosis.
19. 15–30 min for maize meiocyte isolation per batch are acceptable and should not be exceeded. Goal is to capture a close representation of gene expression and small RNAs in the particular meiotic stage, not gene expression that occurs due to mechanical stress or change in physical location. Wheat meiocyte collection was kept under 1 h.
20. We break off maize stalks at one of the lowest internodes, bring them to the laboratory, break or cut off again at the new lowest internode immersing them simultaneously or immediately afterward in a vase with water.
21. In case the meiotic stages from top, center and bottom sampling positions differ, resample the outer ones closer to the center until you identified the range along the main tassel where meiosis is synchronous between the upper florets.
22. Maize meiocytes will sink to the bottom while most of the anther wall pieces are floating on the top. In maize, there will be many little starch granules which will be diluted in a purification step. If there are too many, use fewer anthers for the next batch.
23. To position the tip of the microcapillary, look at it directly, and only then through the ocular lenses.
24. Apply slight air pressure whenever the microcapillary submerges or is in the drop. Otherwise, the microcapillary effect will lead to liquid being sucked into the microcapillary.
25. Important notes on collection:
    - Figure out what magnification works best for you. At 100×, the view area is larger but meiocytes are smaller. 100× is optimal for beginners and for instances when hands are not absolutely calm and steady. At 400×, the view area is

smaller, meiocytes big. The microscope stage has to be moved more but the collection quality will be better.

- Settle on a strategy when collecting, for example starting at the left top, then moving all the way to the right, going slightly down, then back left and so on.
- Check the height of the liquid dome in between and apply more PBS whenever needed.
- After collecting throughout the whole drop, use a 200 μl pipette (at a setting of ~20–50 μl) to cautiously but thoroughly resuspend the drop. Add more PBS to form a nice dome again, collect newly released meiocytes.
- Do not give up quickly—the technique requires some practice and patience (*see* **Note 30**). Also, be aware that different species, inbreds, and stages will result in different sampling efficiency. In 1 h, we routinely collected ~1000–1500 meiocytes in B73, ~3500 in CML228 but only ~500 in Mo17 (*see* Fig. 4b in [9]).

26. First, count single meiocytes for each collection differing in species, inbred line and meiotic stage. Approximate the average number per cluster for future collections. We collected mainly maize zygotene clusters, with ~15 intact meiocytes per cluster in B73, less in Mo17, and more in CML228. CML228 had moderate or broken as well as very big clusters which we then counted as 2 or 3, multiplying the final number with 15.
27. Correlations exist between the length of an anther and the corresponding meiotic stage. For example, in Chinese Spring wheat, anthers of size 0.7, 0.8, 0.9, 1.0, and 1.1 mm predominantly contain meiocytes in leptotene, zygotene, pachytene, diplotene, and diakinesis stages, respectively [10]. The anther size and meiotic stage correlation helps in reducing the time required to obtain high purity meiocytes and ensures the reproducibility of this technique.
28. In wheat, each spikelet contains multiple florets of different meiotic stages, and each floret generally contains three anthers. Within each locule of an anther, meiocytes occur in intact sporogenous archesporial columns (SACs) that are free of somatic cells.
29. In contrast to the maize protocol, where meiocyte clusters are collected when sunk down, the wheat protocol works optimally (i.e., less SACs break) when SACs are brought back up into the dome center by gentle agitation.
30. This might seem difficult in the beginning when collections still result in few meiocytes in a large volume of PBS. Practice and patience with finding the perfect microcapillary will solve this issue. Most important are a perfect microcapillary with a

diameter just slightly bigger than meiocytes, and constant gentle air pressure against the microcapillary effect of constantly sucking in liquid.

31. For lysed tissue, the sample should not exceed 10% of the final volume in TRIzol instead of the here recommended 25% for meiocytes. Collected meiocytes are mostly intact cells and should not release RNA-degrading enzymes as in the case of lysed cells.

32. For optimal reproducibility between samples, use populations of meiocytes from at least 3–5 plants per replicate.

33. We highly recommend TRIzol-based RNA extraction since it consistently yields high quality RNA when compared to alternative RNA extraction kits from various suppliers.

34. Consider that you need to use part of your RNA solution for quality control. Depending on the downstream requirements regarding amount and concentration, decide whether to aim for high concentration (by using only 6 μl for elution) or for higher remaining amount (by using 10–20 μl for elution). A compromise solution is to use 6 μl for elution, then diluting 1 μl of it with 2–4 μl water for use in quality control. If you are not bound by a maximum volume or need of high concentration, elute 2× with 6 μl to achieve the highest overall yield.

35. The RIN of both maize and wheat meiocytes mostly ranged from 2.5 to 5.0. This is not as high as RINs of other samples such as pollen and anthers which showed RINs of 8.0 or above. However, since the RIN is calculated by estimating the size ratios of ribosomal subunits [18], which are biologically purged in meiocyte samples [39], the RIN might not be an appropriate value to estimate the quality of isolated meiocytes RNA.

36. If your lab does not routinely do library preparations, it might be easier and more economical to outsource the library preparation to the sequencing center as well. We successfully used Illumina TruSeq library preparation for single-end reads (50–150 bp for RNA-seq, 36 bp for small RNA-seq). Other kits and sequencing parameters can be used, preferentially with low input requirement but moderate numbers of PCR cycles.

37. Independent of the tool for data processing, always make sure you understand the parameters used, even the default parameters. Document everything, including parameters and versions of software used.

38. There is also an IGV tool called "toTDF." It seems counterintuitive to use the "count" tool; the major difference is the input file format (*see* IGV documentation on https://software.broadinstitute.org/software/igv/igvtools).

39. While we favor the DEseq/DEseq2 package with TMM normalization, simple RPM values facilitates further addition of later data for comparison without the need for a fresh joint normalization run.
40. We found a vast difference in successfully aligned small RNA-seq reads between different alignment algorithms and parameters, ranging from ~40 to over 80% (*see* Table 1 in [3]).
41. If using the miRBase annotations, remove the miR genes before normalization. They are rather meant for use with mRNA-seq data.

Acknowledgments

This work was supported by the National Science Foundation (IOS: 1025881, IOS-1546792), a grant-in-aid fund from the University of Minnesota, and Genome Canada funded Canadian Triticum Applied Genomics^2 (CTAG^2) initiative.

References

1. Caryl AP, Jones GH, Franklin FCH (2003) Dissecting plant meiosis using Arabidopsis thaliana mutants. J Exp Bot 54:25–38
2. Dukowic-Schulze S, Harris A, Li J, Sundararajan A, Mudge J, Retzel E, Pawlowski W, Chen C (2014) Comparative transcriptomics of early meiosis in Arabidopsis and maize. J Genet Genomics 41:139–152
3. Dukowic-Schulze S, Sundararajan A, Ramaraj T, Kianian S, Pawlowski WP, Mudge J, Chen C (2016) Novel meiotic miRNAs and indications for a role of phasiRNAs in meiosis. Front Plant Sci 7:762
4. Sánchez-Morán E, Mercier R, Higgins JD, Armstrong SJ, Jones GH, Franklin FCH (2005) A strategy to investigate the plant meiotic proteome. Cytogenet Genome Res 109:181–189
5. Li J, Yuan J, Ji K, Zhang Y, Wang J, Zhang S, Liu S, Qiu P, Li N, Li M (2016) Small-scale isolation of meiocytes from Arabidopsis thaliana anthers for cytological analyses. Pak J Bot 48:1989–1992
6. Flórez-Zapata NMV, Reyes-Valdés MH, Hernandez-Godínez F, Martinez O (2014) Transcriptomic landscape of prophase I sunflower male meiocytes. Front Plant Sci 5:277
7. Collado-Romero M, Alós E, Prieto P (2014) Unravelling the proteomic profile of rice meiocytes during early meiosis. Front Plant Sci 5:356
8. Chen C, Retzel EF (2013) Analyzing the meiotic transcriptome using isolated meiocytes of Arabidopsis thaliana. In: Pawlowski WP, Grelon M, Armstrong S (eds) Plant meiosis. Humana Press, New York, pp 203–213
9. Dukowic-Schulze S, Sundararajan A, Ramaraj T, Mudge J, Chen C (2014) Sequencing-based large-scale genomics approaches with small numbers of isolated maize meiocytes. Front Plant Sci 5:57
10. Shunmugam ASK, Bollina V, Dukowic-Schulze S, Bhowmik PK, Ambrose C, Higgins JD, Pozniak C, Sharpe AG, Rozwadowski K, Kagale S (2018) MeioCapture: an efficient method for staging and isolation of meiocytes in the prophase I sub-stages of meiosis in wheat. BMC Plant Biol 18:293
11. Chen C, Farmer AD, Langley RJ, Mudge J, Crow JA, May GD, Huntley J, Smith AG, Retzel EF (2010) Meiosis-specific gene discovery in plants: RNA-Seq applied to isolated Arabidopsis male meiocytes. BMC Plant Biol 10:280
12. Chang P, Tseng Y-F, Chen P-Y, Wang C-JR (2018) Using flow cytometry to isolate maize meiocytes for next generation sequencing: a time and labor efficient method. Curr Protoc Plant Biol 3:e20068
13. Barra L, Aiese-Cigliano R, Cremona G, De Luca P, Zoppoli P, Bressan RA, Consiglio FM, Conicella C (2012) Transcription

profiling of laser microdissected microsporocytes in an Arabidopsis mutant (Atmcc1) with enhanced histone acetylation. J Plant Biol 55:281–289
14. Sims J, Chen C, Schlögelhofer P, Kurzbauer M-T Targeted analysis of chromatin events (TACE). Plant meiosis: methods and protocols
15. Hsu SY, Huang YC, Peterson PA (1988) Development pattern of microspores in Zea mays L.—the maturation of upper and lower florets of spikelets among an assortment of genotypes. Maydica 33:77–98
16. Bennett MD, Smith JB (1972) The effects of polyploidy on meiotic duration and pollen development in cereal anthers. Proc R Soc Lond B 181:81–107
17. Nan G-L, Zhai J, Arikit S, Morrow D, Fernandes J, Mai L, Nguyen N, Meyers BC, Walbot V (2017) MS23, a master basic helix-loop-helix factor, regulates the specification and development of the tapetum in maize. Development 144:163–172
18. Schroeder A, Mueller O, Stocker S, Salowsky R, Leiber M, Gassmann M, Lightfoot S, Menzel W, Granzow M, Ragg T (2006) The RIN: an RNA integrity number for assigning integrity values to RNA measurements. BMC Mol Biol 7:3
19. Mills JD, Kawahara Y, Janitz M (2013) Strand-specific RNA-seq provides greater resolution of transcriptome profiling. Curr Genomics 14:173–181
20. Afgan E, Baker D, van den Beek M et al (2016) The Galaxy platform for accessible, reproducible and collaborative biomedical analyses: 2016 update. Nucleic Acids Res 44:W3–W10
21. R Development Core Team (2008) R: a language and environment for statistical computing. R Foundation for Statistical Computing, Vienna, Austria
22. RStudio Team (2015) RStudio: integrated development for R. RStudio, Inc., Boston, MA
23. Love MI, Huber W, Anders S (2014) Moderated estimation of fold change and dispersion for RNA-seq data with DESeq2. Genome Biol 15:550
24. Robinson JT, Thorvaldsdóttir H, Winckler W, Guttman M, Lander ES, Getz G, Mesirov JP (2011) Integrative genomics viewer. Nat Biotechnol 29:24–26
25. Du Z, Zhou X, Ling Y, Zhang Z, Su Z (2010) agriGO: a GO analysis toolkit for the agricultural community. Nucleic Acids Res 38: W64–W70
26. Axtell MJ (2013) ShortStack: comprehensive annotation and quantification of small RNA genes. RNA 19:740–751
27. Pertea M, Pertea GM, Antonescu CM, Chang T-C, Mendell JT, Salzberg SL (2015) StringTie enables improved reconstruction of a transcriptome from RNA-seq reads. Nat Biotechnol 33:290–295
28. Bray NL, Pimentel H, Melsted P, Pachter L (2016) Near-optimal probabilistic RNA-seq quantification. Nat Biotechnol 34:525–527
29. Grabherr MG, Haas BJ, Yassour M et al (2011) Full-length transcriptome assembly from RNA-Seq data without a reference genome. Nat Biotechnol 29:644–652
30. Robinson MD, McCarthy DJ, Smyth GK (2010) edgeR: a Bioconductor package for differential expression analysis of digital gene expression data. Bioinformatics 26:139–140
31. Anders S, Huber W (2012) Differential expression of RNA-Seq data at the gene level – the DESeq package. EMBL, Heidelberg, Germany
32. Lin Y, Golovnina K, Chen Z-X, Lee HN, Negron YLS, Sultana H, Oliver B, Harbison ST (2016) Comparison of normalization and differential expression analyses using RNA-Seq data from 726 individual Drosophila melanogaster. BMC Genomics 17:28
33. Aanes H, Winata C, Moen LF, Østrup O, Mathavan S, Collas P, Rognes T, Aleström P (2014) Normalization of RNA-sequencing data from samples with varying mRNA levels. PLoS One 9:e89158
34. Sahraeian SME, Mohiyuddin M, Sebra R et al (2017) Gaining comprehensive biological insight into the transcriptome by performing a broad-spectrum RNA-seq analysis. Nat Commun 8:59
35. Axtell MJ (2014) Butter: high-precision genomic alignment of small RNA-seq data. bioRxiv 007427
36. Kozomara A, Griffiths-Jones S (2014) miRBase: annotating high confidence microRNAs using deep sequencing data. Nucleic Acids Res 42:D68–D73
37. Smedley D, Haider S, Ballester B, Holland R, London D, Thorisson G, Kasprzyk A (2009) BioMart – biological queries made easy. BMC Genomics 10:22
38. Begcy K, Dresselhaus T (2017) Tracking maize pollen development by the Leaf Collar Method. Plant Reprod 30:171–178
39. Mackenzie A, Heslop-Harrison J, Dickinson HG (1967) Elimination of ribosomes during meiotic prophase. Nature 215:997–999

Chapter 18

How to Study the Proteomes and Phosphoproteomes of Anther and Pollen

Chi Zhang, Xiaobing Feng, Menghui Hu, and Zaibao Zhang

Abstract

Proteomics analysis was a powerful technology for characterizing proteins and protein posttranslational modification (PTMs). Recently, many anther and pollen-related proteomic analyses have been reported, which have expanded our understanding of anther and pollen development and regulation. In this chapter, we describe a detailed, optimized protocol for the separation, digestion, tagging, and subsequent mass spectrometry-based identification and quantification of proteins and phosphoproteins from anther and pollen.

Key words Anther, Liquid chromatography–mass spectrometry (LC-MS/MS), Pollen, Proteome, Phosphoproteome

1 Introduction

In flowering plants, anther is the male reproductive organ that produces the pollen mother (meiotic) cells which generate the mature pollens by meiosis and mitosis [1]. Anther and pollen development have been widely studied based on genetic, molecular and transcriptional methods [1–4]. However, the global proteomic investigations in anthers and pollens are still limited.

Proteomics is the large-scale study and characterization of all proteins (proteome) in a cell, tissue or organism at a given time [5]. The aim of proteomics is to identify, characterize, quantify, and functional investigate of proteome [6]. The most widespread methods for the majority of ongoing proteome projects include gel-based approach (e.g., one-dimensional and two-dimensional polyacrylamide gel electrophoresis) and gel-free approach [7–10]. In recent years, with great advances in proteomic technologies (e.g., gas chromatography–mass spectrometry), many anther and pollen proteomic and phosphoproteomic studies were analyzed, which increased the understanding of physiological functions and signaling networks in anthers and pollens at protein levels [11–15].

Mónica Pradillo and Stefan Heckmann (eds.), *Plant Meiosis: Methods and Protocols*, Methods in Molecular Biology, vol. 2061, https://doi.org/10.1007/978-1-4939-9818-0_18, © Springer Science+Business Media, LLC, part of Springer Nature 2020

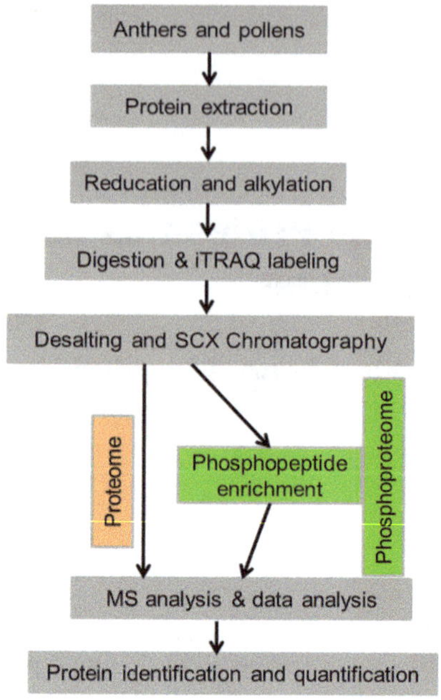

Fig. 1 An overview of anther proteomic and phosphoproteomic strategies

In addition to the standard mass spectrometry (MS)-based protein identification, MS-based quantification of proteins is rapidly developed in proteomic studies. In this chapter, we describe in detail the MS-based quantification of anther and pollen proteomic and phosphoproteomic workflow generally followed in our laboratory. It comprises protein extraction, digestion and isobaric tags for relative and absolute quantification (iTRAQ) labeling, peptide fractionation, enrichment of phosphorylated peptides, and liquid chromatography–mass spectrometry (LC-MS)/MS analysis of the labeled peptides for protein identification and subsequent quantification (Fig. 1).

2 Materials

2.1 Protein Extraction

A leica stereoscopic microscope (Leica Microsystems Ltd., Wetzlar, Germany) with a digital camera.

Potter homogenizer (Fisher Scientific).

Ultrapure water (LiChrosolv®).

10% (w/v) trichloroacetic acid/acetone solution containing 65 mM dithiothreitol (DTT).

SDT buffer: 4% SDS, 100 mM DTT, and 150 mM Tris–HCl, pH 8.0.

BCA protein assay reagent (Promega).

2.2 Protein Reducation, Alkylation, and Digestion

5 mM DTT.

15 mM iodoacetamide.

100 mM ammonium bicarbonate.

Trypsin.

2.3 iTRAQ Labeling

iTRAQ™ reagents (Applied Biosystems).

50 mM glycine.

100% ethanol.

Thermomixer.

2.4 Desalting

Poros R3 microcolumn (Framingham).

ACN.

TFA.

IMAC loading solution: 50% ACN and 0.1% TFA.

2.5 SCX Chromatography

AKTA Purifier system (GE Healthcare).

Buffer A: 10 mM KH_2PO_4 in 25% of ACN, pH 2.7.

Buffer B: 500 mM KCl, 10 mM KH_2PO_4 in 25% of ACN, pH 2.7.

PolySULFOETHYL 4.6 × 100 mm column (5 μm, 200 Å, PolyLC Inc.).

C18 Cartridges (Empore™ SPE Cartridges C18 (standard density), bed I.D. 7 mm, volume 3 mL, Sigma).

2.6 Phosphopeptides Enrichment by TiO_2 Beads

TiO_2 loading buffer: 2% glutamic acid, 65% ACN, 2% TFA.

Washing buffer I: 30% ACN, 3% TFA.

Washing buffer II: 80% ACN, 0.3% TFA.

Elution buffer: 40% ACN, 15% NH_4OH.

2.7 LC-MS/MS Analysis

Q Exactive MS (Thermo Finnigan).

Buffer B: 80% ACN and 0.1% formic acid.

The routine solvents and buffers as well as the routine settings in the MS can be used (*see* **Note 1**).

3 Methods

3.1 Anther Collection and Protein Extraction

1. Anther developmental stages of Arabidopsis and rice are determined according to previous cytological description [16, 17].
2. For quantification analysis, at least 10,000 Arabidopsis anthers and 5000 rice anthers are isolated under a stereoscopic

microscope, immediately frozen in liquid nitrogen and stored at −80 °C until use.

3. Mature pollens are collected with a handheld vacuum cleaner. About 50 mg dry pollens are collected over 4 weeks and stored at −80 °C.
4. Anther and pollen tissues are homogenized in liquid nitrogen with a Potter homogenizer. Then the powders are precipitated in a 10% (w/v) trichloroacetic acid/acetone solution containing 65 mM dithiothreitol (DTT) at −20 °C for 1 h.
5. Spin down for 45 min, 10,000 × *g*, 4 °C.
6. Discard the supernatant, dry pellets in vacuum centrifuge and dissolve samples in 30 μL SDT buffer.
7. Heat samples in boiling water for 3 min, then the suspensions are ultrasonicated (80 W for ten times at 10 s, with a 15 s interval).
8. Incubate the sample at 100 °C for 3 min.
9. Spin down for 10 min, 13,000 × *g*, 25 °C.
10. Determine the protein concentration with the BCA Protein Assay Reagent and stored at −80 °C until use.

3.2 Protein Reducation, Alkylation, and Digestion

1. 200 μg of the protein solution is reduced by 5 mM DTT at 37 °C for 1 h.
2. Then 15 mM iodoacetamide is added for alkylation in dark at room temperature for 1 h. The excess iodoacetamide is quenched by the addition of 5 mM DTT.
3. Eightfold dilution of the solution with 100 mM ammonium bicarbonate.
4. Digest overnight at 37 °C with 2.5% (w/w) trypsin, and the resulting peptides are collected as a filtrate (*see* **Note 2**).

3.3 iTRAQ Labeling

Labeling of peptides with iTRAQ™ reagents. Briefly, one vial of iTRAQ labeling reagent is used for every 100 μg of proteins.

1. Solubilize each iTRAQ™ reagent in 70 μL ethanol (vortex).
2. Add to each peptide sample 5 μL of the different iTRAQ™ reagents, resulting in a final concentration of 25% (v/v) ethanol.
3. Incubate for 1 h in a thermomixer at 500 rpm and room temperature.
4. Quench excess of reagent by adding 50 mM glycine, resulting in a final concentration of 10 mM glycine.
5. Incubate for 30 min in a thermomixer at 500 rpm and room temperature.
6. Mix the samples at equal ratios and dry by centrifugal evaporation.

3.4 Desalting

1. Sample desalting is carried out using Poros R3 microcolumns according to the supplier's recommendations (*see* **Note 3**).
2. Wash the R3 microcolumn with 100% ACN, repeat three times.
3. Wash the R3 microcolumn with 0.1% TFA, repeat three times.
4. Dissolve the labeled peptides with 400 μL 0.1% TFA, and load the sample onto the R3 microcolumn.
5. Wash the R3 microcolumn with 0.1% TFA, repeat three to five times.
6. Elute the sample with IMAC loading solution.
7. Concentrate the solution by vacuum centrifugation and store at −80 °C.

3.5 Strong Cation Exchange (SCX) Chromatography

iTRAQ labeled peptides were fractionated by SCX chromatography using the AKTA Purifier system.

1. The dried peptide mixtures are reconstituted and acidified with 2 mL buffer A.
2. Load the sample onto a PolySULFOETHYL 4.6 × 100 mm column.
3. Elute the peptides at a flow rate of 1 mL/min with a gradient of 0–10% buffer B for 2 min, 10–20% buffer B for 25 min, 20–45% buffer B for 5 min, and 50–100% buffer B for 5 min.
4. Monitor the elution by absorbance at 214 nm, and collect the fractions every 1 min.
5. Combine the collected fractions (about 30 fractions) into ten pools and desalt on C18 Cartridges.
6. Concentrated by vacuum centrifugation and reconstituted in 40 μL of 0.1% TFA.
7. All samples are stored at −80 °C until LC-MS/MS analysis (*see* **Note 4**).

3.6 Enrichment of Phosphorylated Peptides by TiO_2 Beads

1. Phosphopeptides are enriched using TiO_2 microcolumn according to the supplier's recommendations [18].
2. Resuspend the peptides in 500 μL TiO_2 loading buffer.
3. Add TiO_2 beads and agitate for 40 min.
4. Centrifugate for 1 min at 5000 × *g*, resulting with the first set of beads.
5. Mix the supernatant from the first centrifugation with another set of TiO_2 beads, centrifugate for 1 min at 5000 × *g*, resulting with the second set of beads.
6. Combine both sets of beads and wash the beads with 50 μL of washing buffer I three times.

7. Wash the beads with 50 μL of washing buffer II three times to remove the remaining nonadsorbed material.
8. Elute the phosphopeptides with 50 μL of elution buffer.
9. The samples are stored at −80 °C until LC-MS/MS analysis.

3.7 Standard LC-MS/MS Analysis

1. LC-MS/MS analysis is performed using a Q Exactive MS (Thermo Finnigan) equipped with Easy nLC (ThermoFisher).
2. LC systems should be equipped with a trapping column (RP C18) so that the sample is extensively desalted (several minutes with elevated flow rates, e.g., 10 μL/min) to remove the excess of iTRAQ reagent.
3. Separation of labeled peptides by standard "peptide mapping" RP-C18 reversed phase column with a flow rate of 250 nL/min controlled by IntelliFlow technology over 240 min.
4. Elute the peptides with a gradient of 0–60% buffer B from 0 to 200 min, 60–100% buffer B from 200 to 216 min, 100% buffer B from 216 to 240 min.
5. For MS analysis, peptides are analyzed in positive ion mode. MS spectra are acquired using a data-dependent top10 method dynamically choosing the most abundant precursor ions from the survey scan (300–1800 *m/z*) for HCD fragmentation.
6. To analyze the iTRAQ data, Mascot 2.2 is used. For protein identification, the following options are used: Peptide mass tolerance = 20 ppm, MS/MS tolerance = 0.1 Da, Enzyme = Trypsin, Missed cleavage = 2, Fixed modification: Carbamidomethyl (C), iTRAQ8plex (K), iTRAQ8plex (N-term), Variable modification: Oxidation (M), Phosphorylation (S/T/Y), FDR ≤0.01 (*see* **Note 5**).

4 Notes

1. All buffers and solutions should be prepared with p.a. grade water and solvents. All buffers for protein extraction, protein digestion, desalting, and iTRAQ-labeling are prepared fresh prior to use.
2. Trypsin is the preferred endoproteinase for protein digestion. It cleaves proteins specifically at the C-terminal side of arginine and lysine residues. Note that the pH value should be around 8 and is checked before each digestion step.
3. To minimize sample loss, the length of the column is varied depending on the sample amount (approximately 1 μL bed volume of R3 material per 10 μg of sample).
4. The fractionated samples are ready for proteome analysis by LC-MS/MS.

5. The iTRAQ labeling efficiency is checked by searching the fragment spectra with iTRAQ as a variable modification. The labeling efficiency must be >90%.

References

1. Ma H (2005) Molecular genetic analyses of microsporogenesis and microgametogenesis in flowering plants. Annu Rev Plant Biol 56:393–434
2. Chang F, Wang Y, Wang S, Ma H (2011) Molecular control of microsporogenesis in Arabidopsis. Curr Opin Plant Biol 14:66–73
3. Huang MD, Hsing YI, Huang AH (2011) Transcriptomes of the anther sporophyte: availability and uses. Plant Cell Physiol 52:1459–1466
4. Sanders PM, Bui AQ, Weterings K, Mcintire KN, Hsu YC, Pei YL et al (1999) Anther developmental defects in Arabidopsis thaliana male-sterile mutants. Sex Plant Reprod 11:297–322
5. Wilkins MR, Sanchez JC, Gooley AA, Appel RD, Humphery-Smith I, Hochstrasser DF et al (1996) Progress with proteome projects: why all proteins expressed by a genome should be identified and how to do it. Biotechnol Genet Eng Rev 13:19–50
6. Pandey A, Mann M (2000) Proteomics to study genes and genomes. Nature 405:837–846
7. Gevaert K, Van Damme J, Goethals M, Thomas GR, Hoorelbeke B, Demol H et al (2002) Chromatographic isolation of methionine-containing peptides for gel-free proteome analysis: identification of more than 800 Escherichia coli proteins. Mol Cell Proteomics 1:896–903
8. Roe MR, Griffin TJ (2006) Gel-free mass spectrometry-based high throughput proteomics: tools for studying biological response of proteins and proteomes. Proteomics 6:4678–4687
9. Van den Bergh G, Arckens L (2005) Recent advances in 2D electrophoresis: an array of possibilities. Expert Rev Proteomics 2:243–252
10. Vercauteren FG, Bergeron JJ, Vandesande F, Arckens L, Quirion R (2004) Proteomic approaches in brain research and neuropharmacology. Eur J Pharmacol 500:385–398
11. Grobei MA, Qeli E, Brunner E, Rehrauer H, Zhang R, Roschitzki B, Basler K, Ahrens CH, Grossniklaus U (2009) Deterministic protein inference for shotgun proteomics data provides new insights into Arabidopsis pollen development and function. Genome Res 19:1786–1800
12. Mayank P, Grossman J, Wuest S, Boisson-Dernier A, Roschitzki B, Nanni P, Nuhse T, Grossniklaus U (2012) Characterization of the phosphoproteome of mature Arabidopsis pollen. Plant J 72:89–101
13. Ye J, Zhang Z, Long H, Zhang Z, Hong Y, Zhang X et al (2015) Proteomic and phosphoproteomic analyses reveal extensive phosphorylation of regulatory proteins in developing rice anthers. Plant J 84:527–544
14. Ye J, Zhang Z, You C, Zhang X, Lu J, Ma H (2016) Abundant protein phosphorylation potentially regulates Arabidopsis anther development. J Exp Bot 67:4993–5008
15. Zhang Z, Hu M, Feng X, Gong A, Cheng L, Yuan H (2017) Proteomes and phosphoproteomes of anther and pollen: availability and progress. Proteomics 17(20). https://doi.org/10.1002/pmic.201600458
16. Wijeratne AJ, Zhang W, Sun Y, Liu W, Albert R, Zheng Z et al (2007) Differential gene expression in Arabidopsis wild-type and mutant anthers: insights into anther cell differentiation and regulatory networks. Plant J 52:14–29
17. Zhang D, Luo X, Zhu L (2011) Cytological analysis and genetic control of rice anther development. J Genet Genomics 38:379–390
18. Larsen MR, Thingholm TE, Jensen ON, Roepstorff P, Jorgensen TJ (2005) Highly selective enrichment of phosphorylated peptides from peptide mixtures using titanium dioxide microcolumns. Mol Cell Proteomics 4:873–886

Chapter 19

Rice Female Meiosis: Genome-Wide mRNA, Small RNA, and DNA Methylation Analysis During Ovule Development

Helian Liu, Aqin Cao, Liyu Yang, and Jianbo Wang

Abstract

Meiosis is an essential process in sexual life cycle, not only for the genomic stability maintenance but also for the genetic diversity creation through recombination. In rice ovule, megaspore mother cells undergo meiosis to form megaspores; then the functional megaspore performs three rounds of mitoses to form female gametophyte. However, the mechanism of gene expression and regulation in female meiosis process is still poorly understood. As important gene regulatory factors, miRNAs and DNA methylation are widely involved in plant meiosis and ovule development. In order to systematically study the potential mechanism of gene expression and regulation in female meiosis, ovules at megaspore mother cell meiosis stage, functional megaspore mitosis stage, and mature female gametophytes are collected to perform genome-wide RNA sequencing, small RNA sequencing, and bisulfite sequencing. Through bioinformatics analysis, we obtained many differentially expressed genes, miRNAs, and differentially methylated genes related to female meiosis. These data may provide important clues for further revealing the mechanism of female meiosis in rice.

Key words Rice, Female meiosis, Ovule development, RNA sequencing, Small RNA sequencing, Bisulfite sequencing

1 Introduction

Meiosis is the division of a cell in which the number of chromosomes halved when producing mature germ cells in sexual reproductive organisms. In the rice ovule, an archesporial cell differentiates under the nucellus epidermis and develops into a megaspore mother cell (MMC). The MMC undergoes meiosis to form four megaspores, and this process is called megasporogenesis. Subsequently, three megaspores are degraded and the remaining one develops into the functional megaspore, which then forms the female gametophyte with eight nuclei through mitosis [1].

Related studies have shown that the female meiosis is the result of the precise control of many genes [2–5]. Ovule sporophytic tissues act as a place for megasporogenesis to provide mechanical

Mónica Pradillo and Stefan Heckmann (eds.), *Plant Meiosis: Methods and Protocols*, Methods in Molecular Biology, vol. 2061, https://doi.org/10.1007/978-1-4939-9818-0_19,

support and delivery of nutrients, and the change of any related gene may affect the process of the female meiosis and eventually lead to the abnormity of the female reproduction [6]. Therefore, the transcriptome analysis of developing ovules is important for detecting the genes related to female meiosis.

Plant miRNAs are a class of single-stranded noncoding small RNAs widely involved in multiple processes of plant growth and development, including reproductive growth [7–9]. They often act as important negative regulators to alter gene expression by cutting target mRNA or translational repression. At present, the miRNA expression detection methods mainly include northern blot, real-time quantitative PCR, miRNA chip, and miRNA high-throughput sequencing. Among them, the high-throughput sequencing technology greatly improves the efficiency and accuracy of miRNA detection and promotes the miRNA research process [8, 10].

DNA methylation is a common epigenetic phenomenon of eukaryotic genomes and is widely involved in gene expression, silencing and activation of transposons, chromatin remodeling, and maintenance of genomic stability [11, 12]. Currently, whole-genome DNA methylation detection methods mainly include three kinds: methylation-sensitive amplification polymorphism (MSAP), methylated DNA immunoprecipitation sequencing (MeDIP-seq), and whole-genome bisulfite sequencing (WGBS). Among them, WGBS realizes genome-wide single-base resolution and is called the gold standard of DNA methylation detection.

Studies on the gene expression and regulation of plant male meiosis in miRNA or DNA methylation have been performed in many plants [13–15]. However, research on the gene expression and regulation of female meiosis in miRNAs or DNA methylation is scarce. The rapid development of sequencing technology makes it possible to understand the mechanism of female meiosis systematically and deeply. Therefore, genome-wide RNA sequencing, small RNA sequencing, and bisulfite sequencing are performed in our study. All these data provide abundant clues to explore the female meiosis-related genes and their regulatory networks in rice.

2 Materials

2.1 General Materials

1. ddH_2O.
2. Ethanol (100%).
3. dNTP (10 mM).
4. 2× HotStart Taq PCR Master Mix.
5. DL 2000 DNA marker.
6. Agarose.

7. 1× TAE buffer: 0.04 M Tris–acetic acid, 0.001 M EDTA, pH 8.0.
8. 6× loading buffer.
9. Clean bench.
10. Frozen high-speed centrifuge.
11. PCR amplifier.
12. Electrophoresis apparatus.
13. Agilent Bioanalyzer 2100.
14. Molecular Imager Gel Doc.
15. Autoclaves sterilizers.
16. RNase-free PCR tube.
17. RNase-free centrifuge tube.
18. Pipette.
19. Electronic balance.

2.2 Collection of Ovule Tissue

1. 0.1% DEPC (*see* **Note 1**).
2. CPW buffer: 10 mM $CaCl_2$, 0.7 mM KH_2PO_4, 1.6 mM $MgSO_4$, 0.5 mM D-Mannitol, 0.1% DEPC, adjust the pH to 5.8 with KOH or HCl, and sterilize for 15 min at 121 °C (*see* **Note 2**).
3. Dissecting needle.
4. Stereoscopic microscope.

2.3 Extraction of Total RNAs

1. TRIzol reagent.
2. Trichloromethane.
3. Isopropyl alcohol.
4. 0.1% DEPC.
5. 75% ethanol: 100% ethanol, 0.1% DEPC.
6. RNase-free DNase I and 10× DNase I buffer.
7. EDTA.

2.4 Reverse Transcription of mRNA and Small RNA

1. M-MLV RT (200 U).
2. MLV 5× Reaction buffer.
3. RNasin ribonuclease inhibitor.
4. Oligo (dT).
5. Chloroform/isoamyl alcohol (24/1).
6. 3 M NaAc: 40.8 g $NaAc \cdot 3H_2O$, 40 mL 0.1% DEPC, adjust the pH to 5.2 with glacial acetic acid, add 0.1% DEPC to a total volume of 100 mL, sterilize at 121 °C for 15 min, and store at 4 °C.
7. Small RNA-specific reverse transcription primer (10 μM/L).

2.5 RNA-Seq and Small RNA-Seq

1. First strand master mix.
2. SuperScript II reverse transcriptase.
3. Agencourt RNAClean XP Beads.
4. Second strand master mix.
5. End repair mix.
6. A-tailing mix.
7. Adenylate 3′ends DNA.
8. Index adapter.
9. Ligation mix.
10. Uracil-N-glycosylase enzyme.
11. 14–30 ssRNA ladder marker.
12. PAGE gel.
13. Ethidium bromide (10 mg/mL).
14. PCR primer cocktail.
15. Ampure XP beads.
16. TruSeq PECluster kit V3-cBot-HS.
17. Agilent RNA 6000 Nano Kit.
18. ABI StepOnePlus Real-Time PCR System.
19. TruSeq SBS KIT-HS V3.

2.6 DNA Library Preparation

1. Plant genomic DNA kit.
2. Ultraviolet spectrophotometer.
3. AIR DNA fragment kit.
4. End-It™ DNA end-repair kit.
5. Ampure XP magnetic beads.
6. QIAquick PCR purification kit.
7. 10× Klenow buffer.
8. 2 mM dATP.
9. Klenow exo-3′ to 5′ exo minus.
10. 10× DNA Ligation buffer.
11. Methylation adapter.
12. T4 DNA ligase.
13. Zymo EZ DNA methylation lightning kit.
14. KAPA HiFi HotStart Uracil+ Ready Mix (2×).
15. NEXTflex™ Primer Mix.
16. 80% ethanol.
17. Qubit hsDNA.
18. Illumina HiSeq4000 Platform.

2.7 Real-Time Quantitative PCR

1. 2× Taq PCR master mix.
2. SYBR® qPCR mix.
3. 50× ROX.
4. Forward primer (10 μM/L).
5. Reverse primer (10 μM/L).
6. ABI StepOnePlus Real-Time PCR System.

2.8 Bisulfite Sequencing PCR

1. Plant genomic DNA kit.
2. Ultraviolet spectrophotometer.
3. EZ DNA methylation gold kit.
4. Methyl Primer Express v1.0.
5. Forward primer (10 μM/L).
6. Reverse primer (10 μM/L).
7. AxyPrep DNA gel extraction kit.
8. *pEASY*®-T3 cloning kit.
9. Constant temperature oscillator.

3 Methods

3.1 Collection of Ovule Tissue

1. Based on the correspondence between floret morphology and the development of female gametophyte in rice, the florets containing developing female gametophytes at meiosis stage, mitosis stage, and maturation stage are selected.
2. Sterilize the stereomicroscope with 75% ethanol, dry and take a clean slide in the operating area of the stereomicroscope, then place the centrifuge tube with 1 mL of CPW buffer vertically into the ice tray. In addition, the operator needs to wear sterile mask and gloves, and the dissection needle is presterilized and is ensured RNase free (*see* **Note 3**).
3. Place the floret on the slide and find the proper field of vision, securing the floret with one dissecting needle and tearing one of the lemmas to expose the ovary with the other dissecting needle.
4. Immediately afterward, hold the concavity of the remaining one lemma with one of the dissecting needles and dissect gently across the surface of the ovary to completely remove the ovule with the other needle (*see* **Note 4**).
5. Put the ovule into the 1 mL CPW buffer rapidly and collect about 200 ovules; centrifuge at 5729 × *g* rpm for 5 min, 4 °C, aspirate the CPW buffer, and then perform the RNA or DNA extraction.

3.2 Extraction of Total RNAs

All tools used in this process should be RNase-free, and all the solutions should be prepared with RNase-free water (0.1% DEPC).

1. Total RNA is isolated by mixing 0.1 g ground ovule tissue and 1 mL TRIzol reagent and by allowing to stand for 5 min in a 1.5 mL centrifuge tube at room temperature (*see* **Note 5**).
2. Add 0.2 mL trichloromethane, allow to stand for 6 min after shaking for 15 s, and then centrifuge at 12,891 × *g* rpm for 15 min, 4 °C.
3. Transfer the supernatant into another 1.5 mL centrifuge tube and repeat the second step.
4. Transfer the supernatant into another 1.5 mL centrifuge tube and add 0.5 mL isopropyl alcohol (−20 °C precooled); after mixing well the mixture is stored at −20 °C for more than 30 min.
5. Take out the liquid mixture and centrifuge at 8952 × *g*–12,891 × *g* for 10 min at 4 °C.
6. After pouring out the supernatant, the precipitation is washed with 1 mL 75% ethanol and centrifuged at 5729 × *g* for 5 min at 4 °C. (This step needs to be repeated once again.)
7. After pouring out the supernatant, the precipitation is naturally air-dried in the ice box for 5–10 min and dissolved in 30–35 μL 0.1% DEPC.
8. DNase digestion: Mix 16 μL RNA, 2 μL 10× buffer, and 2 μL DNase I in a RNA-free PCR tube and incubate it at 37 °C for 30 min, then add 2 μL EDTA to stop the reaction at 65 °C for 10 min.
9. RNA quality testing: RNA concentration, RIN value, and 28S/18S ratio of total RNA are measured by Agilent Bioanalyzer 2100; finally, qualified RNAs are stored at −80 °C for further use.

3.3 Reverse Transcription of mRNA

1. Mix 12 μL RNA and 1 μL Oligo (dT) in a RNase-free PCR tube at 70 °C for 5 min, then immediately put the tube in ice box for 5 min.
2. Mix gently 5 μL MLV 5× Reaction Buffer, 1 μL M-MLV RT, 1.25 μL dNTP, 0.625 μL RNasin Ribonuclease inhibitor, and 3.75 μL 0.1% DEPC in the tube, and incubate at 42 °C for 60 min, then at 75 °C for 15 min; finally, synthetic cDNAs are stored at −20 °C for further use.

3.4 Reverse Transcription of Small RNA

1. DNase-digested RNAs are transferred to an RNase-free centrifuge tube and expanded to 600 μL by adding 0.1% DEPC, then 600 μL chloroform–isoamyl alcohol (24:1) is added. After considerable oscillation for 15 s, the mixture is incubated in an ice bath for 10 min.

2. After centrifuging at 15,129 × *g*, 4 °C, for 10 min, the supernatant (about 300 μL) is transferred into another new RNase-free centrifuge tube and mixed with 30 μL NaAc (1/10 of the supernatant) and 900 μL 100% ethanol. Store the mixture at −20 °C for more than 2 h.
3. The following procedures are the same as **steps 5–8** of Subheading 3.2.
4. 12 μL RNAs and 1 μL of specific reverse transcription primer of one small RNA with stem-loop structure are mixed in a RNA-free PCR tube at 70 °C for 5 min, then immediately put the tube in the ice box for 5 min.
5. Add 5 μL MLV 5× Reaction Buffer, 1 μL M-MLV RT, 1.25 μL dNTP, 0.625 μL RNasin ribonuclease inhibitor, and 3.75 μL 0.1% DEPC, and mix gently in the tube. The tube is incubated at 42 °C for 60 min, then at 75 °C for 15 min; finally, synthetic cDNA is stored at −20 °C for further use.

3.5 RNA Sequencing and Bioinformatics Analysis

1. The concentration and sample integrity of RNAs are tested by Agilent 2100 Bioanalyzer.
2. Total RNA is heated to 65 °C to open the secondary structure, and mRNAs are enriched by Oligo (dT) magnetic beads, then the enriched mRNAs are fragmented using First Strand Master Mix.
3. First-strand cDNA is synthesized using First Strand Master Mix and Super Script II reverse transcriptase (Reaction condition: 25 °C for 10 min; 42 °C for 50 min; 70 °C for 15 min). Purify the product using Agencourt RNAClean XP Beads, then add Second Strand Master Mix and dNTP mix to synthesize the second-strand cDNA (16 °C for 1 h) (*see* **Note 6**).
4. The purified cDNA is mixed with End Repair Mix and incubated at 30 °C for 30 min. Then combine the Adenylate 3′Ends DNA, Index Adapter, and Ligation Mix in the mixture with A-Tailing and incubate at 30 °C for 10 min.
5. Add uracil-N-glycosylase enzyme to the purified ligation product and incubate at 37 °C for 10 min. Several rounds of PCR amplification (*see* **Note** 7) with PCR Primer Cocktail and PCR Master Mix are performed to enrich the cDNA fragments. Then the PCR products are purified with Ampure XP Beads.
6. Verify the average length and quality of cDNA library by Agilent 2100 Bioanalyzer and RT-qPCR (TaqMan Probe).
7. The qualified libraries are amplified on cBot (TruSeq PE Cluster Kit V3–cBot–HS, Illumina) to generate the cluster on the flowcell, and the amplified flowcell is sequenced pair end on the HiSeq 2000 System (TruSeq SBS KIT-HS V3).

8. The gene expression is evaluated by assigning reads to the reference genomes using ERANGE software (version 4.0, http://woldlab.caltech.edu/gitweb/). The gene expression levels are normalized in terms of FPKM [16], and differential expression analysis is performed using R packages of DESeq with the threshold of foldchange ≥2 and FDR ≤0.001 [17].
9. GO (Gene ontology) annotation (http://www.blast2go.org/) is performed with the Blast2GO program (version 2.3.5). Web gene ontology annotation plot (WEGO) server (http://wego.genomics.org.cn/cgi-bin/wego/index.pl) is utilized for GO classification. For pathway analysis, KEGG (Kyoto Encyclopedia of Genes and Genomes) and MapMan database are employed to seek out the significantly enriched pathways. KEGG pathway analysis is performed using Cytoscape software (version 2.6.2, http://www.cytoscape.org/) with the ClueGO plugin (http://www.ici.upmc.fr/cluego/cluegoDownload.shtml). The MapMan package (http://MapMan.gabipd.org) is performed to visualize more thoroughly the significantly enriched pathways (Fig. 1).

3.6 Small RNA Sequencing and Bioinformatics Analysis

1. The integrity, concentration and NanoDrop to inorganic ions or polycarbonate contamination of total RNAs are tested using Agilent 2100 Bioanalyzer.
2. 5–10 μg RNAs with different fragment sizes are isolated from total RNA using PAGE gel; then small RNAs of size between 18 and 30 nt are recycled.
3. The 5′ linkers and 3′ linkers are added to the corresponding ligation system and incubated at 20 °C for 6 h, and then ligation products are separated and recycled by PAGE gel (The ligation products of 5′ adaptors and 3′ adaptors select 40–60 nt stripe and 60–80 nt stripe, respectively).
4. First Strand Master Mix and SuperScript II reverse transcriptase are added and incubated under the following reaction conditions: 65 °C for 10 min, 48 °C for 3 min, 42 °C for 1 h, 70 °C for 15 min.
5. Several rounds of PCR amplification (*see* **Note 7**) with PCR Primer Cocktail and PCR Mix are performed; then the cDNA fragments are enriched and purified with PAGE gel by selecting ~110 bp PCR product and dissolving the recycled products in EB solution.
6. The average molecule length and quality of cDNA libraries are verified using Agilent 2100 Bioanalyzer and RT-qPCR (EvaGreen).
7. The qualified libraries are amplified on cBot (TruSeq SE Cluster Kit V3–cBot–HS) to generate the cluster on the flowcell,

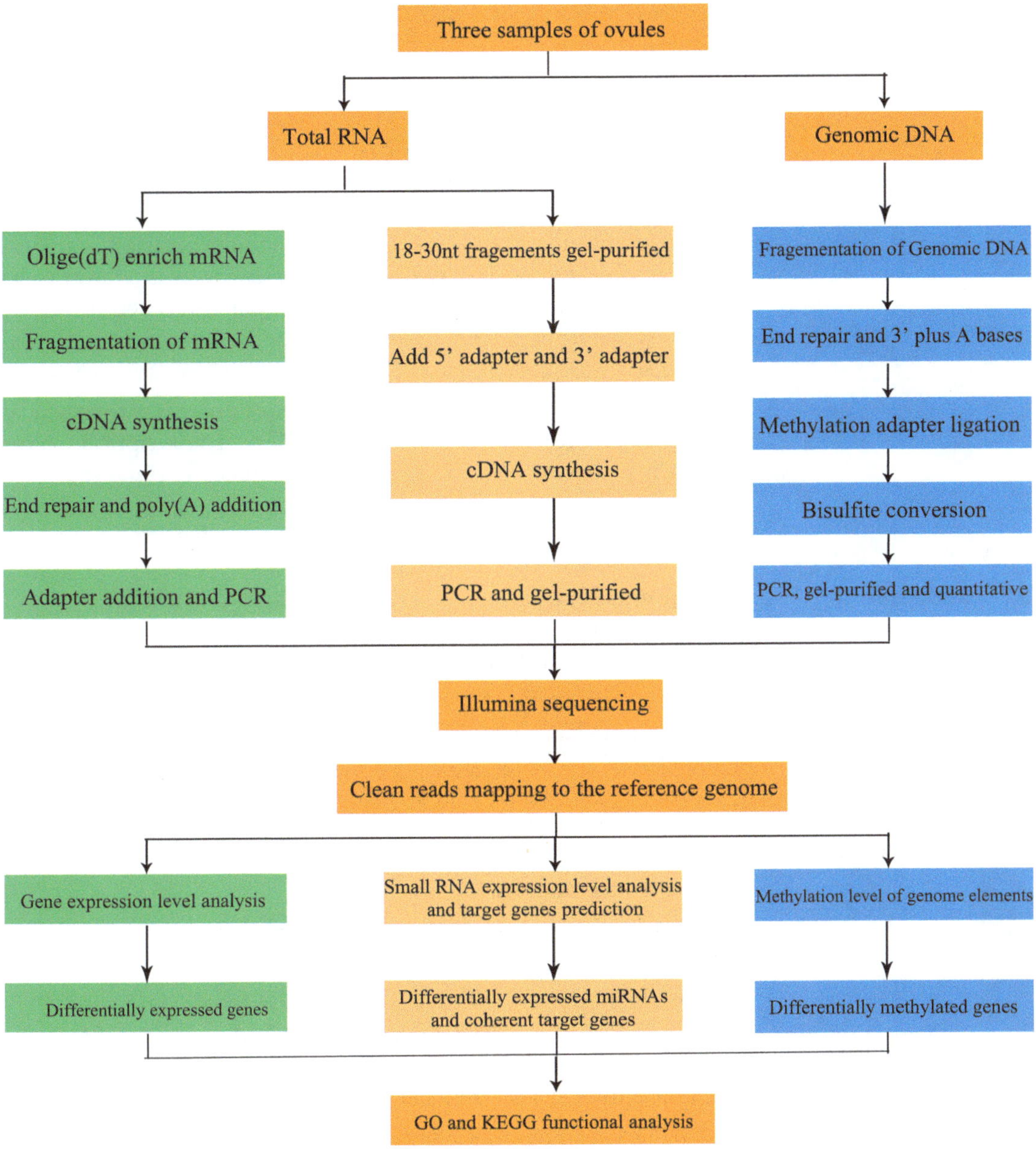

Fig. 1 Workflow of RNA sequencing, small RNA sequencing, and bisulfite sequencing

and the amplified flowcell is sequenced single end on the HiSeq 2000 System (TruSeq SBS KIT-HS V3).

8. The read count of each identified miRNA is normalized to transcripts per million (TPM) using the formula: TPM = actual miRNA count/total count of clean reads × 1,000,000. After normalization, the expression of miRNAs that are not expressed in one of all samples is set to 0.01, and as long as the miRNA expression is <1 in both samples, the differential

expression analysis (log2 Ratio $\geq$ 1 and $P \leq 0.05$) is not performed.

9. miRNA target prediction is executed by aligning the sequences of miRNAs to the genome [18, 19], and two software (psRobot, http://omicslab.genetics.ac.cn/psRobot/ and Target Finder, http://targetfinder.org/) are employed to identify putative miRNA targets, with the default parameters. In addition, combined with information on gene expression, the predicted target gene with opposite expression trends relative to its corresponding miRNA in the same comparison is defined as the coherent target gene. The results are visualized using Cytoscape software (version 2.6.2, http://www.cytoscape.org/).
10. The GO analysis of coherent target genes process is the same as the ninth step in Subheading 3.5. The transcription factor family information of coherent target genes is obtained from the Plant Transcription Factor Database (http://planttfdb.cbi.pku.edu.cn/) (Fig. 1).

3.7 WGBS and Bioinformatics Analysis

1. Genomic DNA is broken into small fragments using The AIR® DNA Fragmentation Kit (Bioo Scientific Corporation) according to the manufacturer's instructions.
2. End repair of DNA fragments is performed using End-It™ DNA End-Repair Kit (Epicentre, USA) according to the manufacturer's instructions. 300–400 bp DNA fragments are screened out using Agencourt Ampure XP magnetic beads, and then QIAquick PCR Purification Kit (QIAGEN, German) is used to purify the DNA fragments according to the manufacturer's instructions.
3. 37 μL End-repaired DNA fragments, 5 μL 10× Klenow buffer, 5 μL 2 mM dATP, and 3 μL Klenow exo-3′ to 5′ exo minus are mix thoroughly to add 3′ end adenine bases to DNA fragments.
4. 31 μL DNA, 4 μL 10× DNA Ligation buffer, 2.5 μL Methylation adapter, and 2.5 μL T4 DNA Ligase are mixed to connect the methylation sequencing adapter.
5. The DNA fragments after adapter addition are subjected to bisulfite conversion using a Zymo EZ DNA Methylation Lightning Kit (Epigenetics, USA) according to the manufacturer's instructions.
6. Reaction system is as follows: 23 μL Bisulfite converted DNA, 25 μL KAPA HiFi HotStart Uracil+ Ready Mix (2×), 2 μL NEXTflex™ Primer Mix.

   The reaction programs: (a) 98 °C for 3 min. (b) 98 °C for 30 s. (c) 65 °C for 30 s. (d) 72 °C for 1 min. (e) 72 °C for 1 min. Twelve cycles of steps (**b**), (**c**), and (**d**) are performed.

7. Add 50 μL AMPure XP magnetic beads, place on the magnetic stand after thoroughly mixing for 5 min, and discard the supernatant. Add 200 μL 80% ethanol, stand for 30 s, suck out the ethanol, and repeat this step once again. Allow to stand at room temperature until the magnetic beads are dry, add 22 μL ddH_2O, remove the PCR tube from the magnetic stand, mix well, and allow to stand for 2 min. Put the PCR tube back on the magnetic stand for 5 min, and then transfer 20 μL supernatant to a new PCR tube. Take 1 μL of the remaining liquid and quantify with Qubit hsDNA.
8. The libraries are numbered and then sequenced with the Illumina 4000.
9. The raw reads filter out low-quality, unknown, and adapter-fouled reads to get clean reads, and then the clean reads are mapped to the rice reference genome (https://phytozome.jgi.doe.gov/pz/portal.html#!info?alias=Org_Osativa) using Bowtie 2 software (http://bowtie-bio.sourceforge.net/bowtie2/index.shtml). The method introduced by [20] is used to identify genome elements and calculate the methylation ratio of each genome element. Significant differences of methylation ratio in the same genome element between the two samples are tested using a chi-square test. The threshold FDR $\leq$0.001 and at least a 2.0-fold methylation ratio change in genome elements are set to screen significant differentially methylated regions (DMRs).
10. A gene is considered as a differentially methylated gene (DMG) if at least one DMR is located on that gene. The GO information of each DMG obtained by the Blast2GO software (version 2.3.5) is uploaded to the WEGO website (http://wego.genomics.org.cn/) and the KEGG website (http://www.genome.jp/kegg/) to perform functional analysis (Fig. 1).

3.8 Real-Time Quantitative PCR (RT-qPCR)

1. A mixture of 1 μL cDNA, 0.5 μL forward primer, 0.5 μL reverse primer, 10 μL 2× Taq PCR Master Mix, and 8 μL ddH_2O is subjected to PCR to validate primers.

   PCR reaction program: (a) 94 °C for 5 min. (b) 94 °C for 30 s. (c) 55 °C for 30 s. (d) 72 °C for 45 s. (e) 72 °C for 10 min. Forty cycles of steps (**b**), (**c**), and (**d**) are performed.
2. Amplification system: 1 μL cDNA, 0.5 μL Forward primer, 0.5 μL Reverse primer, 0.4 μL ROX, 10 μL SYBR qPCR Mix, and 7.6 μL ddH_2O.

   PCR reaction program: (a) 95 °C for 10 min. (b) 95 °C for 15 s. (c) 60 °C for 1 min. Forty cycles of steps (**b**) and (**c**) are performed.
3. The fold changes of nine genes/six miRNAs in different samples are performed by a relative standard curve. The results are

represented as the mean ± SD of three replicates after normalization; *Actin1* and the U6 snRNA are selected as internal control for gene and miRNA expression, respectively.

3.9 Bisulfite Sequencing PCR

1. Total genomic DNA is extracted using a Plant Genomic DNA Kit according to the manufacturer's instructions.
2. The extracted genomic DNA is tested for quality and concentration using agarose gel electrophoresis and UV spectrophotometer.
3. 1 μg genomic DNA is taken for bisulfite conversion using EZ DNA Methylation Gold Kit according to the manufacturer's instructions (*see* **Note 8**).
4. Specific primers are designed for rice target sequences by Methyl Primer Express v1.0, and then amplify the target sequences by PCR.
5. The reaction system of PCR amplification is as follows: 1 μL bisulfite-converted genomic DNA, 0.5 μL Forward primer, 0.5 μL Reverse primer, 5 μL 2× HotStart Taq PCR Master Mix, 3 μL ddH_2O.

   The following PCR amplification program is used: (a) 94 °C for 3 min. (b) 94 °C for 30 s. (c) 52 °C for 30 s. (d) 72 °C for 1 min. (e) 72 °C for 10 min. Thirty-eight cycles of steps (**b**), (**c**), and (**d**) are performed.
6. The PCR product is subjected to agarose gel electrophoresis, and the band corresponding to the length of the target fragments is excised for gel recovery using AxyPrep DNA Gel Extraction Kit according to the manufacturer's instructions.
7. The DNA fragments of recovery are cloned into the *pEASY*®-T3 cloning vector using *pEASY*®-T3 cloning Kit according to the manufacturer's instructions; 15 clones are selected and sequenced.
8. The 15 sequences and the original sequence are imported into the Kismeth website (http://katahdin.mssm.edu/kismeth); we can get the distribution of each methylation site in this sequence.

4 Notes

1. Add 1 mL diethyl pyrocarbonate (DEPC) to 1000 mL of ddH_2O and shake once every 2 h in fume hood, and then autoclave at 121 °C for 20 min after 1 day; cool and store at 4 °C.
2. If ovules are used for RNA extraction, 0.1% DEPC should be added to prevent RNA degradation. DNA extraction does not

need addition of 0.1% DEPC. Transfer CPW buffer to a 1.5 mL centrifuge tube in the laminar flow cabinet; it can be stored for a month at 4 °C.

3. The CPW buffer should always be kept on ice to keep the ovules fresh, but only for 2–3 h, so the collection of ovules must be rapid. Sterilize and remove RNase with DEPC–alcohol wipe gloves every 15 min.
4. Use dissecting needles to rip the ovary gently to maintain the integrity of the ovule.
5. The amount of sample and the amount of TRIzol should be added in proportion; otherwise it will affect the inhibitory effect of endogenous RNA enzyme and result in the degradation of RNA.
6. During the second strand cDNA synthesis, the RNA template is removed and dTTP is replaced by dUTP. The involvement of dUTP ensures that the second strand of cDNA unable to be amplified in subsequent flows for the polymerase is unable to extend across the dUTP site on the template.
7. PCR reaction program: (a) 98 °C for 30 s. (b) 98 °C for 10 s. (c) 72 °C for 15 s. (d) 72 °C for 10 min. 12–15 cycles of steps (**b**) and (**c**) are performed.
8. To optimize conversion efficiency and DNA concentration, we add about 1 μg genomic DNA each time.

Acknowledgments

This work was supported by the State Key Basic Research and Development Plan of China (2013CB126900).

References

1. Yang WC, Shi DQ, Chen YH (2010) Female gametophyte development in flowering plants. Annu Rev Plant Biol 61:89–108
2. Siddiqi I, Ganesh G, Grossniklaus U, Subbiah V (2000) The dyad gene is required for progression through female meiosis in Arabidopsis. Development 127:197–207
3. Barrell PJ, Grossniklaus U (2005) Confocal microscopy of whole ovules for analysis of reproductive development: the elongate1 mutant affects meiosis II. Plant J 43:309–320
4. Panoli AP, Ravi M, Sebastian J, Nishal B, Reddy TV, Marimuthu MPA et al (2006) *AtMND1* is required for hoMogous pairing during meiosis in Arabidopsis. BMC Mol Biol 7:24
5. Schmidt A, Wuest SE, Vijverberg K, Baroux C, Kleen D, Grossniklaus U (2011) Transcriptome analysis of the Arabidopsis megaspore mother cell uncovers the importance of RNA helicases for plant germline development. PLoS Biol 9:e1001155
6. Bencivenga S, Colombo L, Masiero S (2011) Cross talk between the sporophyte and the megagametophyte during ovule development. Sex Plant Reprod 24:113–121
7. Millar AA, Gubler F (2005) The Arabidopsis *GAMYB-like* genes, *MYB33* and *MYB65*, are microRNA-regulated genes that redundantly facilitate anther development. Plant Cell 17:705–721
8. Wei LQ, Yan LF, Wang T (2011) Deep sequencing on genome-wide scale reveals the

unique composition and expression patterns of microRNAs in developing pollen of *Oryza sativa*. Genome Biol 12:R53

9. Fei Q, Yang L, Liang W, Zhang D, Meyers BC (2016) Dynamic changes of small RNAs in rice spikelet development reveal specialized reproductive phasiRNA pathways. J Exp Bot 67:6037–6049
10. Chen ZX, Li FL, Yang SN, Dong YB, Yuan QH, Wang F et al (2013) Identification and functional analysis of flowering related microRNAs in common wild rice (*Oryza rufipogon* Griff.). PLoS One 8:e82844
11. Kim MY, Zilberman D (2014) DNA methylation as a system of plant genomic immunity. Trends Plant Sci 19:320–326
12. Niederhuth CE, Schmitz RJ (2017) Putting DNA methylation in context: from genomes to gene expression in plants. Biochim Biophys Acta 1860:149–156
13. Omidvar V, Mohorianu I, Dalmay T, Zheng Y, Fei Z, Pucci A et al (2017) Transcriptional regulation of male-sterility in 7B-1 male-sterile tomato mutant. PLoS One 12:e0170715
14. Yang X, Zhao Y, Xie D, Sun Y, Zhu X, Esmaeili N et al (2016) Identification and functional analysis of microRNAs involved in the anther development in cotton genic male sterile line Yu98-8A. Int J Mol Sci 17:1677
15. Walker J, Gao H, Zhang J, Aldridge B, Vickers M, Higgins JD et al (2018) Sexual-lineage-specific DNA methylation regulates meiosis in Arabidopsis. Nat Genet 50:130–137
16. Li B, Dewey CN (2011) RSEM: accurate transcript quantification from RNA-Seq data with or without a reference genome. BMC Bioinformatics 12:323
17. Wang L, Feng Z, Wang X, Wang X, Zhang X (2010) DEGseq: an R package for identifying differentially expressed genes from RNA-seq data. Bioinformatics 26:136–138
18. Allen E, Xie Z, Gustafson AM, Carrington JC (2005) MicroRNA-directed phasing during trans-acting siRNA biogenesis in plants. Cell 121:207–221
19. Schwab R, Palatnik JF, Riester M, Schommer C, Schmid M, Weigel D (2005) Specific effects of microRNAs on the plant transcriptome. Dev Cell 8:517–527
20. Li M, Wu H, Luo Z, Xia Y, Guan J, Wang T et al (2012) An atlas of DNA methylomes in porcine adipose and muscle tissues. Nat Commun 3:850

Chapter 20

Quantification of Recombination Rate and Segregation Distortion by Genotyping and Sequencing of Single Pollen Nuclei

Steven Dreissig, Jörg Fuchs, Axel Himmelbach, Martin Mascher, and Andreas Houben

Abstract

Meiosis is a specialized cell division during which homologous chromosomes can exchange genetic material through recombination. This mechanism generates novel allelic combinations, which can be exploited by plant breeders to achieve crop improvement. Pollen grains are the haploid products of meiosis required in fertilization. Here, we describe two approaches to measure meiotic recombination in single haploid pollen nuclei. Pollen nuclei are first separated by fluorescence-activated cell-sorting. Afterwards, the DNA of single pollen nuclei can be amplified by multiple-displacement-amplification using Phi29 DNA polymerase and meiotic recombination events can be measured using KASP markers. Alternatively, the PicoPLEX DNA-seq kit can be used to amplify the DNA of single pollen nuclei followed by library preparation for whole-genome sequencing and subsequent bioinformatic analysis.

Key words Pollen, Single cell, Recombination, Segregation distortion, Flow sorting, Whole-genome amplification

1 Introduction

Meiosis is a fundamental mechanism in eukaryotic reproduction, which generates genetic diversity through genetic exchange between homologous chromosomes [1]. This mechanism is key in genome evolution and adaptation [2]. It is also exploited by plant breeders to combine desirable alleles in a common genetic background to achieve crop improvement [3, 4]. It is therefore of great interest to understand how recombination is regulated and how recombination patterns can be modified or targeted to specific sites. In order to assay meiotic recombination for these purposes, efficient methods are required to measure recombination at high throughput.

Mónica Pradillo and Stefan Heckmann (eds.), *Plant Meiosis: Methods and Protocols*, Methods in Molecular Biology, vol. 2061, https://doi.org/10.1007/978-1-4939-9818-0_20, © Springer Science+Business Media, LLC, part of Springer Nature 2020

In human and animal genetics, sperm genotyping was developed to enable measurements of large numbers of meiotic events [5]. Great improvements regarding the number of sites to be genotyped were recently achieved by using Phi29 DNA polymerase for whole-genome amplification (WGA) [6]. Furthermore, a massive decrease in cost and increase in efficiency was achieved by conducting WGA on microfluidic lab-on-a-chip devices [5–7]. Current microfluidic lab-on-a-chip devices are not applicable for pollen due to the rigid cell wall of pollen. The pollen wall is resistant to most cell lysis protocols which were initially developed for mammalian cells. Additionally, most pollen is considerably larger than mammalian cells which might require special microfluidic devices with larger flow channels and reaction chambers. To overcome this problem we developed an approach for the isolation of individual pollen nuclei via fluorescence-activated cell-sorting (FACS) combined with two different DNA amplification and genotyping methods.

Pollen genotyping, as sperm genotyping in animals, enables direct measurements of meiotic events. The advantages are increased resolution over microscopy-based methods and omitting the necessity to cultivate the offspring for analysis [8–11]. However, there are four major technical challenges associated with single-pollen genotyping, (1) physical isolation of individual pollen grains or pollen nuclei, (2) amplification of the pollen DNA to acquire sufficient amounts of downstream analyses, (3) analysis of the amplified DNA in a cost-efficient way to address the hypothesis of the study, and (4) interpretation of the data in the context of biases and errors which might arise during the previous steps [6].

Here, we describe our approaches to measure meiotic recombination events based on the genotyping of individual pollen nuclei. We present exemplary methods for data analysis and discuss possible errors.

2 Materials

2.1 Isolation of Pollen Nuclei

1. 20 μm CellTrics filter (Sysmex-Partec, 04-0042-2315).
2. 100 μm CellTrics filter (Sysmex-Partec, 04-0042-2318).
3. Pellet pestle (Merck, Z359947).
4. 6 mm metallic beads (Intec GmbH).
5. Galbraith buffer: 45 mM $MgCl_2$, 30 mM sodium citrate, 20 mM MOPS, 0.1% Triton ×-100,
   pH to 7.0 using HCl or KOH depending on the initial pH value [12].
6. MM 400 ball mill (Retsch).

2.2 Fluorescence-Activated Cell Sorting

1. Flow-sorter capable of sorting single cells (e.g., BD FACSAria IIu or BD Influx cell sorter; BD Biosciences).
2. Fluorescent beads (e.g., Alignflow Flow Cytometry Alignment Beads for Blue Lasers, 2.5 μm; ThermoFisher, A16500).
3. Sheath fluid (1× PBS): 137 mM NaCl, 2.7 mM KCl, 10 mM Na_2HPO_4, 1.8 mM KH_2PO_4.
4. Propidium iodide (Invitrogen, P3566).
5. Fluorescence microscope.

2.3 Whole-Genome Amplification by Multiple-Displacement Amplification

1. Illustra GenomiPhi V2 DNA amplification kit (GE Healthcare Life Sciences).
2. Nuclei lysis buffer: 400 mM KOH, 100 mM DTT, 10 mM EDTA [7].
3. Neutralization buffer: 666 mM Tris–HCL, 250 mM HCL [7].
4. 384-microwell plate (Merck, BR781358).
5. Sterile sealing film (Merck, Z369667).
6. Agarose.
7. 1 kb plus DNA ladder (ThermoFisher Scientific, 10787018).
8. Gel electrophoresis chamber.

2.4 Single-Copy PCR

1. PCR: DNA Polymerase (Qiagen), 10× PCR buffer (Qiagen), 5× Q-solution (Qiagen), 2.5 mM dNTP mix (Bioline).
2. Sterile distilled water.
3. 96-well plates.
4. Sealing film.
5. 96-well plate centrifuge.
6. 96-well plate thermal cycler.
7. Agarose (ThermoFisher Scientific, UltraPure Agarose).
8. HDGreen (Intas).
9. Ultralow range DNA ladder (ThermoFisher Scientific, 10597012).

2.5 KASP Marker Genotyping

1. Plate reader capable of detecting FAM (520 nm) and HEX (556 nm) (e.g., QuantStudio 6 (Applied Biosystems)).
2. Enzyme mix (LGC Genomics).
3. Assay mix (LGC Genomics).
4. Microwell plate for 96 (ThermoFisher Scientific, N8010560) or 384 (ThermoFisher Scientific, 4309849) samples.
5. Sealing film suitable for qPCR (Merck, Z734438).

2.6 PicoPLEX Single Nucleus Sequencing

1. PicoPLEX DNA-seq kit (Rubicon Genomics).
2. 384-microwell plate (Merck, BR781358).
3. PBS: 137 mM NaCl, 2.68 mM KCl, 10.14 mM Na_2HPO_4, 1.76 mM KH_2PO_4, adjust pH to 7.4 using HCl.
4. Sterile sealing film (Merck, Z369667).
5. AMPure XP beads (Beckman Coulter Inc.).
6. Tris-EDTA (TE) buffer: 10 mM Tris–HCl, 1 mM EDTA, pH 8.
7. SYBR Gold nucleic acid gel stain (ThermoFisher Scientific, S11494).
8. Qiagen MinElute Kit (Qiagen).
9. Agilent 2100 Bioanalyzer (Agilent Genomics).
10. Illumina HiSeq2500 (Illumina).

3 Methods

3.1 Isolation of Pollen Nuclei and Fluorescence-Activated Cell Sorting

Carry out all procedures at room temperature, unless otherwise specified. Before starting the experiment, consider which whole-genome amplification approach will be used and prepare a 384-microwell plate accordingly (*see* Subheading 3.2.).

1. Collect 10–20 mature anthers from a barley spike of an F_1 hybrid between cultivar "Morex" and cultivar "Barke" in a 1.5 mL reaction tube using forceps (*see* **Note 1**). Pollen can be disrupted either by filter bursting [13] or crushing using metallic beads [14].
2. For filter bursting, add 500 μL of Galbraith buffer and vortex for 30 s. Transfer the liquid including all anthers onto a 100 μm CellTrics-filter placed inside a 1.5 mL reaction tube and let it pass through. Rinse the filter with additional 500 μL of Galbraith buffer. Subsequently, pipette the filtrate onto a 20 μm CellTrics-filter placed inside a new 1.5 mL reaction tube. Use a plastic pestle to burst the pollen by gentle grinding. Stain the nuclei suspension with propidium iodide (final concentration: 50 μg/mL) (*see* **Note 2**).
3. If pollen grains are to be disrupted by metallic beads, add 300 μL of sterile, double-distilled H_2O into the 1.5 mL reaction tube containing the anthers and vortex for 30 s. Place the tube into a thermomixer at maximum speed for 10 min.
4. Centrifuge for 5 min at 18533,8 × *g* in a microfuge where after all anthers are removed from the tube using forceps. Centrifuge again for 5 min at 18533,8 × *g* in a microfuge, remove the supernatant by pipetting, and resuspend the pollen pellet in 100 μL of Galbraith buffer.

5. Transfer the pollen suspension into a 2.0 mL reaction tube containing two 6 mm metallic beads. Centrifuge for 5 min at 18533,8 × *g* in a microfuge. Subsequently, insert the tube into an appropriate adapter for a ball mill and homogenize the pollen-metallic bead suspensions at 30 Hz for 40 s.
6. Afterward, add another 500 μL of Galbraith buffer and filter the suspensions through a 20 μm CellTrics-filter. Stain the filtrate with propidium iodide (final concentration: 50 μg/mL).
7. Before attempting to sort single pollen nuclei into microwell plates setup the flow sorter as recommended by the supplier. Depending on the sorter used this includes fluidics start-up, laser alignment, optimizing the droplet break-off, optimizing the side streams and determining the accurate drop delay. For precise sorting a sort mode ensuring accurate single cell sorting should be chosen (e.g., "1.0 drop single" on the BD Influx). It is recommended to control the spatial deposition of droplets (*see* **Note 3**) and the precision of the sorting process (*see* **Note 4**) before the final sort.
8. Prepare the desired number of 384-microwell plates according to the selected WGA approach. Carry out all procedures with reagents and microwell plates (kept) on ice.
9. For multiple-displacement-amplification and KASP genotyping, prepare the desired number of 384-microwell plates by dispensing 2 μL of lysis mix (0.5 μL steril H_2O, 0.5 μL lysis buffer, 1 μL sample buffer (Genomiphi V2 kit) per sample) into the desired number of wells under sterile conditions and cover with sealing film. Briefly centrifuge the microwell plates to bring the liquid to the bottom of the well.
10. For single-nucleus whole-genome sequencing by PicoPLEX, add 2.5 μL of 1× PBS into the desired number of wells under sterile conditions and cover with sealing film. Briefly centrifuge to bring the liquid to the bottom of the well.
11. Analyze the nuclei suspension using the propidium iodide fluorescence signal as trigger and display the events in a dot plot: PI fluorescence intensity versus side scatter (Fig. 1a). Gate the corresponding region of pollen nuclei and display them in a histogram representing the PI fluorescence intensity and define the sort gate (Fig. 1a).
12. Sort single pollen nuclei into individual wells of a 384-microwell plate containing either 2 μL of lysis mix or 2.5 μL of 1× PBS for WGA followed by KASP genotyping or PicoPLEX whole-genome sequencing, respectively (*see* **Note 5**). Cover plate with sealing film and centrifuge at 2755 × *g* for 1 min in a microfuge.

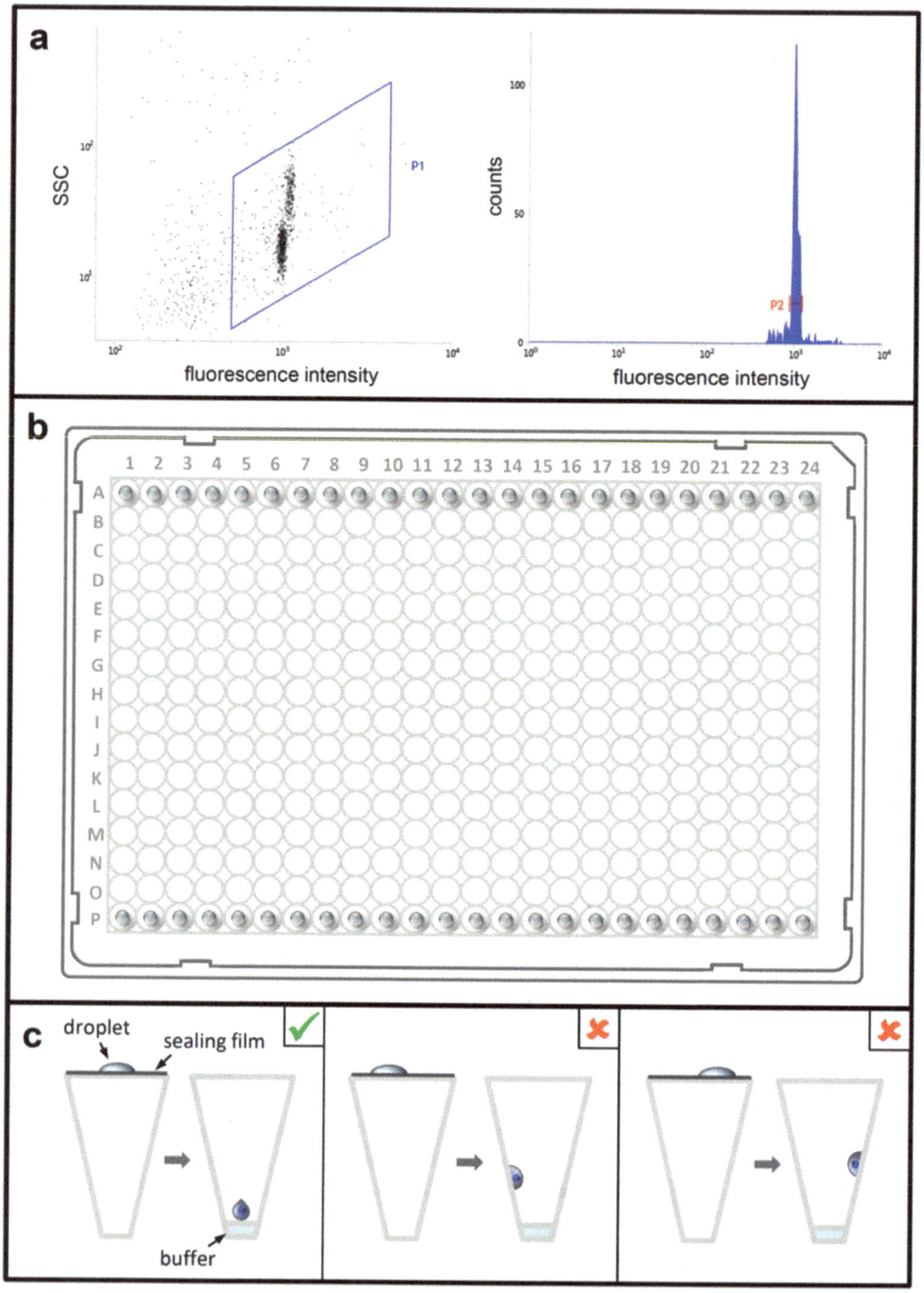

Fig. 1 Single pollen nuclei sorting (**a**) Dot plot (fluorescence intensity versus side scatter) of flow cytometric analysis of isolated pollen nuclei of barley (left). P1 indicates the population of nuclei displayed in the histogram (right). P2 represents the gate used for sorting. (**b**) Scheme of a 384-microwell plate with droplets sorted onto the surface of a sealing film covering the plate in row A and P to control the spatial deposition of the droplets. (**c**) Side views of individual wells illustrating the correct positioning of droplets and potential consequences of a wrong alignment

3.2 Single-Nucleus Whole-Genome Amplification and Genotyping

Following the successful sorting of single pollen nuclei, WGA is conducted to provide sufficient amounts of DNA for genotyping. A variety of approaches are currently available for this purpose. Here, we present two protocols which were tested by us, namely multiple-displacement amplification in combination with KASP markers and PicoPLEX single-nucleus whole-genome sequencing.

Carry out all procedures with reagents and microwell plates kept on ice. All centrifugation steps are to be carried out at 4 °C.

3.2.1 Multiple-Displacement Amplification Followed by KASP Marker Genotyping

1. Incubate the 384-microwell plate containing nuclei in nuclei lysis mix (0.5 μL steril H_2O, 0.5 μL lysis buffer, 1 μL sample buffer (Genomiphi V2 kit) per sample) at 65 °C for 3 min, then cool down to 4 °C and keep on ice.
2. Carefully remove the sealing film and add 0.5 μL of neutralization buffer into each well. Cover again using the same sealing film and centrifuge at 2755 × *g* for 1 min in a microfuge. Leave the plate on ice while preparing the amplification master mix.
3. Prepare the amplification master mix (3.5 μL sample buffer, 4.5 μL reaction buffer, 0.5 μL enzyme mix per sample) of the illustra GenomiPhi V2 DNA amplification kit for the desired number of samples, remove sealing film, and dispense it into each well. Cover the plate again using sealing film and centrifuge at 2755 × *g* for 1 min in a microfuge.
4. Incubate the plate at 30 °C for 8 h followed by 10 min at 65 °C to stop the amplification reaction (*see* **Note 6**).
5. Load 1 μL of each reaction on a 0.8% agarose gel and conduct gel electrophoresis to visualize a DNA "smear" indicative of successful DNA amplification (Fig. 2).
6. Dilute each amplified DNA sample in a 1:50 ratio in sterile double-distilled H_2O.
7. To test for amplification success carry out PCR using the barley-specific primers P1 to P8 and cereba2 (*see* **Note 7**, Table 1) according to the following thermal cycling conditions: 3 min at 95 °C; 30 s at 95 °C; 30 s at 65 °C reduced by 1 °C for 9 cycles; 30 s at 72 °C; 25 cycles at final annealing temperature of 55 °C. Prepare the following master mix for each primer pair and use 5 μL of single-nucleus DNA per reaction: 1× PCR buffer, 0.2 mM dNTPs, 1× Q-solution, 0.3 μM of each primer and 0.02 units of Taq DNA polymerase.
8. Perform agarose gel electrophoresis (2% agarose) to visualize PCR products ranging between 100 bp and 200 bp and count the number of positive reactions per single-nucleus sample to estimate WGA efficiency (*see* **Note 7**).

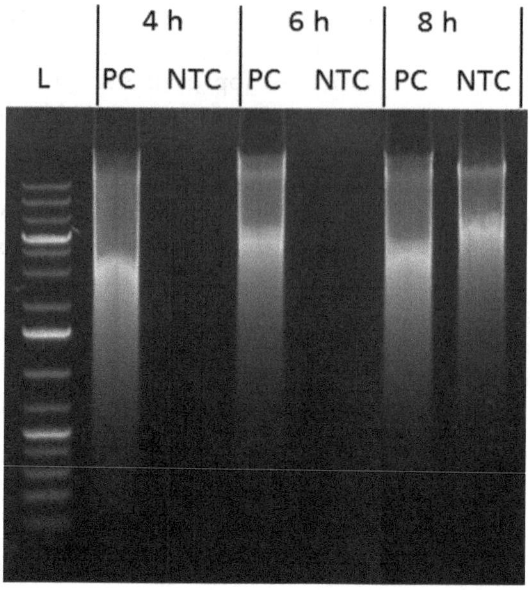

Fig. 2 Multiple-displacement amplification incubation time series. This figure shows the products of multiple-displacement amplification (MDA) after 4 h, 6 h, and 8 h for a positive control (PC) and a no-template control (NTC). The DNA smear indicates successful amplification of the template DNA after 4 h, 6 h, and 8 h. As for the no-template control, a DNA smear after 8 h indicates unspecific amplification due to contamination and/or primer dimers. These results were found to vary between different lot numbers of the GenomiPhi V2 kit. We therefore advise to test each kit by conducting an incubation time series to determine the threshold for unspecific amplification. L = GeneRuler 1 kb plus DNA ladder

9. Select samples showing successful PCR amplification of at least three different primer pairs to achieve a high KASP call rate (*see* **Note 8**).
10. Perform KASP reactions using markers A to X (Table 2) according to the following thermal cycling conditions: 15 min at 94 °C; 20 s at 94 °C and 1 min at 65 °C for 10 cycles (decrease annealing temperature by 0.8 °C per cycle); 20 s at 94 °C; and 1 min at 57 °C for 40 cycles. Prepare the following master mix for each KASP marker and use 5 μL of amplified single-nucleus DNA per reaction: 1× KASP enzyme mix, 0.14 μL KASP marker mix per reaction. Include positive controls (genomic DNA of barley cultivar "Morex" and "Barke") and a negative control containing no DNA but H_2O.
11. Perform end-point fluorescence readout in an appropriate plate reader or quantitative-PCR device at 30 °C for 1 min. Measure fluorescence at 520 nm (FAM) and 556 nm (HEX).
12. Plot FAM fluorescence values against HEX fluorescence values for each sample (including controls) in Microsoft Excel and

Table 1
List of primers used to evaluate WGA performance

Primer name	Sequence 5′–3′	Product size (bp)
P1-1F	GCAGCTTCACCTCATGTTCA	100
P1-1R	GCAGACCAACCTGACCCTAC	
P2-1F	GGCACTCGCTCACATTCTTA	87
P2-1R	GGGAAGGCATGAACAACACT	
P3-1F	GAAAGCAGACGCAGACACAG	92
P3-1R	AGCAGCAGCAGCAACTAATG	
P4-1F	TGCTTCGTATGTGGACTGAACT	96
P4-1R	GCAACATCCCTGTCATCAATAA	
P5-1F	CGATATGGCAGATGGGGTAT	109
P5-1R	CAGATGGCTGGACTTTGACA	
P6-1F	CCTCGTCCTCCAATCAGTCT	136
P6-1R	CACCCGAACCCTAGATGTGT	
P7-1F	CCCACACTGCCATACTCTCA	106
P7-1R	CCTTTTTGCAGCGTTTTCTC	
P8-2F	AGAAGAACTCGCCGTACAAGAT	117
P8-2R	GAGGACCCCTTAATCCAGAACT	
cereba2-F	CACATGGATACAACCGCAAC	114
cereba2-R	CAGGAACATAGCGACCATCA	

manually assign the allelic state of each sample by referencing against the positive controls (*see* **Note 9**, Fig. 3).

13. Repeat allele calling for all KASP markers and summarize data in a table for further analysis.

3.2.2 PicoPLEX Single-Nucleus Sequencing

For single-nucleus sequencing to measure meiotic recombination across the whole genome of individual pollen nuclei, a commercial WGA and library preparation kit (PicoPLEX single-cell DNA seq, Rubicon Genomics) is employed, following essentially the manufacturer's instructions. Briefly, genomic DNA is isolated from single pollen and used as a template for several rounds of quasi-random priming followed by linear amplification. The products are amplified exponentially to yield Illumina NGS-ready libraries.

1. Single pollen nuclei sorted into a 384-microwell plate and suspended in 2.5 μL of 1× PBS are provided with 2.5 μL of Cell Extraction Buffer without touching the nuclei with the pipette tip. Prepare a no-template control (2.5 μL of 1× PBS).

Table 2
Gene-based KASP markers

KASP marker	Gene	Chromosome	Physical map position (Mbp)	Polymorphism [Morex/Barke]
A	AK252051.1	3H	2.94996	TAACCACCAGCAAGCAAACAAGCCCTCTGTTTGCATGAGTGCGCAATATAACTTACATCAAGAAACCCCACAAACTGAGGAAC**[A/G]**AACATCATTTTGTACAGAGACAGCACATGCCACTCTCACATCTGTGATAAATATGCAAAGTGAGATGAAATAATTAGCACACAAAAACCCTTGTGACTGACCTCACCCATGGCACCCCGT
B	MLOC_62156.1	3H	4.50244	AGATAACCTMAAATCAGCTCTACAACAATTTGGTGCCCTGGAGGGCTAGAATCAATCACC**[A/G]**ACTTGTCACAGYCGGCAGCGGAATAAAGTAGGTGTGTGGATGCACACTTTGCTATGGGGT
C	MLOC_36532.1	3H	8.01784	GCCGCCAGAATGATAGTATGCCGACGGAGGAGCTTGTGGAAACTCAGGTGGAAGGTAGAAGTCAAAGAAGAATAGGCCATCTTGGTATGGTGTGCCACTTGCTCCAATAATCACAGCTCT**[A/C]**ATGAGATCCATACGATCTTCAAATACCCTCACATAAATATAATCTGGTAGGTTCTTCTCAAGTATGCCCCATTCTTGCTGAACTTTTTTACCCACTTCCTTCCACCGGTACCATGAGCT
D	AK372389	3H	13.6276	CGGGAGAATGAACAGAGCAATCACGTGCGATATGGCTTGCAGAATACCGTCCAATGGATGGAATGCTAGTCCAGCAAATGGTGATAGGGTATTCTCCTTGTTGTAAATGTGGTGGGTTGC**[A/G]**TGTAGGTGCTTATACAG TGGTTTTATGTCATGCAACTCTCTGTGCATCCAGTAAATTCCAAACTCCACAAAGGTGAGATATAAAGCCACATATACAAAATACATCGGCCAACCAACTTCA
E	AK248960.1	3H	47.14852	GAAGGTAGAGGTCCCCCTGTCGAGGACCCACAGCGCAAACATGTTCTGCGACTGGCGCTTGACCCTTATGAGGTCGCCGGATGGAGTCTCCACCAGGCAATTGAACTTCTGGTGCTGRTA**[A/G]**TGGACGAGGTTGTGCATATGCGGTTCCACCTGGGGGAAATCCTCATCTGGCCATGAACGGGGAGGCTCTGCCCACACCTCCACCGTACCATATACATCAATGGTGTAAAGTTGCCTTCTG
F	AK250083.1	3H	50.33532	CCGCAGCCGCCGTGCCACCGGGATGACCCTCGCCTTCCTCACCGCTGCTGCTAACGAGCT**[C/G]**GGCTGCCCCCTTCCTTGCTGACAAGGCCATATTTGAAGCTGACAATCAGCAACACTTGAC

G	AK365312	3H	89.0754	GATATGAGGAGAGGATCAGGCTGGTCATGTTCAACACCCATTTTGGAGCCCCCCAC AAGC**[A/G]**CATCCATCCAGAGTCTTTCTACACAAACTCATGGCCAGAGTGCTTC TGCCAGATGAATGC
H	AK359814	3H	117.73428	TACAACCGAGATCCTGCAAGCAGGATGCCTCTAGCTAGCACAACCGGATTATTG ATCATCATACAGACATCCATAACTTAAGAAACTTTACAGAACCAAATATCAGCAG CTAATAATATA**[A/G]**TTAAGCGCAACCACGACGTTATTATTTTATACAGTAATAAAA TACAAAAAAGGAACGCAGCGCACGAACTTAGCAAGAGCTACATGCATGTAGTAGAA GCAGAAACCCGACTTCTACTTC
I	MLOC_54754.1	3H	349.5932	ACTTCTCGATCTGGCTGTCCATGACGACCAGCGCCGAGGAGTCGCCGACGTAGTA GGGGTGGTTCCCTGACCAAACATCCACCGAGTACTCCGGCTTCGTGCCCCCCG TCACCAGCACCA**[A/G]**CTCACCGTTGCAGTACACCTTCGCGTCCTCGAARATC TCCGGGTGCAGCCCCTTCTTCCGCATCGAGCACCTCATCCTCTGGGAGGGGAC GAGGGAGCGGAGGGTACGGTTTGAATTTCT
J	AK367893	3H	407.21724	CCATCAGAAGGGYGCATTGTTGTGGATGAAGTCTGCTGCTGTMTACCTGTCTCTGT ACTT**[A/G]**TTGTTCTGTCTACGTTGTAAGCATATGTTGTTACGCCCGTTTCGAGCT GCTTGTGATGAT
K	MLOC_39198.3	3H	407.95256	TCCGACTGAGTATTTGGGAGTCAAATTGGTGTCTTTCGGCTTTGCTGGACAAGGAA GAGC**[A/C]**ATCATGCGAGGTCCAATGGTTTCAGGAGTCATTAATCAGTTGCTGA CCACGACTGACTGG
L	AK252342.1	3H	423.49472	TGTTGCCTATGAAACATTTGGTTTAACCACAGGTTTCATGTCAAGGACACGGTGTA TCCT**[A/C]**TCCTCTATGTAACATCCTATTAGTGTTCTAGTCTGGAATTTGTTAGC TTGAAGCATTCTT
M	MLOC_2480.1	3H	429.59304	AGAAACTCACAGGCAAGCTATTGTTTCTGCTGTAACGAGTGGGAGCAGACAATAC TTTCT**[T/C]**GGTACCGACAGTGCTCCCCATGATAAACGGATGAAGGAGTGCTCC TGTGGATGTGCTGGA
N	MLOC_10890.2	3H	441.08536	GTCTTCAACATTGAGCATCCACGTGTTACCGGAAGACTGGTCGATGGTGGTGTTAGT ATT**[A/G]**TCCCTACCATATTGTTCAGCCTCGCTTGCTCCTGAAGAACTCTCGGGG CCTTGATGGCTG

(continued)

Table 2
(continued)

KASP marker	Gene	Chromosome	Physical map position (Mbp)	Polymorphism [Morex/Barke]
O	AK363304	3H	445.6994	CAGGTCTCATCTACCAAATTTGAAGTACCTCCTTCAATGATCCCAGACATGTAACCAC TTTCAAATTTAGCCTTGTTGGCAAAATATTTCTAGTGAACCTTGTATATTC TGATTTCTGCC[**A/G**]AAAGAATTATGCATGAAGAAGCACAATTCTTAGTCTACAGC TTGTGGTTCTGCATTTGAATCTCAAAACTGCCTCAAAAGTTATGCAGAGAAGTC AAGTTCAGCTTCCACAAAGAGAAGA
P	AK357864	3H	459.52416	TCACAACTAATTGCTGAATTCACCAATCGTAATCTGGCTGGTATTTAACATGCTGTT GTA[**A/G**]AATAGTTGATCGCGACTTCACATGCTTCTGCGCCCGAAGACATGTATGT CACGGATGTGT
Q	AK355032	3H	465.49724	GTCGTTGCCCAGGAGCATGGCTATGGCGTTGTTGAAGACGGACGAGTCGGCCGT GTTCCC[**A/G**]GAGAAGAACGCGTCTGGCCTCAAGTTTTCAGGATGAGACGAAGT GTCCCCTGCCCCGGGC
R	AK251764.1	3H	468.14092	GCCATATATAATTCTGCACATTCCTTTCTGATATACAGGAGAGCTGGGAGTGAGAC AATG[**A/G**]AATCGAAGATATCTACTGTGGTGTTTCTGATCTCAAGCATATAGTAC ACCCGATGGTCCT
S	AK365737	3H	484.32664	GCTACCAGGTCCTGCATATCTTGGGCTCAGCAAATCATAGTTCTGAGAATAAAGTGC ACC[**A/G**]GTCTCGAATAAGTGTGGGCTTGTCCGCAACTTTTCTGAATC TGGTGAAAGTTTTTTTTTT
T	AK354719	3H	484.62176	GCGGGCTCTCCGGCGCTGGAGCGGGAGGAGCAGGCTATAAATAAGGGGGGTGGGG CGCCCACAAGACAGAACCACACACCACAAGCAAGAGTGCACAAAAGCATACTG ACCGCAGAGACC[**A/G**]CAGTGTTGAGATAGGATGGCGTCGGAGATGAGCAAGG ACGTGAAGTTCTCCGACGAGGAAGTCACCTCGCACCCGCGCGTCGGCGAAGGTG AGGAGCAAACGGTTGGGCCGGCGCGGCAG

U	MLOC_13043.1	3H	487.0162	TGGTGTCAGTTGGGTAGAGCTTAAGGGTCAAGTGCATCAGTTCTGCACAGGAGAT AAATC[**T/C**]CATCCAAGGCAAGGAGAGATCGATGAGTACTTGGAAGAAATGAT GGCCAAAATCAGGCAA
V	AK367638	3H	492.51312	CTTATTGACTGTAATGAGGGGTGTCCTCTGACTGTTTGAGGGTATGGAGAAACTGC GAGC[**T/C**]AAGGAAGCAGGAGGCCGCGGAAGTGGCGAGGCTGTCTGCGATGC TGTTGGTGAGGCACGG
W	AK370453	3H	506.33044	GCGTCGACGCTCATGTCGAACACTTCGTTTCTGTGCAGCTCAGGCGCAGTGTAGT AACCT[**A/G**]AATCACAGGAGAGTGCGATTAAGTATATGTCAGCGAAGCATTATCT
X	AK355583	3H	534.45488	GTGATGTCAAGACCATTATTTTTGCTAGTTTTACATAGTACAGTAGCAAGGGAAAC TTC[**A/C**]AATATGCAGCAGCTTCACCCAAGTGGGTTCTTCAGCGAATATATTTAG CCCAGGAATAGT

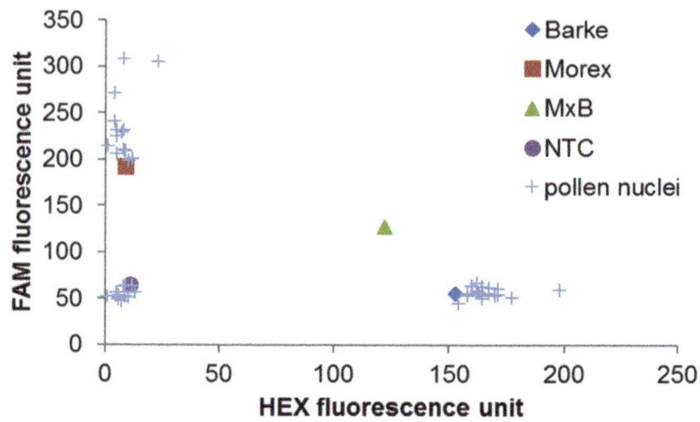

Fig. 3 KASP scatter plot after multiple-displacement amplification. This figure shows a scatter plot of KASP marker C (Table 2). FAM and HEX fluorescence values of pollen nuclei are plotted to visualize allele-specific clusters in agreement with positive controls (genomic DNA of the two parental genotypes "Barke" and "Morex"). Importantly, no heterozygous samples, as shown by a genomic DNA control from a "Morex" × "Barke" hybrid, are expected. The occurrence of heterozygous samples would indicate the presence of a DNA contamination or multiple pollen nuclei in the same reaction tube

2. The Cell Extraction Master Mix (4.8 μL of Extraction Enzyme Dilution Buffer and 0.2 μL of Cell Extraction Enzyme per reaction) is prepared for the chosen number of samples. Mix gently and incubate on ice until used.
3. For lysis, add 5 μL of Cell Extraction Master Mix without touching the nuclei with the pipette tip and mix gently (final volume of 10 μL). Seal the microwell plate with sealing film and pulse-spin to collect the liquid at the bottom of the well.
4. Perform lysis using a thermal cycler with a heated lid (10 min at 75 °C, 4 min at 95 °C and hold at 22 °C). Pulse-spin the microwell plate.
5. The Pre-Amplification Master Mix (4.8 μL of Pre-Amp Buffer and 0.2 μL of Pre-Amp Enzyme per reaction) is prepared for the chosen number of samples. Mix gently and incubate on ice until used.
6. Remove the sealing film from the plate, add 5 μL of Pre-Amplification Master Mix, seal the plate with sealing film, pulse-spin and amplify the DNA using a thermal cycler with heated lid (1 cycle: 2 min at 95 °C; 12 cycles: 15 s at 95 °C, 50 s at 15 °C, 40 s at 25 °C, 30 s at 35 °C, 40 s at 65 °C, and 40 s at 75 °C; hold reactions at 4 °C). Pulse-spin the microwell plate (total volume of 15 μL).
7. The Amplification Master Mix (25 μL of Amplification Buffer, 0.5 μL of Amplification Enzyme and 4.5 μL of nuclease-free water per reaction) is prepared for the chosen number of samples. Mix gently and incubate on ice until used.

8. Remove the sealing film from the plate, add 30 μL of Pre-Amplification Master Mix and 5 μL of a specific index of the Dual Index Plate. Follow the index handling instructions of the manufacturer (Rubicon Genomics).
9. Seal the plate with sealing film, pulse-spin and amplify the DNA using a thermal cycler with a heated lid (1 cycle: 4 min at 95 °C; 4 cycles: 20 s at 95 °C, 25 s at 63 °C, 40 s at 72 °C; 7 cycles: 20 s at 95 °C, 55 s at 72 °C. Hold reactions at 4 °C). Pulse-spin the microwell plate (final volume of 50 μL).
10. Pool 10 μL aliquots of each sample and store the remainder of the library at −20 °C. If less than 10 samples are processed, adjust the volume accordingly to obtain 100 μL.
11. Typically, a 100 μL aliquot of the pooled library is mixed thoroughly by pipetting with an equal volume of AMPure XP beads (RT, homogenous suspension). Incubate at RT for 5 min.
12. Pulse-spin the pool and collect the beads using a magnetic rack (2 min). Discard the supernatant.
13. Add 300 μL of freshly prepared 80% (v/v) ethanol to the beads.
14. While placed in the magnetic rack, rotate each tube 90° and wait until beads form a pellet. Repeat three more times.
15. With the sample in the magnetic rack, discard the supernatant.
16. Repeat **steps 13–15**.
17. Pulse-spin and return the tube to the magnet for the removal of residual liquid.
18. Dry the beads (cap open) for 3–6 min in a heating block (37 °C). Do not overdry the beads.
19. Elute the DNA by adding 30 μL of TE buffer pH 8.0. Resuspend the beads completely by pipetting and incubate for 3 min.
20. Pulse-spin and place the tubes into the magnetic rack (3 min).
21. Transfer the supernatant containing the library pool into a fresh tube.
22. Perform a standard size-fractionation using a SYBR-GOLD-stained 2% agarose gel [15].
23. Excise the region of interest (350–1000 bp) and extract the DNA using the Qiagen MinElute kit [15]. Elute the final library by the addition of 25 μL of buffer EB (preheated to 50 °C) to the column.
24. Characterize sequencing library on an Agilent 2100 Bioanalyzer and quantify by Real-Time PCR [15, 16].
25. Add 8 % PhiX DNA and sequence the library (e.g., Illumina HiSeq2500 device, rapid run, 1 lane, cBot clustering,

2 × 100 cycles paired-end, dual indexing with 8 cycles per index) (http://www.illumina.com).

26. For sequence read mapping, trim Illumina adapters using Cutadapt [17].
27. Align trimmed reads to the barley cultivar "Morex" reference genome sequence assembly [18] using BWA-MEM with default parameters.
28. Convert resulting SAM files to BAM format using SAMtools.
29. Analyze data for optical and PCR duplicates using Novosort (http://www.novocraft.com/products/novosort/).
30. Perform genotype calling at single-nucleotide polymorphism (SNP) sites using SAMtools (SNPs were previously ascertained in a "Morex" × "Barke" RIL population using the POPSEQ method [16]).
31. Import resulting VCF files into the R statistical environment (https://www.r-project.org/contributors.html).
32. Derive consensus genotypes by aggregating information in 1 Mb bins using functions of the "data.table" R package (https://cran.r-project.org/web/packages/data.table/index.html).
33. The resulting genotype file containing allele information at 1 Mb resolution is then used to analyze recombination.

3.3 Calculating Recombination Rate and Segregation Distortion

Using the above generated genotype file, recombination rate and segregation distortion can be calculated using Microsoft Excel or any other program suitable for manipulating text files and conducting basic mathematical operations. The following protocol is for the use of Microsoft Excel.

1. Import genotype file (derived from KASP genotyping or PicoPLEX sequencing) into Microsoft Excel using the "Data" > "From Text" tab. The genotype matrix should be structured in the following way: (1) column contains chromosome number; (2) column contains physical position of a marker; (3) column until last column, depending on the number of samples, contains genotype information (e.g., "0" or "2") for each marker.
2. Sort data according to chromosome number (1. column) and marker physical position (2. column) using the "Sort" and "Custom list" function.
3. In an empty column, use the "=average()" formula across multiple markers (e.g., 5 markers separated by 1 Mb each, start in the row containing the 5. marker) within the same sample. Copy this formula into the below cell and continue to the end of the chromosome. The resulting column shows

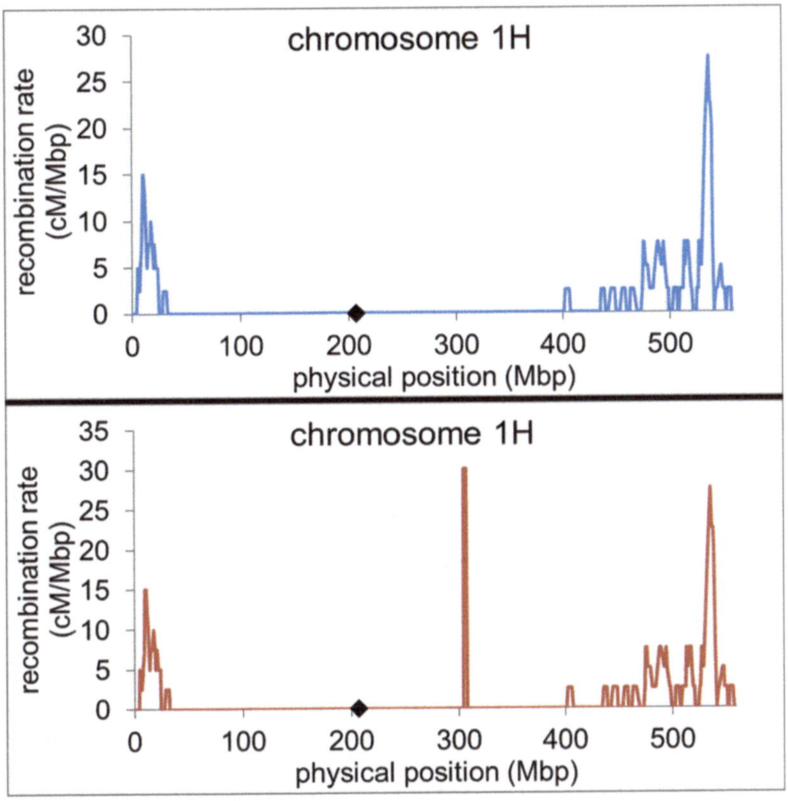

Fig. 4 Visualization of recombination rate. This figure shows an example of how recombination rate (cM/Mbp) can be visualized along chromosome 1H. The lower panel (red) shows an example of an error in genotyping or reference genome assembly, which results in a peak in the low-recombining region of barley chromosome 1H. The upper panel (blue) shows recombination rate (cM/Mbp) after manual removal of the marker responsible for the unexpected peak in the lower panel

recombination events as values between 0 and 2 in sliding windows. Carry out the same procedure for all samples using a new empty column for each sample (*see* **Note 10**).

4. In a new empty column, count the number of recombination events at each marker position using the following formula across all samples: =countifs(*range[first sample to last sample]*, ">0", *range[first sample to last sample]*, "<2").
5. Calculate genetic distance (cM) within each sliding window representing a certain physical distance using the following formula: cM = (number of recombination events/total number of samples) × 100. To visualize recombination rate (cM/Mb), plot cM values at each marker position against the physical position for all markers of a chromosome (*see* **Note 11**, Fig. 4).

6. To measure segregation distortion, calculate the frequency of each parental allele at each marker position using the following formula: =countif()/*total number of samples*. Depending on marker density, calculate average allele frequency in sliding windows of 10 Mb.
7. Apply chi^2-test to determine significantly distorted segregation ratios using the following formula in each sliding window: =chiqu.test(*range[observed value for each allele]*, *range[expected value for each allele]*). Segregation distortion is shown by *P* values <0.05.

4 Notes

1. It is of advantage to use mature anthers which are yellow and readily release pollen by vortexing. Note that the amount of pollen which can be gathered per anther decreases as anthers are older or under abiotic or biotic stress.
2. After vortexing, the liquid should appear yellow to brown which indicates pollen grains were released from the anthers. If the liquid remains rather clear, the anthers were too young and did not release pollen. In this case, pollen can be released from anthers by squeezing entire anthers against the 100 μm CellTrics-filter. However, this is likely to result in tapetum nuclei being isolated which are seen as an additional 2C peak in the flow-cytometry graph.
3. The correct spatial deposition of droplets is controlled by sorting 25 fluorescent beads per well onto a 384-microwell plate covered with sealing film. The position of the formed drops on top of the sealing film is evaluated visually. If necessary adjust the drop charge or the tray control settings until the droplet hits the center of the wells targeted. Validate this at least for the top and the bottom row of the microwell plate you want to use (Fig. 1b). A precise alignment is very important to ensure a sorting of individual pollen nuclei directly into the low amount of liquid at the bottom of each well (Fig. 1c). Samples may be lost by nuclei being attached to the side of the reaction tube.
4. The precision of the sorting processes (an accurate drop delay) is controlled by sorting a defined number of reference particles (e.g., fluorescent beads in a grid pattern onto a microscopic slide placed on top of the microwell plate and subsequent evaluation using a fluorescence microscope). If deviations from the expected number of sorted particles are observed the drop delay has to be adjusted new.
5. The 384-microwell plate is prepared under sterile conditions to avoid nonspecific DNA amplification. Use a UV-sterilized laminar flow hood and UV-sterilize all reagents (except for WGA kit components), reaction tubes, and pipettes.

6. The optimal incubation time varies between 2 h to 8 h. We found differences between lot numbers of the GenomiPhi V2 kit concerning the time required for a no-template control to show a similar amount of DNA amplification compared to a corresponding positive control containing genomic DNA. To achieve optimal DNA amplification of single pollen nuclei for KASP genotyping, we recommend a time series (e.g., 2 h, 4 h, 6 h, and 8 h incubation at 30 °C) to determine after how many hours a no-template control shows similar amplification as the corresponding positive control. Then select an incubation time where DNA yield is maximal in single-nuclei samples and approximately half in a no-template control. However, this should only be done if barley-specific KASP markers are used as these markers do not hybridize to nonspecific DNA contamination (e.g., bacterial or human).
7. Primer pairs P1 to P8 amplify single-copy sequences evenly distributed along chromosome 3H of barley which were validated by single-copy FISH [19]. WGA efficiency is then estimated by counting the number of successful single-copy sequence PCR reactions. We found these PCR products are only present after successful GenomiPhi-based whole-genome-amplification. In addition, we recommend using a high-copy PCR target (i.e., centromeric retrotransposon *cereba*, primer pair cereba2) as a control to test if a nucleus was successfully sorted into the reaction tube. This sequence is amplified even if the WGA has failed.
8. We found a positive correlation between the number of single-copy PCR products and KASP marker call rate ($r = 0.51$, $r^2 = 0.26$, [11]).
9. Allele calling can be done automatically using appropriate software. However, we found it of advantage to manually inspect all plots as most software shows poor performance in terms of handling amplification data derived from WGA samples. Presumably, this is caused by varying degrees of amplification in each sample resulting in copy-number differences.
10. The size of the sliding window depends on the marker density and can be adjusted. If recombination is measured in marker intervals separated by, for example, 100 Mbp, the sliding window approach is not required.
11. Plotting recombination rate (cM/Mbp) along the chromosome can be used to identify genotyping errors or markers which were assigned to a false physical position. In both cases, a single marker results in a drastic increase in recombination frequency. This is seen as a peak in the cM/Mbp plot which is multiple times higher than the average recombination rate or recombination rate in subtelomeric regions (Fig. 4).

Acknowledgments

We thankfully acknowledge Petr Cápal for his advice on single-cell whole-genome-amplification. Furthermore, we like to acknowledge Nils Stein (IPK, Gatersleben) for providing "Morex" × "Barke" F1 seeds and Sandra Dreisslein as well as Ines Walde (IPK, Gatersleben) for their excellent technical assistance. This work was supported by IPK Gatersleben.

References

1. Villeneuve AM, Hillers KJ (2001) Whence meiosis? Cell 106:647–650
2. Stapley J, Feulner PGD, Johnston SE et al (2017) Variation in recombination frequency and distribution across eukaryotes: patterns and processes. Philos Trans R Soc Lond Ser B Biol Sci 372:20160455
3. Ramsay L, Colas I, Waugh R (2014) Modulation of meiotic recombination. In: Kumlehn J, Stein N (eds) Biotechnological approaches to barley improvement. Springer, Berlin Heidelberg, pp 311–329
4. Choi K (2017) Advances towards controlling meiotic recombination for plant breeding. Mol Cells 40:814–822
5. Wang J, Fan HC, Behr B et al (2012) Genome-wide single-cell analysis of recombination activity and de novo mutation rates in human sperm. Cell 150:402–412
6. Gawad C, Koh W, Quake SR (2016) Single-cell genome sequencing: current state of the science. Nat Rev Genet 17:175–188
7. Gole J, Gore A, Richards A et al (2013) Massively parallel polymerase cloning and genome sequencing of single cells using nanoliter microwells. Nat Biotechnol 31:1126–1132
8. Li X, Li L, Yan J (2015) Dissecting meiotic recombination based on tetrad analysis by single-microspore sequencing in maize. Nat Commun 6:6648
9. Li X, Meng D, Chen S et al (2017) Single nucleus sequencing reveals spermatid chromosome fragmentation as a possible cause of maize haploid induction. Nat Commun 8:991
10. Dreissig S, Fuchs J, Himmelbach A et al (2017) Sequencing of single pollen nuclei reveals meiotic recombination events at megabase resolution and circumvents segregation distortion caused by postmeiotic processes. Front Plant Sci 8:1620
11. Dreissig S, Fuchs J, Cápal P et al (2015) Measuring meiotic crossovers via multi-locus genotyping of single pollen grains in barley. PLoS One 10:e0137677
12. Galbraith DW, Harkins KR, Maddox JM et al (1983) Rapid flow cytometric analysis of the cell cycle in intact plant tissues. Science 220:1049–1051
13. Kron P, Husband BC (2012) Using flow cytometry to estimate pollen DNA content: improved methodology and applications. Ann Bot 110:1067–1078
14. De Storme N, Geelen D (2011) The Arabidopsis mutant jason produces unreduced first division restitution male gametes through a parallel/fused spindle mechanism in meiosis II. Plant Physiol 155:1403–1415
15. Himmelbach A, Knauft M, Stein N (2014) Plant sequence capture optimised for Illumina sequencing. Bio-protocol 4:e1166
16. Mascher M, Muehlbauer GJ, Rokhsar DS et al (2013) Anchoring and ordering NGS contig assemblies by population sequencing (POPSEQ). Plant J 76:718–727
17. Martin M (2011) Cutadapt removes adapter sequences from high-throughput sequencing reads. EMBnetjournal 17:10–12
18. Mascher M, Gundlach H, Himmelbach A et al (2017) A chromosome conformation capture ordered sequence of the barley genome. Nature 544:427–433
19. Aliyeva-Schnorr L, Beier S, Karafiátová M et al (2015) Cytogenetic mapping with centromeric bacterial artificial chromosomes contigs shows that this recombination-poor region comprises more than half of barley chromosome 3H. Plant J 84:385–394

Part IV

Further Approaches to Study Plants Meiosis and Beyond

Chapter 21

Identifying and Isolating Meiotic Mutants in a Polyploid Brassica Crop

Marina Pfalz, Adrián Gonzalo, Nicolas Christophorou, Aurélien Blary, Aurélie Berard, Nadia Bessoltane, Emilie Montes, Lydia Jaffrelo, Charles Poncet, Marie-Christine Le Paslier, Nathalie Nesi, Delphine Charif, and Eric Jenczewski

Abstract

This chapter provides a detailed description of TILLING and CRISPR-Cas9 approaches for the purpose of studying genes/factors involved in meiotic recombination in the polyploid species *B. napus*. The TILLING approach involves the screening and identification of EMS-mutagenized M2 *B. napus* plants. The strategy for high-throughput plant pooling, the set up for microfluidic PCR and sequencing is provided and the parameters for the analysis of sequence results and the detection of mutants are explained. The CRISPR-Cas system relies on the optimal design of guide RNAs and their efficient expression. The procedure for the generation and detection of knockout mutants is described with the aims to simultaneously target homologous genes.

Key words *Brassica napus*, Gene editing, Meiotic recombination, Microfluidic PCR, Polyploidy, Reverse genetics, Targeting induced local lesions in genomes

1 Introduction

Meiosis is one of the fundamental factors that generates genetic diversity and ensures fertility. Until recently, the study of meiosis has focused mainly on diploid plant species such as *A. thaliana* or rice (*Oryza sativa*) while polyploids received far less attention. This is a major gap because polyploid species are commonplace among plants, in particular among crops, and because polyploid research provides unique opportunities to explore new avenues/properties for already known meiotic genes.

This chapter describes the setup of reverse genetics approaches to study the function of meiotic genes in polyploid species. Although this work is focused on *Brassica napus* (AACC, $2n = 38$), the methods and strategies we describe here can be

Mónica Pradillo and Stefan Heckmann (eds.), *Plant Meiosis: Methods and Protocols*, Methods in Molecular Biology, vol. 2061, https://doi.org/10.1007/978-1-4939-9818-0_21,

used in a broader range of polyploid species; some of these approaches are indeed based on previous work with other polyploid crops [1]. Our main objective in this chapter is to describe how the specific difficulties linked to the presence of meiotic recombination gene duplicates can be overcome. These approaches can also be used in diploid plant, when it comes to study gene families.

We present two methods where the genes of interest are targeted in a specific or nonspecific way. The first method, TILLING (targeting induced local lesions in genomes), is based on ethyl methanesulfonate (EMS)-mutagenesis and leads to random single base substitutions (mainly GC → AT transitions). The generality and efficiency of chemical mutagenesis makes TILLING applicable to all plants. However, EMS treatment also causes a high density of off-target mutations that, at least conceptually, could be mistaken for mutations in target genes during phenotypic analysis [2, 3]. Another limit of TILLING in polyploids is that mutations affecting the different copies of a gene must be combined by crossing to produce a defective phenotype. The second method, CRISPR-Cas9 (clustered regularly interspaced short palindromic repeats, CRISPR associated protein9) gene editing, induces site-specific double strand breaks through the concerted guidance of the cas9 nuclease to the gene of interest by single guide RNAs (sgRNAs) [4]. The nonhomologous end joining pathway (NHEJ) generates insertions or deletions during the repair of these double strand breaks. One major advantage of this technology is that it can induce mutations simultaneously in multicopy genes [5, 6]. This is especially important when these genes are functionally redundant and where an effect can be observed only when all copies are depleted. However, this technology relies on transgenesis, a limiting step in some plant species, and the efficiency of gRNAS to guide the Cas9 nuclease to the gene of interest, which is difficult to predict. We will show how we address these constraints in *B. napus*.

2 Materials

2.1 *Agrobacterium Tumefaciens*

1. The strain C58C1 carrying the virulence plasmid pMP90 [7] is used to transform *B. napus*. This strain confers a rifampicin resistance gene on the chromosome and a gene for resistance to gentamycin on the virulence plasmid.
2. Agrobacteria were made competent for electroporation to receive the expression plasmid. The agrobacterial cultures are carried out on LB medium [8].
3. During the genetic transformation step, the bacterial cultures are carried out in liquid Min A medium [9] supplemented with 50 mg/L rifampicin, 50 mg/L gentamycin and 75 mg/L spectinomycin.

2.2 *Brassica napus* Transformation

1. Cotyledonary *B. napus* cv. *Westar* and cv. *Drakkar* are used for organogenesis (*see* **Note 1**).
2. Germination takes place on Murashige and Skoog (MS) medium [10] with 3% sucrose and 0.6% Plant Agar pH 5.8 adjusted with KOH 1 M.
3. For the organogenesis steps, the recommended culture medium is the Murashige & Skoog (MS) medium [10] with 3% sucrose and 0.6% Plant Agar supplemented with 2 mg/L 6-benzylaminopurine pH 5.8 adjusted with KOH 1 M.
4. Seedlings conversion steps are conducted on Gamborg B5 medium [11] with 1% sucrose and 0.6% Plant Agar supplemented with 5 mg/L indole butyric acid pH 5.8 adjusted with KOH 1 M.

3 Methods

3.1 Identification of EMS Mutations Using a TILLING-by-Sequencing Approach

Classical TILLING approaches use mismatch-specific endonucleases to identify mutations in target genes [2]. The advent of next generation sequencing (NGS) technologies has provided fast and cost-effective ways to identify such mutations at higher efficiencies [1, 12]. The later approach is described here.

1. Mutagenized population in *Brassica napus*.

   Development of the RapTILL population was described in detail in [3]. The complete population consists of 9970 EMS-mutagenized M2 seed lots that are stored under controlled conditions (5% RH, 8 °C) at INRA Rennes (France). Current (conservative) estimates indicate an average mutation density of approximately 1 mutation per 50 kb, or almost 24,000 mutations per genome [3].
2. DNA extraction.

   For a subset of 1024 M2 families, 5–7 seeds per family are sown in individual pots. Leaf material is collected on 3-to-4-week-old plants as a mixture of 5–21 leaf discs (∅ = 5 mm) per M2 family. After sample freeze-drying and grinding, DNA extractions are performed using the sbeadex livestock kit on oKtopure™ robot platform (LGC Genomics) and then tested for quality and quantity. All samples that are below a concentration of 20 ng/μL are removed from the analysis or extracted again.
3. Pooling in 3D.

   We use 3D pools to screen the 1024 M2 families for mutations in a more efficient way. Each pool is prepared by combining equal amounts of genomic DNAs from eight M2 families (Fig. 1). Pooling is performed in such a way that each M2 family is the unique intersection of three different pools.

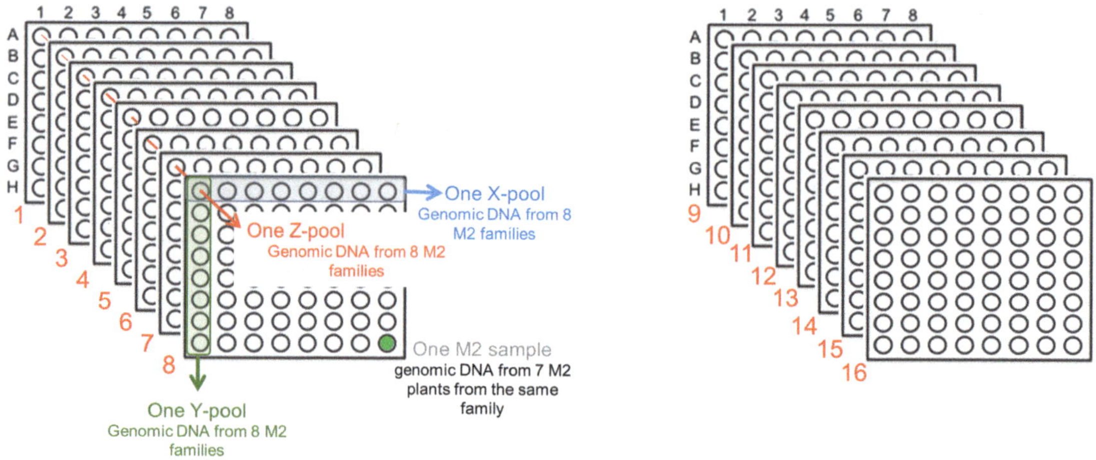

Fig. 1 Design of the 3D pooling to screen 1024 EMS-mutagenized *Brassica napus* M2 families The figure shows 16 plates of 64 wells (8 × 8). Each well contains a bulk of genomic DNAs extracted from 5–7 sister plants of the same M2 family. Arrows and shaded areas indicate which wells are pooled together to produce a triplet of X, Y and Z pools

The three dimensions are designated as *X*, *Y* and *Z*, respectively. As a result, 128 X-pools, 128 Y-pools and 128 Z-pools are produced with a DNA concentration varying from 45 to 60 ng/μL (Quant-iT™ PicoGreen™ quantification) (*see* **Note 2**).

4. PCR amplicons, rationale and design.

The approach is based on PCR amplification of the target region and sequencing performed on each of the 384 pools. Overlapping amplicons are used to either cover the complete sequence of a gene or target specific regions where a high density of nonsense mutations are expected to occur due to EMS-mutagenesis (*see* **Note 3**). The genomic and coding sequences for target genes can be retrieved from the Genoscope *Brassica napus* Genome Browser available at:

http://www.genoscope.cns.fr/brassicanapus/cgi-bin/gbrowse/colza/.

(a) Target region(s).

Amplification regions are selected with the help of the web-based CODDLE (for Codon Optimized to Discover Deleterious Lesions) software available at http://blocks.fhcrc.org/proweb/coddle/. We provide protein sequence alignments encompassing a wide range of eukaryotes (*see* **Note 4**), including the target gene, to CODDLE in order to identify the most deleterious mutations (based on SIFT scores).

(b) PCR amplifications—primer design and testing.

For every target gene, PCR primer pairs are designed from the genomic sequences using Oligo6. The criteria used to retain a primer pair are as follows: (1) amplicons were 500–600 bp long (*see* **Note 5**); (2) the PCR can be performed with an annealing temperature of 60 °C (recommended by Fluidigm protocol: minimum primer T_m = 59 °C and maximum primer T_m = 61 °C); (3) the two homoeologous copies of a gene are equally amplified (*see* **Note 6**). To check whether this last criterion is met, we sequence (Sanger) the amplicons. Mixed bases in the chromatogram are considered as indicative of the presence of the two homeologous copies, providing they strictly overlap with known homoeoSNP (single nucleotide polymorphism distinguishing the homoeologs) positions (*see* **Note 7**).

5. Microfluidic PCRs and construction of libraries.

   We use 48 × 48 Access Array™ Integrated Fluidic Circuits (IFC) to perform 2304 (i.e., 48 unique DNA pools × 48 primer pairs) parallel PCR reactions and library preparations. Every reaction combines both the specific amplicon production and a barcoding step (using 48 different Access Array barcodes library for Illumina sequencing). All amplicons (48) of a DNA pool share the same barcode. As microfluidic chip output, 48 unique tagged pools of amplicon libraries are produced, one for each DNA pool. In total, eight 48 × 48 Access Array chips are used to perform the 384 barcoded amplicons libraries (8 × 48 DNA pools with the 48 primer pairs and the 384 barcodes available) (*see* **Note 2**).

   B. napus has a 2.34–2.56 pg/2C genome [13]. We estimate that the number of template DNA molecules representing a haploid genome (thus an allele) is ~300. Depending on the number of mutant alleles present in a M2 family (50% are expected, based on strict Mendelian segregation), the number of template DNA molecules representing an individual mutant allele may vary from a minimum of 3 (only one mutant allele in a single pool) to a maximum of 39 (only one WT allele in a single pool), with an average of 21. These numbers are important to set up a similar experiment with another species and determine the number of plants per M2 families and the number of M2 families per pool (*see* **Note 8**).

6. Sequencing.

   All amplicons are sequenced on one single MiSeq flowcell using a 600 cycles PE mode; this produced 50 million of paired-end reads (each being 300 bp long). In our experience, ~75% of reads are generally above the Q30 threshold (excepted for four pools). We then apply a 10% spike-in with PhiX for optimizing low complexity libraries sequencing (such as an amplicons library). As a general rule, a total of 35 million PE reads can be used for subsequent steps.

7. Screening of candidate mutations.

   Each of the 384 (128 X, 128 Y, and 128 Z) samples is processed independently as described below:

   (a) We use bwa mem to align reads against a previously determined (Sanger sequencing of amplicons) sequence of the target genes.

   (b) We "clean up" the alignments by removing: (1) reads that are not in proper pair and (2) pairs of reads with a mapping quality below 20.

   (c) We then detect all nucleotide variations in each sample (Sample: e.g., X1, Y15, Z12, ...) and at each position in the gene (e.g., MSH4). For this, we run GATK DepthOfCoverage (—printBaseCounts—minBaseQuality 20—minMappingQuality 20) on the filtered alignment files to have bases count (nA, nC, nG, nT, nN) per positions in the gene. We report all nucleotide variations (as ReferenceBase-MajoritaryBase-SecondMajoritaryBase) and calculate its frequency (SecondMajoritaryBase/MajoritaryBase).

   (d) As a "visual" control, we draw a graph giving the number of reads covering every position along the genes. The A and C homoeologous copies are represented separately (*see* **Note 9**).

8. We then look for mutations for a given gene across the 1024 M2 families.

   (a) This approach is based on the principle that real mutations must be common to three intersecting pools (Fig. 1) and specific to these three pools (*see* **Note 10**).

   (b) For each M2 family, we select all nucleotide variation occurring in the three samples with a min Frequency ≥ 0.01 in each of the three samples. This allows discarding sequencing errors (expected at a rate of 0.002) without being too conservative (the expected frequency is 6.25%; *see* **Note 3**) and remove nucleotide variation occurring in more than one M2 family.

   (c) We also only retain nucleotide variations that correspond to GC $\rightarrow$ AT transitions as these are the most frequent alterations induced by EMS mutagenesis.

   (d) Sanger sequencing validation of candidate mutations.

      Copy-specific primers are used to amplify and sequence the regions surrounding the most promising candidate mutations. For this, we used the genomic

DNA of the M2 families in which the candidate mutations are detected, as template for PCR amplification.

(e) Identification of mutant plants for specific copies.

According to Mendelian rules, half the seeds from a M2 family in which a candidate mutation is detected should carry the identified mutation. ~10 seeds are sown for each M2 family in which a candidate mutation is detected. The plants are genotyped either by sequencing directly the copy-specific PCR amplicons obtained as in h, or after developing (derived) cleaved amplified polymorphic sequences (CAPS or dCAPS) markers that distinguish the wild type and mutant alleles. We use the web-based tool dCAPS Finder 2.0 (http://helix.wustl.edu/dcaps/) to develop these markers. Plants homozygous for the mutations affecting one or the other of the two homoeologs are crossed to produce F_1 hybrids combining both mutations at the heterozygous state.

9. Production of mutant plants defective for all copies.

The F_1 s can then be selfed to produce a full set of segregating F2 plants, among which double A/C mutant (i.e., homozygous for the mutations affecting the two homoeologs) and their WT siblings (homozygous for the WT alleles at the two homoeologous loci) plants can be identified using the same markers as in *i* (*see* **Note 11**).

3.2 Gene Editing by CRISPR-Cas9

The CRISPR-Cas9 system has been already successfully applied to generate mutants in oilseed rape [5, 6]. A big advantage of this approach is that it allows cotargeting of homoeologs with one sgRNA in polyploid species.

3.2.1 Design of Single Guide RNA (CRISPOR)

Our goal is to design sgRNAs that induce simultaneously mutations in the open reading frame of homoeologous gene copies. For this, we use the CRISPOR (http://crispor.tefor.net) program. We select series of <1000 bp genomic sequences centered on exons and paste one sequence after the other into the window of the CRISPOR website. The reference genome of *B. napus* cv. *Darmor bzh* (v. 4.1; 14) is chosen to screen for putative off-targets. If required, additional genomes can be uploaded by the CRISPOR support team. We also select the PAM (protospacer adjacent motif) for the Cas9 from *Streptococcus pyogenes* that corresponds to the pDe-Cas9 vector we use [14]. We select gRNAs from the CRISPOR output table using the following criteria: (1) the position to be targeted must be located within an exon (*see* **Note 12**); (2) the specificity score must be as high as possible; (3) the homoeologous gene copy should be listed as an off-target showing an identical sequence to the sgRNA (*see* **Note 13**); off-targets in other genes than the homoeolog should contain more than 4 mismatches to the sgRNA; (4) when possible predicted sgRNA efficiencies are taken into account. Additional criteria can then be applied to avoid poly-T sequences (no more than 3Ts), choose a GC content between 45% and 60% along the length of the guide and between 50% and 83% among the six nucleotides upstream of the PAM, or select guides starting with a G (or replacing the first base by a G) when using U6 promoters.

3.2.2 Validation of the sgRNA Sequence in the Genotypes Before Transformation

Copy-specific primers are used to amplify and sequence the regions flanking the sgRNAs in cv. *Drakkar* and cv. *Westar*, and verify that they are identical to the reference genome.

3.2.3 Vector Construction

Due to varying efficiencies of sgRNAs, we introduce three sgRNAs that target separate regions within the same gene into the pDe-Cas9 vector [15]. The sgRNAs are tandemly arrayed, with each sgRNA expression cassette containing the U6 promoter from *Arabidopsis thaliana* or *Brassica napus* (*see* **Notes 14** and **15**), the sequence targeting the gene(s) of interest (sgRNA), the sgRNA scaffold and the U6 terminator sequence (Fig. 2). Both ends of the cassette are flanked by a restriction site (Fig. 2) to facilitate replacing a gRNA by another one in case the sgRNA proved to be inefficient in generating mutations. Therefore, the

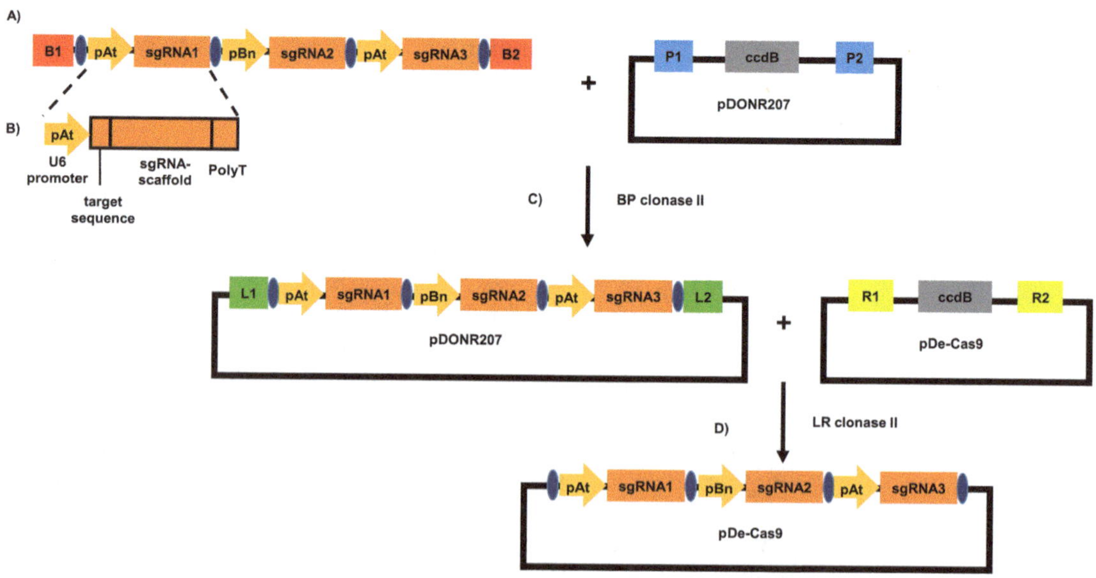

Fig. 2 Schematic presentation for the construction of a three sgRNAs-containing CRISPR-Cas9 vector. (**a**) Three sgRNA cassettes are tandemly arranged and the cassettes are enclosed by restrictions enzymes sites (dark blue). The complete construct is flanked by Gateway-vector specific primers and is synthesized by a company. (**b**) Each cassette consists of an U6 promoter either from *A. thaliana* or *B. napus*, the 20 bp target sequence, the sgRNA scaffold and U6 terminator. (**c**, **d**) Two sequential rounds of cloning lead to the insertion of the sgRNA construct into the binary vector pDe-Cas9

restriction sites should be unique within the construct and not present in the vector (check also gRNA sequence for restriction sites). Finally, the complete construct is flanked by Gateway-vector specific sequences to allow recombination to the vector. A company (GeneArt) can synthesize such a construct directly. It is first introduced by a Gateway cloning reaction into pDONR207 and then subcloned into pDe-Cas9 for *Agrobacterium tumefaciens*-mediated transformation.

3.2.4 B. napus *Transformation*

1. *A. tumefaciens* transformation.

   The binary vector pDe-Cas9 [15] is electroporated into *Agrobacterium tumefaciens* C58C1::pMP90 competent strain. The vector construct contains the SpecR gene that confers resistance to spectinomycin.

2. *A. tumefaciens* preculture.

   A single agrobacterium colony (taken from a culture in a petri dish on LB medium [7] is grown at 28 °C in 10 mL MinA liquid medium [9] supplemented with 50 mg/L rifampicin, 50 mg/L gentamycin and 75 mg/L spectinomycin. The liquid cultures are shaked at 250 rpm for 48 h.

3. *A. tumefaciens* final culture.

   The day before transformation, 10 mL MinA liquid medium is inoculated with the preculture (0.5%) without antibiotics. The suspension was grown overnight at 28 °C under constant shaking (at 210 rpm). Absorbance at 600 nm is then measured. The suspension is used at an absorbance of about 0.100.

4. Preparation of plant material.

   Seeds of *B. napus* cv. *Westar* and *Drakkar* are surface sterilized in 10% sodium hypochlorite solution added a drop of tween 20 for 20 min just after being dusted with a 70% alcohol solution for 2 min. The seeds are washed in sterile distilled water three times for 5 min and planted at a density of 20 seeds per plate. Seeds are stratified 48 h at 4 °C before being placed at 22–24 °C in a 16 h light/8 h dark photoperiod at a light intensity of 220 μmol m^2 s^{-1}. After 5 days, cotyledons are excised to eliminate apical meristem. Excised cotyledons are placed on MS medium, 3% sucrose and 0.6% Plant Agar supplemented with 2 mg/L 6-benzylaminopurine, pH 5.8. The cotyledons are plated at a density of 10 cotyledons per plate (Fig. 3).

5. Transformation procedure.

   The genetic transformation procedure has been adapted from previous protocols [11, 16, 17].

   The agrobacteria suspension is used to cover the base of a 5 cm petri dish. Individual excised cotyledons are taken from

Fig. 3 Stages of *B. napus* genetic transformation. (**a**) seedling, scale bar 2.5 cm; (**b**) seedlings at 5 days of development, scale bar 1 cm; (**c**) agrobacteria culture; (**d**) cotyledonary petioles inoculation, scale bar 1 cm; (**e**) culturing explants, scale bar 1 cm; (**f**) organogenesis in cultured cotyledonary petioles, scale bar 1 cm; (**g**) Potentially transformed plantlets from regenerated shoots, scale bar 1 cm

the plates described above and the cut surface of their petioles is immersed into the bacterial suspension for a few seconds. Excised cotyledons are immediately returned to the same MS plates from which they have been taken (Fig. 3). The cotyledons are cocultivated with the *Agrobacterium* for 60 h in the dark at 22–24 °C.

After cocultivation, the cotyledons are transferred to the regeneration and selection medium comprising MS medium supplemented with 400 mg/L ticarcillin disodium/clavulanate potassium to slow down growth of agrobacteria and 25 mg/L paromomycin to select potential transformants (*see* **Notes 16** and **17**). Plating density is maintained at ten explants per plate. Higher densities reduce regeneration frequency.

In parallel, two controls are prepared: the first one uses non inoculated explants with selection antibiotic to check the selection efficiency and the second one uses noninoculated explants without selection antibiotic for the organogenesis control.

6. Selection and plant regeneration.

   Plants are maintained on the regeneration and selection medium, under light and temperature conditions specified above, for 2 weeks during three subcultures. At the second and third subculture shoots usually appear on callus formation (Fig. 3). They are isolated on a rooting medium Gamborg B5 medium [10], supplemented with 500 mg/L ticarcillin disodium/clavulanate potassium and 50 mg/L paromomycin to increase the pressure of selection. Two or three subcultures are generally necessary at this stage for rooting.

(a) Selection of transgenic plants.

The presence of the transformation cassette is tested in the primary transformants by PCR using *Cas-9* and *NPTII*-specific primers. We retain all plants that carry the cassette as well as a few plants that do not carry it; these latter plants can be used as control afterward.

(b) Detection of mutations.

The primary transformants (T_0) that contain the T-DNA are analyzed further by amplifying the region targeted by the guide with homoeolog-specific primers and subsequent sequencing. Analysis of the sequencing chromatograms is used to evaluate the mutation efficiency of each of the transgenics and each of the sgRNAs. The T_0 plants (that are chimeric) in which mutations are already detected at this generation provide the best chance of detecting mutations in the next generations (*see* **Note 18**). Plants are also screened for large deletion between sgRNAs by PCR. Therefore, primer pairs are chosen which amplify the regions spanning the targets of two successive gRNAs and the fragment in between.

(c) Production of mutant plants defective for all copies.

T_0 mutated lines are self-pollinated by covering the inflorescences with selfing bags at the beginning of flowering. In parallel T_0 lines obtained from cv. *Westar* and cv. *Drakkar* are crossed to each other.

The progenies (T_1 lines and F_1 hybrids) are genotyped with copy specific primers for the presence of mutations in the gRNA target-sites as well as for the presence/absence of the T-DNA. Plants missing the T-DNA but showing mutations on both homoeologous copies are selected for cytological and genetic evaluation of cross over rate in this or the next generation.

4 Notes

1. In *B. napus*, some genotypes are more sensitive to transformation than others. Indeed, the transformation efficiency rate primarily depends on organogenesis ability and susceptibility to *A. tumefaciens* and reaches around 4–5% for cv. *Westar* (spring type) and cv. *Drakkar* (winter type).
2. The number of pools is adjusted to equal the total number of *Access Array Barcodes Library for Illumina sequencing* (384).
3. Previous results [3] clearly demonstrate that the effect of missense mutations on protein function is difficult to predict, even if these mutations target highly conserved amino acids located

in essential domains. Given that the effect of some mutations (e.g., altering anticrossover proteins) can only be assessed once recombination frequency has been assessed, missense mutations should not be primarily considered.

4. To improve predictions, the orthologs must be as diverse as possible and involve species for which functional characterization/validation exists. Based on this, we always include sequences from *Arabidopsis thaliana*, *Oryza sativa*, *Mus musculus*, *Caenorhabditis elegans*, and *Saccharomyces cerevisiae* in the alignments.

5. A reading length of 300 bases is usually achieved for all reads. PCR fragments of 500–600 bp in length can thus be completely sequenced using paired-end sequencing. Longer amplicons can also be designed, which are anchored within exons and cover introns in their central part.

6. We have strong evidence that "template switches" do occur during simultaneous PCR amplification of the two homoeologous copies (*see* also [18]); as a consequence, the SNPs that differentiate the homoeologs can be detected as mutations because of the mapping of chimeric reads derived from the "template switched" amplicons (Fig. 4). This is not necessarily a problem for all sequence changes occurring at the exact position on the homoeoSNP can simply be filtered out. Development of copy-specific primers must be weighed against the time required and difficulty to develop them and the reduction in the total number of primers that can be used in total.

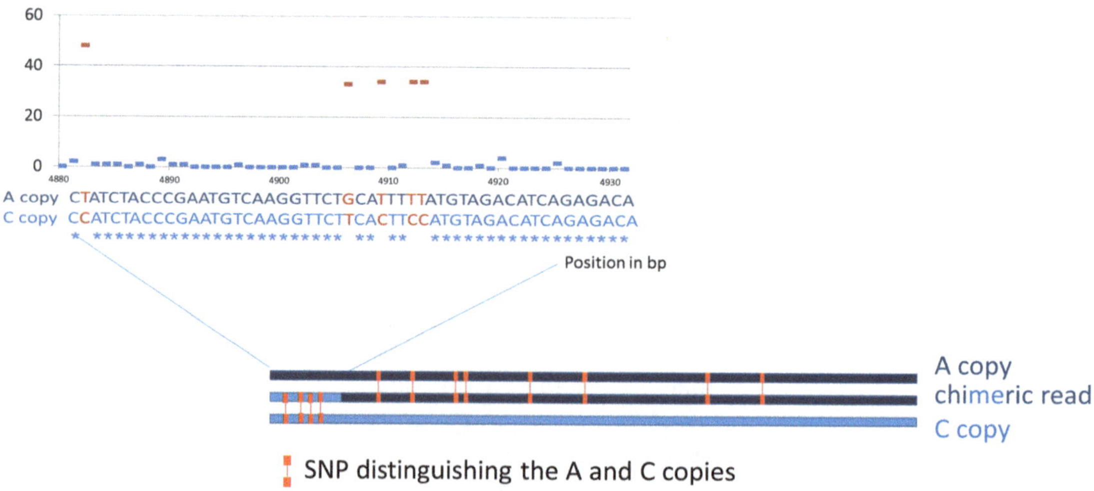

Fig. 4 Detection of homoeoSNPs as recurrent and frequent mutations. The figure on top shows the number of pools in which a mutation has been detected (*Y* axis) for each position along the gene sequence (*X* axis).The position of SNPs that distinguish the A and C copies (homoeoSNP) is indicated in red. In this example, the chimeric read mapped to the A copy, to which it was more closely related (i.e., higher number of common homoeoSNPs). The SNP specific to the C copy were thus detected as candidate mutations

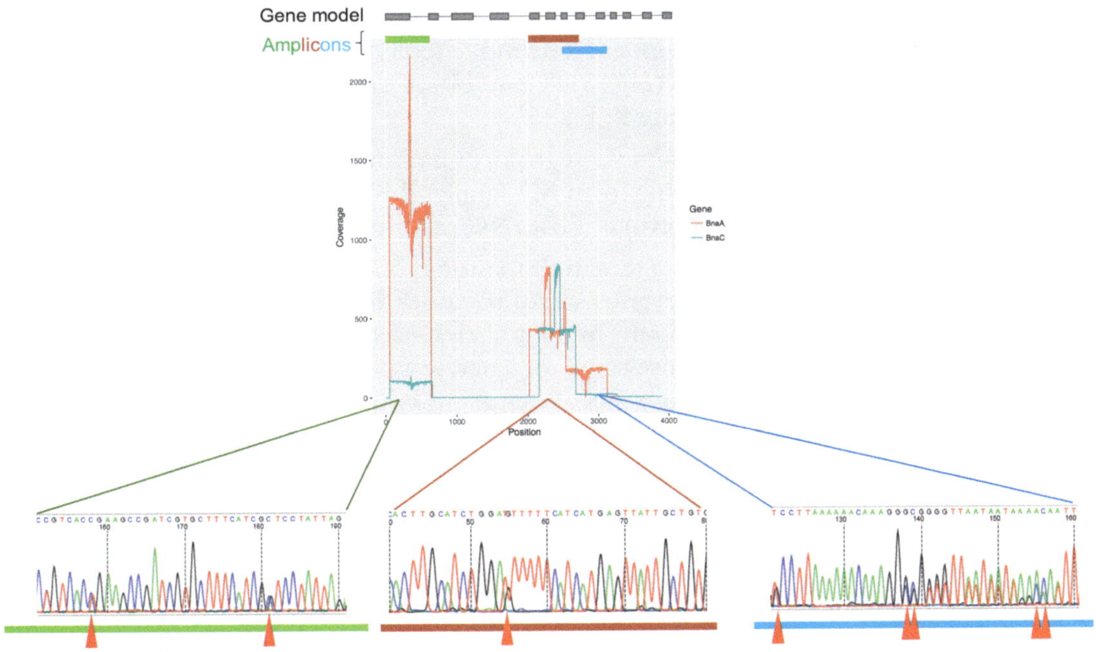

Fig. 5 Variation of coverage between homoeologous copies. The figure on top shows the exon–intron structure of the gene (in gray) and the position of three amplicons (in green, red, and blue, respectively). The figure at the bottom shows part of the chromatogram obtained by sequencing (Sanger) the corresponding amplicons; mixed bases (red arrow heads) indicate that the two homeologous copies were equally amplified. The figure in-between shows the number of reads covering every position along two homoeologous genes (A copy in red; C copy in blue) for three amplicons; the C copy of the gene is far less covered than the A copy for the green and blue Access Array™ Integrated Fluidic Circuit amplicons, respectively

7. We have examples of simultaneous PCR amplification of the two homoeologous copies during the testing of the primers that turned to be copy-specific afterward (Fig. 5). Initial Sanger sequencing of amplicons thus represents a first, necessary but not sufficient step.
8. In a pilot experiment, DNA from two plants carrying known mutations was mixed to wild type DNA to assess the limit of resolution of our pooling design. Mutation detection was always efficient at a dilution of 6.25% corresponding to the expected Mendelian ratio of a mutation segregating within a single M2 family (6 mutant alleles out of 96; considering 6 plants per family and 8 families per pool). Detection of less frequent mutations was more erratic.
9. These graphs are useful to estimate the accuracy of calls by visual inspection. When coverage is low, sampling can make estimation of nucleotide frequency inaccurate, leading to errors in the identification of the most frequent nucleotides at a given position. When coverage is high, nucleotide frequencies are correctly estimated, which improves the chance of identifying

real mutations. Note that these graphs can also reveal unexpected bias in the simultaneous amplification of the two homoeologs (Fig. 5). These biases cannot be necessarily anticipated (*see* **Note 8**; Fig. 5).

10. We assume here that a mutation at any given sequence position is specific to an M2 family; this is supported by evidence from our previous TILLING experiments.
11. As an alternative to using one single mutant allele (at the homozygous stage) to knock out the A or the C copy, a pair of different mutant alleles, each at the hemizygous stage, can also be selected to generate a plant deficient for this copy. This strategy relies on crossing two different F_1 s that altogether combine four different mutant alleles: two for the A- and two for the C-copy, respectively. Therefore, this approach largely reduces the risk that an off-target mutation could be mistaken for mutations in target genes during phenotypic analysis because WT alleles should be present for all loci, except the target genes.
12. It is necessary to verify that the sequence of the target gene does not contain alternative ATG in frame (EST databases) a bit downstream the mutation, which could lead to shorter but still functional mRNAs.
13. When one wants to cotarget the homoeologous copies of a gene with the same sgRNA, the specificity score has to be balanced with the presence/absence of SNPs between homoeologs in the target region. In situations where the two constraints cannot be addressed at the same time, the later (identity between homoeologs) should be preferred.
14. Although the U6 promoter from *A. thaliana* works well for generating mutations in *B. napus*, species-specific promoter might increase the mutation rate. The integration of several gRNA constructs in one vector allowed to combine U6 promoters from *A. thaliana* and *B. napus*.
15. SgRNA are transcribed by Polymerase III and put under regulation of U6 promoters which are recognized by polymerase III. To obtain an optimal expression of the gRNA the guide sequence has to start with an "G" for U6 promoter [5′-G-$N_{(19)}$-3′]. If the favorite gRNA starts with another nucleotide than a "G," an additional "G" can be inserted upstream of the guide sequence [5′-G-$N_{(20)}$-3′].
16. The nature and concentration of the gelling agent has an impact on in vitro regeneration and the selection of potential transformants. For example, if hygromycin is used to select for potential transformants, then Gelrite is required as the gelling agent. However, a prolonged culture on Gelrite adversely affects the quality of the buds (glassy shoots). It also affects

seedling conversion and plant fertility. If kanamycin or paromomycin are used, then agar is required as gelling agent. However, batches of agar are all different. If the agar concentration is not suitable, then the organogenesis can be blocked. The concentration can vary from 6 to 8 g/L depending on the batch. Experimenter should be acutely aware of this.

17. Both kanamycin and paromomycin are effective at the same concentration with the same number of subcultures. However, selection of transformants with kanamycin requires subcultures of 21 days instead of 15 days for paromomycin and the plants obtained are less fertile.

18. Detection depends on the proportion of mutated cells at collection time and mutations might occur later during plant development. Therefore, it is advisable to produce T_1 progenies for all transgenic plants.

Acknowledgments

This work has benefited from a French State grant (reference ANR-10-LABX-0040-SPS) managed by the French National Research Agency under an Investments for the Future program (reference n°ANR-11-IDEX-0003-02). It has also been funded through the ANR project ANR-14-CE19-0004—CROC and with the support of INRA BAP division (Appel à Manifestation d'intérêt 2012; HyperRec). We also want to acknowledge CEA-IbFJ-Genoscope that gave EPGV group access to their Illumina sequencing resources as well as Isabelle Le Clainche, Aurélie Chauveau, Elodie Marquand and Aurélie Canaguier. A.B. was funded by a "Young Scientist Contracts" (CJS) from INRA. A.G. was funded by the Marie-Curie "COMREC" network FP7 ITN-606956.

References

1. Tsai H, Howell T, Nitcher R, Missirian V, Watson B, Ngo KJ, Lieberman M, Fass J, Uauy C, Tran RK, Khan AA, Filkov V, Tai TH, Dubcovsky J, Comai L (2011) Discovery of rare mutations in populations: TILLING by sequencing. Plant Physiol 156:1257–1268
2. Henikoff S, Till BJ, Comai L (2004) TILLING. Traditional mutagenesis meets functional genomics. Plant Physiol 135:630–636
3. Blary A, Gonzalo A, Eber F, Bérard A, Bergès H, Bessoltane N, Charif D, Charpentier C, Cromer L, Fourment J, Genevriez C, Le Paslier MC, Lode M, Lucas MO, Nesi N, Lloyd A, Chèvre AM, Jenczewski E (2018) FANCM limits meiotic crossovers in Brassica crops. Front Plant Sci 9:368
4. Schiml S and Puchta H (2016) Revolutionizing plant biology: multiple ways of genome engineering by CRISPR/Cas Plant Methods 12:8
5. Yang H, Wu JJ, Tang R, Liu KD, Dai C (2017) CRISPR/Cas9-mediated genome editing efficiently creates specific mutations at multiple loci using one sgRNA in *Brassica napus*. Sci Rep 7:7489
6. Braatz J, Harloff HJ, Mascher M, Stein N, Himmelbach A, Jung C (2017) CRISPR-Cas9 target mutagenesis leads to simultaneous modification of different homoeologous gene copies in polyploidy oilseed rape (*Brassica napus*). Plant Physiol 174:935–942

7. Koncz C, Schell J (1986) The promoter of TL-DNA gene 5 controls the tissue-specific expression of chimaeric genes carried by a novel type of agrobacterium binary vector. Mol Gen Genet 204:383–396
8. Bertani G (1951) Studies on lysogenesis I.: the mode of phage liberation by lysogenic Escherichia coli. J Bacteriol 62:293–300
9. Sparrow PAC, Dale PJ, Snape JW, Irwin JA (2006) The use of phenotypic markers to identify Brassica genotypes for routine high throughput agrobacterium-mediated transformation. Acta Hort (706):239–246
10. Murashige T, Skoog F (1962) A revised medium for rapid growth and bioassays with tobacco tissue cultures. Physiol Plant 15:473–497
11. Gamborg OL, Miller RA, Ojima K (1968) Nutrient requirements of suspension cultures of soybean root cells. Exp Cell Res 50:151–158
12. Gilchrist EJ, Sidebottom CHD, Koh CS, MacInnes T, Sharpe AG, Haughn GW (2013) A mutant *Brassica napus* (canola) population for the identification of new genetic diversity via TILLING and next generation sequencing. PLoS One 8:e84303
13. Arumuganathan K, Earle ED (1991) Nuclear DNA content of some important plant species. Plant Mol Biol Rep 9:208–218
14. Chalhoub B, Denoeud F, Liu S, Parkin IA, Tang H, Wang X et al (2014) Early allopolyploid evolution in the post-Neolithic *Brassica napus* oilseed genome. Science 345:950
15. Fauser F, Schiml S, Puchta H (2014) Both CRISPR/Cas-based nucleases and nickases can be used efficiently for genome engineering in Arabidopsis thaliana. Plant J 79:348–359
16. Moloney MM, Walker JM, Sharma KK (1989) High-efficiency transformation of *Brassica napus* using agrobacterium vectors. Plant Cell Rep 8:238–242
17. Sparrow PAC, Dale J, Irwin JA (2004) The use of phenotypic markers to identify *Brassica oleracea* genotypes for routine high-throughput agrobacterium-mediated transformation. Plant Cell Rep 23:64–70
18. Cronn R, Cedroni M, Haselkorn T, Grover C, Wendel J (2002) PCR-mediated recombination in amplification products derived from polyploid cotton. Theor Appl Genet 104:482–489

Chapter 22

A Cytological Analysis of Wheat Meiosis Targeted by Virus-Induced Gene Silencing (VIGS)

Stuart Desjardins, Kostya Kanyuka, and James D. Higgins

Abstract

Virus-induced gene silencing (VIGS) is a rapid and cost-effective reverse genetic technology that can be used to assess gene function in wheat. This chapter contains a detailed description of how to target wheat meiotic genes by VIGS. The timing of this technique is critical and has been optimized to silence meiotic genes at peak expression, evidenced by silencing of *Triticum aestivum disrupted meiotic cDNA1* (*TaDMC1*). We also describe cytological techniques that have been adapted for the preparation and analysis of meiocytes in wheat, including fluorescent in situ hybridization (FISH) with directly labeled, synthetic oligonucleotide probes, and immunolocalization on spread material.

Key words VIGS, Meiosis, Cytology, Wheat, FISH, Chromosomes, Immunolocalization

1 Introduction

The model diploid plant *Arabidopsis thaliana* has enabled the identification and characterization of over eighty genes required to ensure accurate homologous recombination, synaptonemal complex formation, and correct chromosome segregation during meiosis [1, 2]. As sequencing technologies have improved over the past 5 years, a considerable number of genomes are now available on open databases, such as Ensembl (http://plants.ensembl.org/index.html) and Phytozome (http://phytozome.jgi.doe.gov), so that orthologous genes in wild species and crops may be identified. In particular, completion of the draft allohexaploid bread wheat (*Triticum aestivum*) genome ($2n = 6x = 42$, AABBDD) has provided opportunities for a bioinformatic analysis to be performed [3], and consequently key meiotic genes that are essential for mediating crossover formation can be identified. Approaches and technologies to investigate gene function by mutagenesis have concurrently been developed in bread wheat including the extensive panel of TILLING lines [4] and gene targeting approaches including CRISPR/Cas (Subheading 3 in [5]). These techniques

Mónica Pradillo and Stefan Heckmann (eds.), *Plant Meiosis: Methods and Protocols*, Methods in Molecular Biology, vol. 2061, https://doi.org/10.1007/978-1-4939-9818-0_22, © Springer Science+Business Media, LLC, part of Springer Nature 2020

are extremely effective at creating null mutants, and the TILLING lines may also produce hypomorphic phenotypes. The major limiting factor is the time and cost required to produce such knockouts, which limits the number of genes that can be screened. An alternative approach for investigating meiotic orthologous gene function in wheat is to utilize virus-induced gene silencing (VIGS) [6–9]. This provides a rapid, efficient approach for determining gene function in wheat, so further analyses using TILLING or CRISPR/Cas may be employed on promising targets. Therefore, in this chapter we describe the methodology of how to generate VIGS knockdowns of meiotic genes in wheat along with the approaches used to analyze the phenotype cytologically.

2 Materials

2.1 Plant Material and Cultivation

1. Hexaploid bread wheat (*Triticum aestivum*) cultivar Bobwhite is routinely used for VIGS experiments due to its susceptibility to barley stripe mosaic virus (BSMV) infection, but many other genotypes can also be utilised.
2. Tobacco (*Nicotiana benthamiana*) is used as an intermediate host.
3. The plants are grown in a Level 3 biological containment facility under controlled environmental growth room conditions: photoperiod 16 h, temperature 20 °C (night) and 23 °C (day), light intensity (at the soil level) ~180 μmol m^{-2} s^{-1}, and relative humidity ~60%.

2.2 Virus Induced Gene Silencing (VIGS)

2.2.1 BSMV VIGS Binary Vectors Are Described in [10]

1. pCaBS-α: plasmid for BSMV RNA alpha; contains kanamycin resistance gene for selection in bacteria.
2. pCaBS-β: plasmid for BSMV RNA beta; contains kanamycin resistance gene for selection in bacteria.
3. pCa-γbLIC: plasmid for BSMV RNA gamma; contains a ligation independent cloning (LIC) site for insertion of wheat gene fragments for VIGS, and kanamycin resistance gene for selection in bacteria.

2.2.2 VIGS Construct Design

1. Software si-Fi version siFi21 (downloadable from http://labtools.ipk-gatersleben.de/).
2. An entire wheat cDNA annotation TGACv1 or IWGSC RefSeq v1.0 in FASTA format (downloadable from Ensembl Plants http://plants.ensembl.org/Triticum_aestivum/Info/Index, or from the Wheat Portal at URGI https://wheat-urgi.versailles.inra.fr/Seq-Repository/Annotations, respectively).

2.2.3 PCR and Purification of PCR Products

1. PCR of target sequence for VIGS: ~100 ng of template DNA, 1 U of high-fidelity DNA polymerase (e.g., Phusion DNA Polymerase), 1× reaction buffer, 0.4 mM dNTPs, and 0.4 μM primers containing the following 5′ sequence extensions: 5′-AAGGAAGTTTAA-3′ (F-primer) and 5′-AACCACC ACCACCGT-3′ (R-primer).
2. DNA gel extraction kit (e.g., QIAquick Gel Extraction Kit (Qiagen) or similar).

2.2.4 Ligation-Independent Cloning

1. T4 DNA polymerase mixture for vector: 200 ng of *Apa*I-digested pCa-γbLIC DNA, 5 mM dTTP, 1× BSA, 1× NEBuffer 2.1, 1.2 U of T4 DNA polymerase. For 20 μL: add 5 μL of ApaI-digested pCa-γbLIC, 1 μL of 100 mM dTTP, 0.2 μL of 100× BSA, 2 μL of 10× NEBuffer 2.1, 0.4 μL of 3000 U/mL T4 DNA polymerase, and make up to 20 μL with ddH_20.
2. T4 DNA polymerase mixture for insert: 200–250 ng of PCR product, 5 mM dATP, 1× BSA, 1× NEBuffer 2.1, 0.6U of T4 DNA polymerase. For 10 μL: add *x* μL of PCR product, 0.5 μL of 100 mM dATP, 0.1 μL of 100× BSA, 0.2 μL of T4 DNA polymerase, and make up to 10 μL with ddH_2O.

2.2.5 Generation of Recombinant BSMV RNA Gamma Constructs

1. Chemically competent *E. coli* strain JM109 or DH5α.
2. Super Optimal broth with Catabolite repression (S.O.C.) medium: 0.5% yeast extract, 2% tryptone, 10 mM NaCl, 2.5 mM KCl, 10 mM $MgCl_2$, 10 mM $MgSO_4$, 20 mM glucose. Add glucose after autoclaving broth with the remaining ingredients and sterilize the final solution by passing it through a 0.2 μm filter.
3. Lysogeny Broth (LB) agar plates supplemented with kanamycin: agar (15 g/L), NaCl (10 g/L), tryptone (10 g/L), yeast extract (5 g/L), kanamycin (50 μg/mL), pH to 7.0 with 5 N NaOH.
4. Liquid LB medium supplemented with kanamycin: NaCl (10 g/L), tryptone (10 g/L), yeast extract (5 g/L), kanamycin (50 μg/mL), pH to 7.0 with 5 N NaOH.
5. Colony-PCR: 1 U of *Taq* DNA polymerase, 1× reaction buffer, 0.4 mM dNTPs, and 0.4 μM primers (2235.F and 2615.R).
6. Colony-PCR/sequencing primers: 2235.F (5′-GATCAACTG CCAATCGTGAGTA-3′) and 2615.R (5′-CCAATTCAGG CATCGTTTTC-3′).
7. Miniprep plasmid DNA Kit.

2.2.6 Transformation of BSMV Vectors into Agrobacterium tumefaciens

1. Electrocompetent *A. tumefaciens* strain GV3101 (or EHA105).
2. Electroporator and cuvettes.
3. LB agar plates supplemented with kanamycin (50 μg/mL) and gentamycin (25 μg/mL).

2.2.7 Agroinfiltration of Nicotiana benthamiana *Seedlings*

1. Liquid LB medium supplemented with kanamycin (50 μg/mL) and gentamycin (25 μg/mL).
2. Infiltration buffer: 10 mM $MgCl_2$, 10 mM 2-(*N*-morpholino) ethanesulfonic acid (MES; pH 5.6), 150 μM acetosyringone.
3. Young 25–30-day-old *N. benthamiana* seedlings.
4. 1-mL needleless plastic syringes.

2.2.8 Inoculation of Wheat Plants with BSMV

1. 26–30 days old (4–4.5 leaf stage) wheat cultivar Bobwhite plants (Fig. 1).
2. Prechilled mortars and pestles.
3. Distilled or reverse osmosis water, or 10 mM potassium phosphate buffer pH 7.

Fig. 1 Bread wheat cultivar *Bobwhite* at the 4–4.5 leaf stage (~28 days post-sowing) ready for inoculation with BSMV. First symptoms of viral infection (yellow spots or stripes) will appear on upper leaves 5–11 days postinoculation, and target genes are silenced at ~14 days postinoculation. Photograph courtesy of Wing Sham Lee

4. Celite 545 AW (Sigma-Aldrich) abrasive.
5. Suitably sized plastic bags or incubation boxes large enough to accommodate individual or groups of inoculated plants.

2.3 Preparing Meiotic Spreads

1. Fixative solution: three parts of absolute ethanol to one part of acetic acid fixative (v/v). Prepare fresh.
2. Citrate buffer (0.01 M): 445 μL of 0.1 M sodium citrate, 555 μL of 0.1 M citric acid, made up to 10 mL with ddH_2O, pH 4.5.
3. Enzyme solution: 1% cellulose, 1% pectolyase in 0.01 M citrate buffer, pH 4.5. Store in 333 μL aliquots at −20 °C. Working concentration: mix 333 μL of enzyme solution with 667 μL of 0.01 M citrate buffer, pH 4.5.
4. Acetic acid (70%).
5. VECTASHIELD® Mounting Medium with DAPI (Vector Laboratories).

2.4 FISH

1. Denaturation mixture: 50% formamide, 10% dextran sulfate, 2× SSC, made up to 40 μL per slide. For 40 μL: add 20 μL of 100% formamide, 8 μL of 50% dextran sulfate, 4 μL of 20× SSC, and make up to 40 μL with ddH_2O.
2. Saline–sodium citrate (SSC; 20×): 3 M sodium chloride, 0.3 M trisodium citrate, pH 7. For 1 L: dissolve 175.3 g of NaCl and 88.2 g of $Na_3C_6H_5O_7$ in 1 L of ddH_2O, adjust pH with a few drops of 14 N HCl. Sterilize by autoclaving. Dilute in ddH_2O to prepare 2× SSC.
3. Oligonucleotide probes directly labeled with fluorophores:

   Oligo-pSc119.2-2 [11]:

   5′-[A488]TTCCACGATTGACGATTCCGGGGGTGCGTTT ACGTGTCCGTCGTC-3′.

   Oligo-pTa794–1:

   5′-[T×Rd]YRGAGTTCTGATGGGATCCGGTGCTTTAGT GYTGSTATGATCGCA-3′.

   Oligo-pTa71-2 [11]:

   5′-[A647]GGGCAAAACCACGTACGTGGCACACGCCGC GTA-3′.
4. Probe mixture: 10% dextran sulfate, 2× SSC, 50 ng of each probe. For 40 μL: add 8 μL of 50% dextran sulfate, 4 μL of 20× SSC, 2 μL of 25 ng/μL probe(s), and make up to 40 μL with ddH_2O (*see* **Note 1**).
5. 2T: 2× SSC, 0.2% Tween 20.
6. VECTASHIELD® Mounting Medium with DAPI.

2.5 Immuno-localization on Spread Material

1. Citrate buffer: 10 mM trisodium citrate, pH 7. For 500 mL: dissolve 1.47 g of trisodium citrate in 500 mL of ddH_2O, adjust pH to 7 using citric acid granules.
2. Phosphate-buffered saline (PBS; 1×): 1 preprepared tablet of PBS (100×) per 100 mL of ddH_2O. Prepare 500 mL.
3. PBST: 1× PBS, 0.1% Triton X-100. For 500 mL: add 0.5 mL of Triton X-100 to 499.5 mL of 1× PBS.
4. Blocking solution: 1× PBS, 3% Bovine Serum Albumen (BSA). For 10 mL: dissolve 0.3 g of BSA in 10 mL of 1× PBS.
5. Primary antibodies: make up to the preferred dilution (e.g., 1:200, 1:500) in 1× PBS containing 3% BSA. We typically use 1:500 for linear axis (e.g., ASY1, ASY3) and synaptonemal complex proteins (e.g., ZYP1), but 1:200 for focal recombination machinery proteins (e.g., MLH3, HEI10).
6. Secondary antibodies: make up to the preferred dilution (e.g., 1:200) in 1× PBST containing 3% BSA (e.g., goat anti-rabbit Alexa Fluor 488, goat anti-rat Alexa Fluor 594, goat anti-guinea pig Alexa Fluor 647 (Invitrogen)).
7. VECTASHIELD® Mounting Medium with DAPI.

3 Methods

3.1 VIGS

1. Conduct an in silico analysis of the whole coding sequence (CDS) of the target meiotic gene using si-Fi21 and select 250–400 bp regions with good predicted silencing efficiency (*see* **Notes 2** and **3**).
2. Design primers to the selected region and amplify the target gene sequence by RT-PCR from wheat spike RNA or, if contained within a single exon, by PCR from total genomic DNA. Ensure ligation independent cloning (LIC) adaptor sequences are appended to the 5′ ends of the primers.
3. Resolve the amplicon by gel electrophoresis, excise the band and purify using a commercially available extraction kit.
4. Meanwhile, linearize 1 μg of pCa-γbLIC plasmid with *Apa*I at 25 °C for 2 h in a 25 μL reaction. Stop the reaction by heat inactivating the restriction enzyme at 65 °C for 20 min.
5. In separate tubes, treat ~200 ng of the *Apa*I-linearized pCa-γb-LIC vector and ~200–250 ng of the gel-purified PCR product with T4 DNA polymerase in 20 μL and 10 μL reactions, respectively, and incubate at 23 °C for 30 min (*see* **Note 4**). Stop the reaction by heat-inactivating the T4 DNA polymerase at 75 °C for 15 min.
6. To clone the insert into the vector, add 2 μL of treated vector (~20 ng) to 10 μL treated insert (200–250 ng), incubate at

65 °C for 2 min, and then leave at RT for 15 min to allow the annealing of complementary ends.

7. Transform the recombined plasmid into chemically competent *E. coli* cells, plate out and grow at 37 °C under kanamycin selection (50 μg/mL).
8. Confirm inserts in transformants by colony PCR with a pair of flanking primers, 2235.F and 2615.R, and verify by sequencing from isolated plasmid DNA.
9. Once verified, transform the pCa-γbLIC VIGS construct into electrocompetent *A. tumefaciens* cells with a 2.2 kV pulse on an electroporator, plate out and grow at 28 °C under gentamycin (25 μg/mL) and kanamycin selection (50 μg/mL).
10. In addition to BSMV RNAγ, α, and β genomic segments are also required for successful infection. It is therefore necessary to generate additional *A. tumefaciens* transformants with pCaBS-α and pCaBS-β (as in **step 9**).
11. Grow 5 mL cultures of *A. tumefaciens* overnight at 28 °C with constant shaking (220 rpm) under gentamycin (25 μg/mL) and kanamycin (50 μg/mL) selection. Three different transformants are grown, one containing the pCa-γbLIC VIGS construct, one containing pCaBS-α and one containing pCasBS-β.
12. Spin the overnight cultures down for 20 min at 2500 × *g*, discard supernatant, and resuspend the pellets in the infiltration buffer to a final OD_{600} of 1.5. Incubate suspension at RT, without shaking, for a minimum of 3 h.
13. Mix the α, β, and γ strains together at a 1:1:1 ratio, and then pressure infiltrate 0.5–1 mL of the suspension directly into the underside of *N. benthamiana* leaves using a needleless 1 mL syringe (*see* **Note 5**).
14. 3–5 days later collect directly infiltrated leaves from infected *N. benthamiana* plants, and grind them using a cold pestle and mortar with 3 mL of potassium phosphate buffer (or water) per g of leaf tissue.
15. To inoculate wheat plants dip forefinger into the sap inoculum, and gently rub each leaf between forefinger and thumb, 3–6 times. Let the plants absorb the virus for 5–10 min, mist with water, cover with plastic bags or place inside incubation boxes and keep under low light overnight to allow the plants to recover from the inoculation stress. Next day, return plants to standard growth conditions.
16. To knockdown meiotic genes wheat plants are inoculated 14 days prior to harvesting the anthers for cytological analysis, when the effects of VIGS are typically at its strongest. When using wheat cultivar Bobwhite inoculate the plants at the

~4–4.5 leaf stage (~28 days post-sowing; Fig. 1). Early anthers are ready for sampling by fixation from ~14 days post inoculation.

3.2 Preparing Meiotic Spreads

1. At the booting stage (~42 days post-sowing), harvest immature wheat spikes, and place them on a moist 9 cm Ø filter paper in a 9 cm Ø petri dish.
2. Dissect early anthers (0.5–1.5 mm length) from the spike with watchmaker's forceps and a fine mounted needle, and place them in 1 mL of fixative solution.
3. Replace the fixative solution after 1 h and again at the end of the day. Leave at RT for at least 24 h, and store at 4 °C for up to 3 months.
4. Transfer the anthers to a watch glass, remove the fixative solution, and wash with 2 mL of 0.01 M citrate buffer (3 × 5 min).
5. Remove the final wash, add 1 mL of enzyme solution, place a lid on the watch glass, and incubate at 37 °C for 60 min.
6. After an hour has passed stop the reaction by removing the enzyme solution and replacing it with 2 mL of ice-cold sterile distilled water. Keep the anthers on ice until ready to use.
7. Using a calibrated eyepiece graticule sort the anthers by size, typically into 0.1 mm piles.
8. Transfer ~6 anthers of the same size onto a slide, with a minimum amount of water, and quickly macerate them with a mounted needle.
9. Add 10 μL of 70% acetic acid to the material, mix into a suspension, place on a 45 °C hot plate, spread horizontally with a mounted needle, add a further 10 μL of 70% acetic acid, spread again, allow suspension to collect into a single droplet, and leave for 45 s.
10. Take the slide off the hot block, and add 200 μL of fixative solution in a ring around the material. Then pick the slide up, hold it at a 45° angle over a waste container, and add a further 200 μL of fixative solution directly over the entirety of the spread area.
11. Dry the preparation by blowing the back of the slide with a commercial hair dryer.
12. Mount the preparation in 10 μL of DAPI in Vectashield and apply a cover glass (24 × 40 mm; No. 1) to visualize immediately (Fig. 2a), or leave unmounted for downstream applications (*see* Subheadings 3.3 and 3.4). Unmounted slides can be kept for several months at −20 °C.

3.3 FISH

The meiotic spreads prepared in Subheading 3.2 can be used for FISH with oligonucleotide probes directly labeled with fluorophores.

1. Dehydrate the preparations by washing the slides in 100% ethanol (2 × 5 min) in a Coplin jar, remove the slides and allow to air-dry.
2. Add 40 μL of denaturation mixture onto the spread area, cover with a plastic coverslip (24 × 30 mm), and incubate on a 62 °C hot plate for 4 min to denature the chromosomal DNA.
3. Immediately snap chill the slides in a Coplin jar containing ice-cold 2× SSC, to prevent the DNA from reannealing, transfer to a 4 °C cold room and float off the coverslips.
4. Without delay remove slides from the ice-cold 2× SSC solution, drain off any excess, pipette 40 μL of probe mixture onto the spread area, cover with a plastic coverslip (24 × 30 mm), and incubate them overnight in a humid chamber at 37 °C.
5. The next day take the slides out of the 37 °C incubator, and float off the plastic coverslips in a Coplin jar containing 2× SSC.
6. Wash the slides in 2× SSC in a Coplin jar at 40 °C (3 × 5 min), and then in 2T in a Coplin jar at RT (2 × 2 min).
7. After completing the posthybridization washes remove the slides from the Coplin jar, drain off any excess 2T, mount in 10 μL of DAPI in Vectashield and apply a glass coverslip (24 × 40 mm; No. 1).
8. View the FISH preparations on a fluorescence microscope with filters for DAPI, FITC, Texas Red, and Cy5, and one that is equipped with an image capture and analysis system (Fig. 2b).

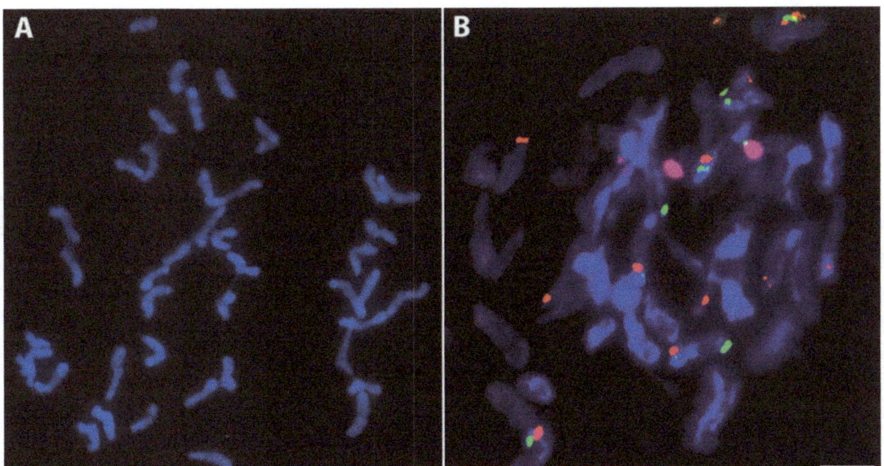

Fig. 2 Meiotic metaphase I chromosome spreads in cultivar *Bobwhite* BSMV:*asTaDMC1* VIGS knockdown lines. (**a**) DAPI only. (**b**) Probed by FISH with directly labeled synthetic oligonucleotide probes: pSc119.2-2 (green), pTa794-1 (red), and pTa71-2 (purple). The meiotic phenotype is variable between cells, but in the most severe knockdowns the majority of homologous chromosomes remain unpaired, visible as univalents. Scale bar = 10 μm

3.4 Immunolocalization on Spread Material

The meiotic spreads prepared in Subheading 3.2 are used for immunolocalization of meiotic proteins, adapted from [12].

1. Place the slides in a staining rack and, in a suitable container, wash in 1× PBS at RT (3 × 5 min).
2. While on the final wash, microwave citrate buffer on maximum power until boiling (typically ~5 min) and, while still bubbling, submerge the slides in it and leave for 45 s.
3. After 45 s has elapsed remove the slides, and wash immediately in 1× PBST at RT for 5 min.
4. Drain off any excess 1× PBST, add 50 μL of blocking solution to the spread area, cover with Parafilm, and incubate at RT for 30 min in a humid chamber.
5. Carefully remove Parafilm with tweezers, drain off any excess blocking solution, add 50 μL of primary antibodies at the desired concentration, cover with Parafilm, and incubate at 37 °C for 30 min (or overnight at 4 °C) in a humid chamber.
6. Carefully remove Parafilm and wash slides in a Coplin jar in 1× PBST at RT (2 × 5 min).
7. Drain off any excess 1× PBST, add 50 μL of secondary antibodies at the desired concentration, and incubate at 37 °C for 30 min in a humid chamber.
8. Carefully remove Parafilm and wash slides in a Coplin jar in 1× PBST at RT (2 × 5 min).
9. Drain off any excess 1× PBST, mount in 10 μL of DAPI in Vectashield, and apply a glass coverslip (24 × 40 mm; No. 1).
10. View the immunocytological preparations on a fluorescence microscope with filters for DAPI, FITC, Texas Red, and Cy5, and one that is equipped with an image capture and analysis system (Fig. 3).

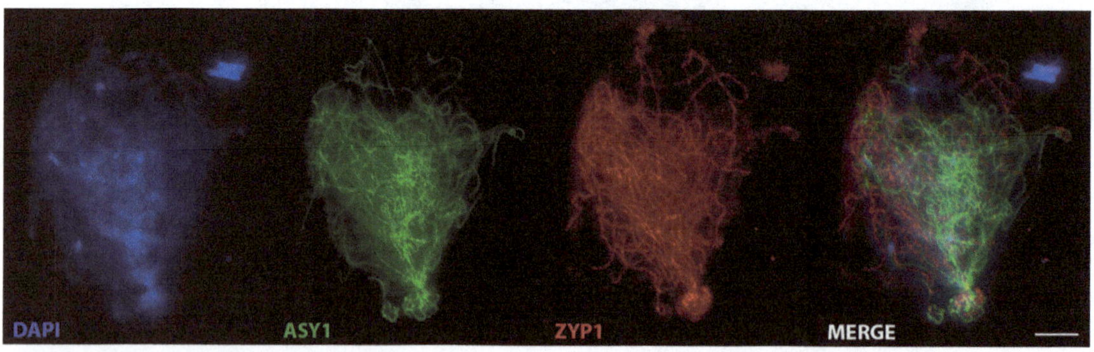

Fig. 3 Dual immunolocalization of ASY1 (green) and ZYP1 (red) to a late prophase I nucleus in a cultivar *Bobwhite* BSMV:*asTaDMC1* VIGS knockdown line, showing lengths of unaligned chromosomes that lack ZYP1. Scale bar = 10 μm

4 Notes

1. Formamide is deliberately omitted from the probe mixture as it can cause instability of probe–target hybrid molecules involving short oligonucleotides. It is present in the denaturation mixture as it is necessary to reduce the melting temperature of the chromosomal DNA, but must be washed away in ice-cold 2× SSC prior to the addition of the probes.
2. It is optimal to select and make VIGS constructs from at least two nonoverlapping fragments for each gene target. These provide independent tests of each other, and similar results with both constructs increases the confidence that the observed phenotype is due to specific silencing of the intended target gene rather than any off-target or aberrant effects.
3. In each experiment at least one negative control VIGS construct containing a 250–400 nt gene fragment of non–plant origin gene, such as the *Aequorea victoria Green Fluorescent Protein* gene (*GFP*), should also be included.
4. The $3' \rightarrow 5'$ exonuclease activity of the T4 DNA polymerase will generate complementary overhanging ends between the vector and the insert.
5. Unlike wheat, *N. benthamiana* is fully susceptible to *A. tumefaciens* infection and is used to accumulate high levels of BSMV. Sap extracted from its leaves is then used to infect wheat plants.

Acknowledgments

This work has been supported by the UK Research and Innovation (UKRI) Biotechnology and Biological Sciences Research Council (BBSRC) grant BB/N002628/1. Technical assistance was provided by Daisy Ogle, Inna Guterman, and Wing-Sham Lee.

References

1. Mercier R, Mézard C, Jenczewski E, Macaisne N, Grelon M (2015) The molecular biology of meiosis in plants. Annu Rev Plant Biol 66:297–327
2. Osman K, Higgins JD, Sanchez-Moran E, Armstrong SJ, Franklin FC (2011) Pathways to meiotic recombination in *Arabidopsis thaliana*. New Phytol 190:523–544
3. International Wheat Genome Sequencing Consortium (IWGSC) (2014) A chromosome-based draft sequence of the hexaploid bread wheat (*Triticum aestivum*) genome. Science 345:1251788
4. Krasileva KV, Vasquez-Gross HA, Howell T, Bailey P, Paraiso F, Clissold L, Simmonds J, Ramirez-Gonzalez RH, Wang X, Borrill P, Fosker C, Ayling S, Phillips AL, Uauy C, Dubcovsky J (2017) Uncovering hidden variation in polyploid wheat. Proc Natl Acad Sci U S A 114:E913–E921
5. Bhalla PL, Sharma A, Singh MB (2017) Enabling molecular technologies for trait

improvement in wheat. Methods Mol Biol 1679:3–24

6. Bennypaul HS, Mutti JS, Rustgi S, Kumar N, Okubara PA, Gill KS (2012) Virus-induced gene silencing (VIGS) of genes expressed in root, leaf, and meiotic tissues of wheat. Funct Integr Genomics 12:143–156
7. Bhullar R, Nagarajan R, Bennypaul H, Sidhu GK, Sidhu G, Rustgi S, von Wettstein D, Gill KS (2014) Silencing of a metaphase I-specific gene results in a phenotype similar to that of the pairing homeologous 1 (Ph1) gene mutations. Proc Natl Acad Sci U S A 111:14187–14192
8. Lee W-S, Hammond-Kosack KE, Kanyuka K (2012) Barley stripe mosaic virus-mediated tools for investigating gene function in cereal plants and their pathogens: virus-induced gene silencing, host-mediated gene silencing, and virus-mediated overexpression of heterologous protein. Plant Physiol 160:582–590
9. Lee W-S, Rudd JJ, Kanyuka K (2015) Virus induced gene silencing (VIGS) for functional analysis of wheat genes involved in *Zymoseptoria tritici* susceptibility and resistance. Fungal Genet Biol 79:84–88
10. Yuan C, Li C, Yan L, Jackson AO, Liu Z, Han C, Yu J, Li D (2011) A high throughput barley stripe mosaic virus vector for virus-induced gene silencing in monocots and dicots. PLoS One 6:e26468
11. Tang Z, Yang Z, Fu S (2014) Oligonucleotides replacing the roles of repetitive sequences pAs1, pSc119.2, pTa-535, pTa71, CCS1, and pAWRC.1 for FISH analysis. J Appl Genetics 55:313–318
12. Chelysheva L, Grandont L, Vrielynck N, le Guin S, Mercier R, Grelon M (2010) An easy protocol for studying chromatin and recombination protein dynamics during *Arabidopsis thaliana* meiosis: immunodetection of cohesins, histones and MLH1. Cytogenet Genome Res 129:143–153

Chapter 23

Induction and Characterization of Diploid Pollen Grains in *Arabidopsis thaliana*

Nico De Storme and Danny Geelen

Abstract

Polyploidization or whole genome duplication (WGD) is one of the main forces driving plant genome evolution and biodiversity with major implications for plant breeding and crop improvement. In nature, de novo formation of polyploid plant genomes most likely occurs through a modification of the sexual reproductive pathway. By interfering with reproductive genome stability, for example, via induction of meiotic restitution, diploid or polyploid gametes are ectopically formed that may participate in fertilization to yield polyploid offspring. This mechanism of WGD is generally referred to as sexual polyploidization. Considering the central role of sexual polyploidization in speciation, genome evolution and crop breeding, we provide here a set of methodologies to induce and characterize $2n$ pollen grain formation in plants. Using *Arabidopsis thaliana* as a model, we outline two different methods, that is, one chemical and one environmental, to induce male meiotic restitution and high frequency $2n$ pollen grain formation. In addition, we provide a set of simple and straightforward techniques to characterize alterations in male meiotic cell division and gametophytic ploidy stability underpinning $2n$ pollen formation. This comprehensive toolbox is applicable in a broad range of plant species to enable quick induction and assessment of $2n$ gamete formation during plant male reproduction.

Key words $2n$ pollen, Meiotic restitution, Temperature, GA and *Arabidopsis thaliana*

1 Introduction

Genome comparison and phylogenetic studies have revealed that most if not all plant genomes have experienced one or more rounds of whole genome duplication (WGD) during evolution [1]. As such, polyploidization is considered one of the major drivers of plant genome evolution, speciation and diversification [2, 3]. Contrary to diploid cells, which only contain two copies of the basic chromosome complement, polyploid cells contain three or more copies of each chromosome, providing them much genomic flexibility and plasticity on which selection can act [4]. Polyploidy can be induced during the vegetative phase of the plant's life cycle, that is, through endomitosis or endoreduplication, as a cellular mechanism to alter the transcriptomic or metabolic state of a single cell,

Mónica Pradillo and Stefan Heckmann (eds.), *Plant Meiosis: Methods and Protocols*, Methods in Molecular Biology, vol. 2061, https://doi.org/10.1007/978-1-4939-9818-0_23, © Springer Science+Business Media, LLC, part of Springer Nature 2020

tissue or organ. For example, in many fruit-bearing species, pericarp cells undergo several rounds of genome duplication as a cellular mechanism to increase fruit size and metabolic activity [5]. However, as these polyploidy events are programmed to only occur in specific somatic cell types, corresponding WGD events are not transferred to the offspring.

Sexual polyploidization events, on the other hand, specifically occur during the reproductive phase of the plant's life cycle and hence lead to the production of functional diploid and/or polyploid gametes [6]. The involvement of these gametes in pollination and fertilization leads to the ad hoc formation of triploid or polyploid progeny, thereby forming a mechanistic basis for the de novo establishment of stable polyploid karyotypes. Interestingly, under standard growth conditions, sexual polyploidization is very rare. The process of meiotic cell division and spore development is strictly regulated to consolidate gametophytic haploidy and to guarantee genomic integrity during germ line development [7]. However, specific (a)biotic stresses or (bio)chemical agents can interfere with strict regulation of gametophytic ploidy stability, causing ectopic events of meiotic restitution or genome doubling in the germ line to yield functional diploid or polyploid gametes [8]. Because of this, stress-induced $2n$ gamete formation and associated induction of sexual polyploidization is considered an important mechanism driving plant genome change and evolution under environmentally challenging conditions [9].

Several studies have already reported on conditions or treatments that cause ectopic induction of $2n$ gamete formation in plants via meiotic restitution [10–14]. However, in most of these studies treatments are laborious, time-consuming, nonefficient, or ambiguous. Here, using *Arabidopsis thaliana* as a model, we outline two simple methods to induce male meiotic restitution (Fig. 1), and additionally provide a set of techniques to detect and quantify resulting diploid and polyploid pollen grains. As these methods are applicable for a broad range of flowering plant species, they constitute a basic toolbox for fundamental research on reproductive genome integrity, sexual polyploidization, and plant genome evolution. Also, in the context of applied crop breeding, proposed methods may form an efficient tool to produce de novo synthetic polyploids, that is, via the implication of meiotic restitution and $2n$ gamete formation in hybridization or backcrossing programs [15].

2 Materials

2.1 Solutions

1. Kl germination medium: For 1 L of liquid Kl medium, add 2.15 g of Murashige and Skoog (1/2 MS) basal salt mixture, 10 g sucrose, 0.5 g of 2-(N-morpholino)ethanesulfonic acid (MES) and 0.1 g myoinositol to a beaker containing 0.9 L of

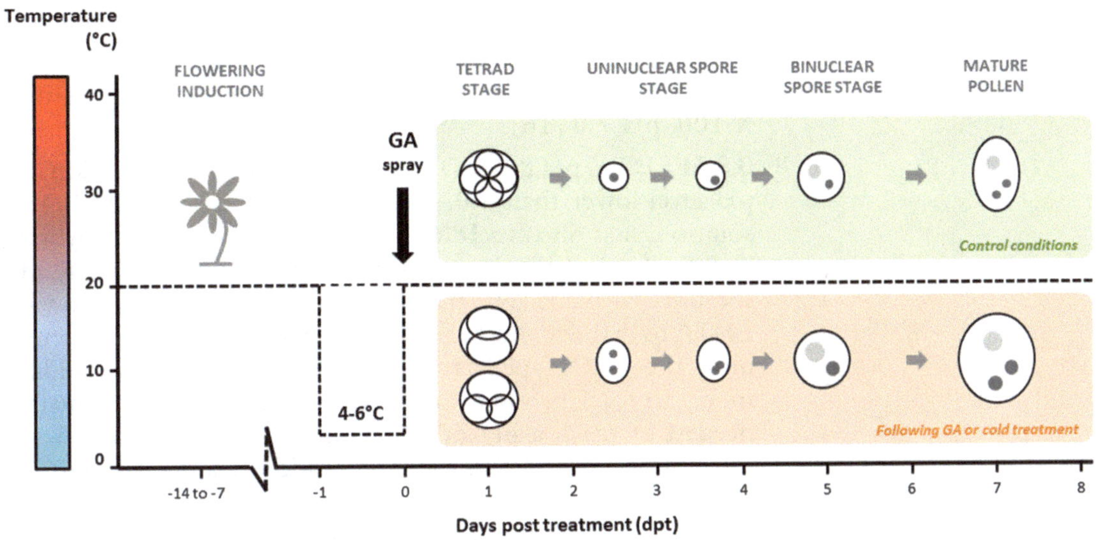

Fig. 1 Schematic overview of the cold shock and 100 μM GA treatment to induce male meiotic restitution and associated formation of diploid and polyploid pollen grain formation in *A. thaliana*. The process of male gamete development under control or following cold or GA treatment is schematically shown in the green and orange boxes, respectively

distilled water and thoroughly stir to dissolve. Adjust pH to 5.7 using 1 M KOH and add distilled water to a final volume of 1 L. To prepare plates with solid K1 medium, add 8 g of agar to 1 L of liquid K1 medium and dissolve by autoclaving the solution (20 min at 121 °C). Once the solution has sufficiently cooled down for handling, pour the agar-dissolved K1 medium in petri dishes and allow solidifying under sterile conditions (laminar flow). Once closed, plates can be stored at 4–5 °C.

2. 100 μM gibberellic acid (GA): 100 μM GA in distilled H_2O.
3. 45% lacto-propionic orcein (LPO) working solution: Dissolve 5% (m/v) synthetic orcein in a 1:1 buffer of propionic acid and lactic acid to generate the LPO stock solution. Mix the solution for several hours at RT using a magnetic stirrer, before filtering through a 20 μm mesh and bottling. Prepare the working solution by diluting the LPO stock solution to a final concentration of 45% (v/v) with distilled water. Filter the working solution again through a 20 μm mesh before use to eliminate unwanted aggregates.
4. Fixative: three parts of absolute ethanol to one part of acetic acid (v/v).
5. Isotonic solution (Phosphate buffered saline - PBS; 1×): 137 mM NaCl, 2.7 mM KCl, 10 mM Na_2HPO_4, 1.8 mM KH_2PO_4, pH 7.4.

6. Isoton® II Diluent: commercially available diluent solution (Beckman Coulter).
7. Pollen Isolation Buffer (PIB; 1×): 0.05 M $NaPO_4$, 0.5% Triton X-100, pH 7.0 [16].
8. 0.5 M EDTA, pH 8.0: EDTA salt does not go into solution at a pH level lower than 8.0. Hence, to produce a 0.5 M EDTA solution, first dissolve 18.61 g EDTA disodium salt dehydrate in 80 mL of distilled water (in a 100 mL bottle). Next, add sodium hydroxide (NaOH) pellets to increase the pH allowing EDTA to dissolve. Add a few pellets at a time and wait patiently until the NaOH pellets have fully dissolved before adding more. In total, around 2 g of NaOH pellets is approximately needed to reach a pH of 8.0. Finally, if necessary, upon full dissolution of EDTA in the buffer, add extra distilled H_20 to reach a final volume of 100 mL.
9. DAPI staining solution: 2 μg/mL (w/v) 4′,6-diamidino-2-phenylindole (DAPI) in distilled water.

2.2 Tools and Equipment

1. ARAsystem flats and trays for growing *Arabidopsis thaliana*.
2. Standard potting soil (Saniflor/Van Israel).
3. Wuxal® Super nutrient solution (8N-8P-6K + TE).
4. Microscopy slides and microscopy cover slips (18 mm × 18 mm).
5. Blunt brass rod or dentist hook.
6. A hot plate at 45 °C.
7. Stereomicroscope.
8. Light microscope (×40, ×100 and ×200 magnification) equipped with color camera.
9. Fluorescence microscope equipped for DAPI analysis (ex./em. maxima: 358/461 nm).
10. Small tweezers.
11. Fine handling needles.
12. Nail polish.
13. Laminar flow(s).
14. Eppendorf centrifuge.
15. Closed cabinet in fume hood.
16. Beckman Coulter Multisizer 3 particle analyzer.
17. Standard Multisizer Cups (25 mL—Beckman Accuvette ST).

3 Methods

3.1 Induction of Male Meiotic Restitution via Cold Shock

Meiotic cell division is tightly regulated to consolidate ploidy reduction and the associated formation of haploid spores. However, under low temperature stress, this tight regulation is impeded and male meiosis produces a subset of dyads and triads that contain unreduced spores [8, 11, 14]. The method to induce high level $2n$ pollen grain formation in Arabidopsis presented here is based on a cold shock treatment (Fig. 1). By briefly exposing flowering plants to cold (for 24 h at 4–6 °C), male meiocytes experience partial loss of MII cell wall formation and form binuclear and polynuclear spores that, upon nuclear fusion before pollen mitosis I (PMI), respectively develop into diploid and polyploid pollen grains [17].

1. Gas-sterilize *Arabidopsis thaliana* seeds for 3–5 h in a closed chamber using chloric gas (by putting 4 mL hydrochloride in 100 mL of bleach—under fume hood) and carefully sow seeds in vitro in petri dishes containing solid K1 germination medium.
2. Incubate plates with Arabidopsis seeds for 3 days at 4–5 °C in the dark to stimulate stratification. After that transfer plates to a growth chamber with standard growth conditions (18–22 °C, 12 h–12 h light–dark photoperiod, and light intensity between 120 and 150 μmol/m^2 s).
3. Following in vitro germination (5–7 days under light), transfer Arabidopsis seedlings to soil and maintain plants under standard growth conditions, 18–22 °C, 70–80% relative humidity (RH), and 12 h–12 h light–dark photoperiod regime with a light intensity ranging between 120–150 μmol/m^2 s, to sustain vegetative development and growth.
4. After 4–5 weeks of vegetative growth, change photoperiod settings to a 16 h–8 h light–dark regime to induce flowering and reproductive development. Settings of all other parameters remain unchanged; an ambient temperature of 18–22 °C and 70–80% RH.
5. Ten to twenty days following flower induction, select green and healthy plants with well-developed inflorescences and transfer to a growth cabin for cold treatment to induce diploid ($2n$) pollen grain formation. In order to maximize the level of male meiotic restitution and to minimize adverse effects on plant growth and development, treat flowering Arabidopsis plants for 24 h at a temperature of 4–6 °C (*see* **Note 1**). During cold shock treatment, settings for both light (photoperiod of 16 h–8 h light–dark, light intensity of 120–150 μmol/m^2 s) and relative humidity (70–80%) remain unchanged.

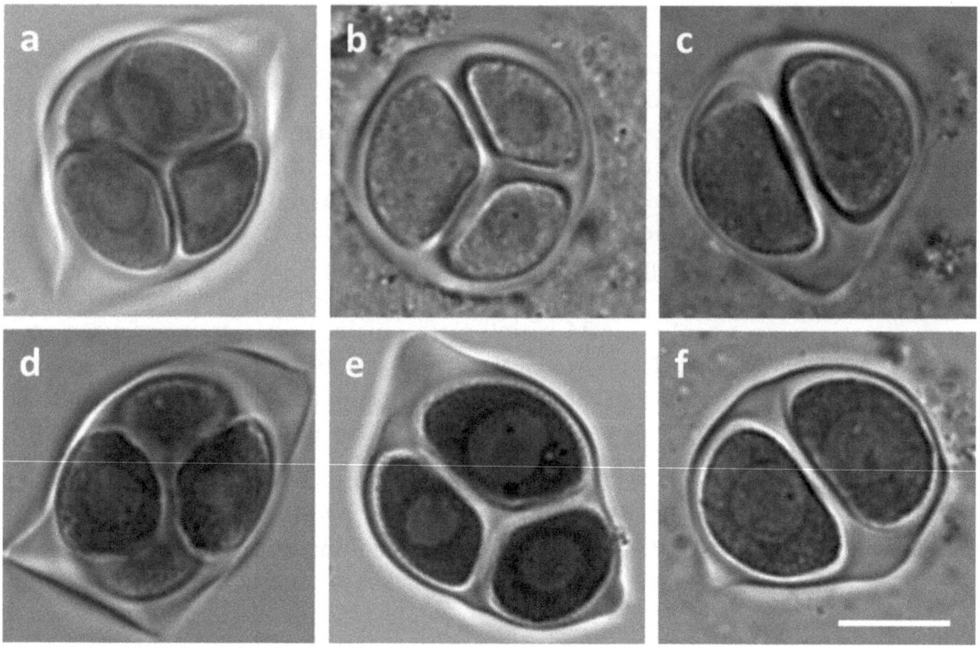

Fig. 2 Lactopropionic orcein (LPO) staining of Arabidopsis male meiocytes at the end of MII (i.e., tetrad stage). Balanced tetrads isolated from wild type *A. thaliana* Col-0 2× plants (**a**, **b**). Under normal conditions meiotic tetrads contain four equally sized spores that are spatially organized in a tetrahedral configuration. Meiotically restituted triads (**c**, **d**) and dyads (**e**, **f**) isolated from the Arabidopsis *jason* mutant, which shows male meiotic restitution due to the ectopic formation of tripolar and parallel spindles respectively during MII [18]. Scale bar: 10 μm

6. Following 24 h of cold treatment, transfer Arabidopsis plants back to standard growth conditions; ambient temperature of 18–22 °C, photoperiod of 16 h–8 h light–dark, light intensity of 120–150 μmol/m^2 s and 70–80% RH. At this moment, induction of male meiotic restitution can be validated and quantified using orcein-based "tetrad-stage analysis" (*see* Subheading 3.3; Fig. 2).

During the first 5–6 days following cold treatment, dinuclear and polynuclear spores formed via cold-induced male meiotic restitution proceed through gametophytic development. At 3–4 days posttreatment (dpt), these spores show nuclear fusion before pollen mitosis I to yield normally configured diploid and polyploid binucleate microspores with one generative and one vegetative nucleus.

Binucleate-stage spores formed upon cold treatment continue gametophytic development to form functional diploid or polyploid pollen grains 7–9 dpt (*see* **Note 2**). This pollen can then be used in selfing or hybridization assays to generate triploid or polyploid seed and progeny plants.

3.2 Induction of Male Meiotic Restitution via Gibberellic Acid

Meiotic nonreduction in Arabidopsis male sporogenesis can also be induced via chemical treatment; namely via exogenous application of GA to the inflorescences (Fig. 1). GA-treated male meiocytes show partial loss of cell wall formation at the end of MII, yielding meiotically restituted dyads and triads with binuclear and polynuclear spores that respectively develop into functional diploid and polyploid pollen grains [19, 20].

1. Germinate *Arabidopsis thaliana* seeds and grow plants in a similar way as outlined in Subheading 3.1, **steps 1–4**.
2. Approximately 10–20 days following flower induction, select green and healthy plants with well-developed inflorescences and exogenously treat them with GA by spraying GA solution (using manual spraying device) until drop off on both the inflorescences and the vegetative parts of the plant. Depending on the growth habitus and developmental size of the plants, 5–10 mL of GA solution is used per plant.
3. [Optionally] To enhance the level of male meiotic restitution, plants are treated a second time or even repeatedly. However, do not perform more than one GA treatment per day.
4. During and following GA treatment, maintain *A. thaliana* plants under standard growth conditions, that is, ambient temperature of 18–22 °C, photoperiod of 16 h–8 h light–dark, light intensity of 120–150 μmol/m^2 s and RH of 70–80%.
5. Approximately 24–48 h following exogenous GA treatment, meiotic nonreduction in male sporogenesis can be validated and quantified using orcein-based "tetrad-stage analysis" (*see* Subheading 3.3; Fig. 2).

From day 2 to day 7 following GA treatment, dinuclear and polynuclear spores, formed via loss of meiotic cell wall formation, proceed through male gametophytic development. At 3–4 dpt, they show nuclear fusion before pollen mitosis I (PMI) after which resulting binucleate spores perform a second mitosis at 5–6 dpt to yield functional trinucleate pollen grains with one vegetative nucleus and two sperm cells.

Mature diploid pollen grains are formed 7–9 days post GA treatment. These pollen grains can be used in selfing or hybridization assays to generate triploid or polyploid seed and progeny plants (*see* **Note 2**).

3.3 Assessing Male Meiotic Outcome via Tetrad Analysis

Visualization of the male meiotic outcome, at the tetrad stage, in *A. thaliana* and other plant species provides indications on the stability of the meiotic cell division and inherent processes, such as homologous recombination, chromosome segregation, and reductional cell division. At the end of male meiosis II, resulting microspores are transiently enclosed within a rigid callosic structure, facilitating quantitative assessment of meiotic chromosome segregation and gametophytic ploidy. Under normal conditions, male meiosis in most diploid plant species (or in stable polyploids with an even chromosome number) consistently generates balanced tetrads, that is, as typically characterized by the presence of four uniformly sized spores that are organized in a tetrahedral manner (*see* **Note 3**). In the case of ectopic induction of meiotic nonreduction, male meiosis yields restituted figures, such as monads, dyads or triads. The lacto-propionic orcein (LPO) staining procedure allows microscopic visualization and detection of the male meiotic outcome (tetrad stage), and provides an easy and straightforward toolkit for the characterization and quantitative analysis of male meiotic restitution in Arabidopsis and other plant species (Fig. 2).

1. Select immature flower buds ranging in overall length between 0.1 and 2.0 mm (male meiosis occurs in buds of approx. 0.5 mm) and fixate by putting them intact in fixative (*see* **Note 4**).
2. Fix the buds at RT for the first 12 h with occasional agitation, after which they can be stored permanently at 4–5 °C.
3. Transfer a single fixed flower bud to a clean microscope glass slide and add a drop (5–10 μL) of 70% ethanol onto the slide via pipetting.
4. Immediately add a single drop (5–10 μL) of 45% LPO working solution to the intact bud and gently tap the flower bud with a blunt brass rod or dentist hook to release male meiocytes into the staining solution.
5. If possible, carefully remove bud and anther wall debris from the sample.
6. Add a cover slip on the sample and gently press on the cover slip to squash the male meiocytes out of the anther locule.
7. (Optionally) Heat the slide preparation on a hot plate (45 °C) for 5–10 min to improve LPO chromosome staining and to reduce background staining of the cytoplasm (*see* **Note 5**).
8. Gently squash the sample by putting pressure on the coverslip with a double layer of filter paper to absorb excess of staining solution.
9. Asses slide preparations either directly microscopically, or preserve them semipermanently by sealing coverslip edges with nail varnish and storage at low temperature (4–6 °C) (*see* **Note 6**).
10. Analyze slides at ×100 magnification to assess the developmental state of the selected flower bud (i.e., premeiotic, postmeiotic, MI, MII, or tetrad stage).

11. Assess slides containing tetrad-stage male meiocytes further using higher magnification (×200–1000) for image acquisition and for quantitative analysis of dyads, triads and tetrads, that is, as a measure for the level of male meiotic restitution.

3.4 Screening for Diploid and Polyploid Male Spores via Pollen Size Analysis

In most plant species, the gametophytic ploidy level of mature pollen grains is closely correlated to their respective size [21]. For example, in *A. thaliana*, diploid (2×) pollen is on average 24% larger compared to their haploid (1×) counterparts [22], and similar observations have been made for other species, including *Solanum* [23], *Hordeum* [24], and *Avena* [25]. Because of this strict ploidy-size correlation, the male gametophytic ploidy level can indirectly be estimated via assessment of pollen size or volume (Fig. 3). Here, we outline two easy and straightforward techniques, that is, one image-based and one impedance-based, to assess pollen grain size in *A. thaliana* as a proxy for estimating gametophytic ploidy and quantifying 2*n* pollen grain formation [22] (*see* **Note 7**).

3.4.1 Isolation of Mature Pollen Grains

1. Select 5–20 mature and open flowers at anthesis from green and healthy plants and collect them in a 1.5 mL Eppendorf tube (*see* **Note 8**).
2. Add 500 μL of 1× PIB or isotonic solution to the Eppendorf tube and gently vortex the sample for 5 min.
3. Manually remove the flower material and large debris from the tube using small tweezers.
4. Centrifuge the pollen solution for 5 min at 22,000 × *g* using a microfuge.
5. Remove the supernatant and resuspend the pollen pellet in 15 μL of isotonic solution.
6. Resulting pollen samples are then used for pollen size analysis; either via 2D microscopy-based image (*see* Subheading 3.4.2) analysis or alternatively via 3D impedance-based assessment of pollen volume (*see* Subheading 3.4.3).

3.4.2 Image-Based Pollen Size Analysis

1. Pipet approximately 8–10 μL of fresh pollen suspension on a clean microscopy slide and gently cover by a 8 × 8 mm cover slip (*see* **Note 9**).
2. (Optional) Cover slip edges are sealed using nail polish to avoid dehydration of the pollen sample. Sealed microscopic slides can be stored at 2–4 °C for maximum 2 days.
3. Perform microscopic analysis of pollen grains either at low magnification (100×), that is, to capture large numbers of pollen grains per image (for automated image analysis), or at higher magnifications (200–1000×) to visualize single pollen grains at high resolution for accurate pollen size analysis via manual assessment (diameter or area measurements).

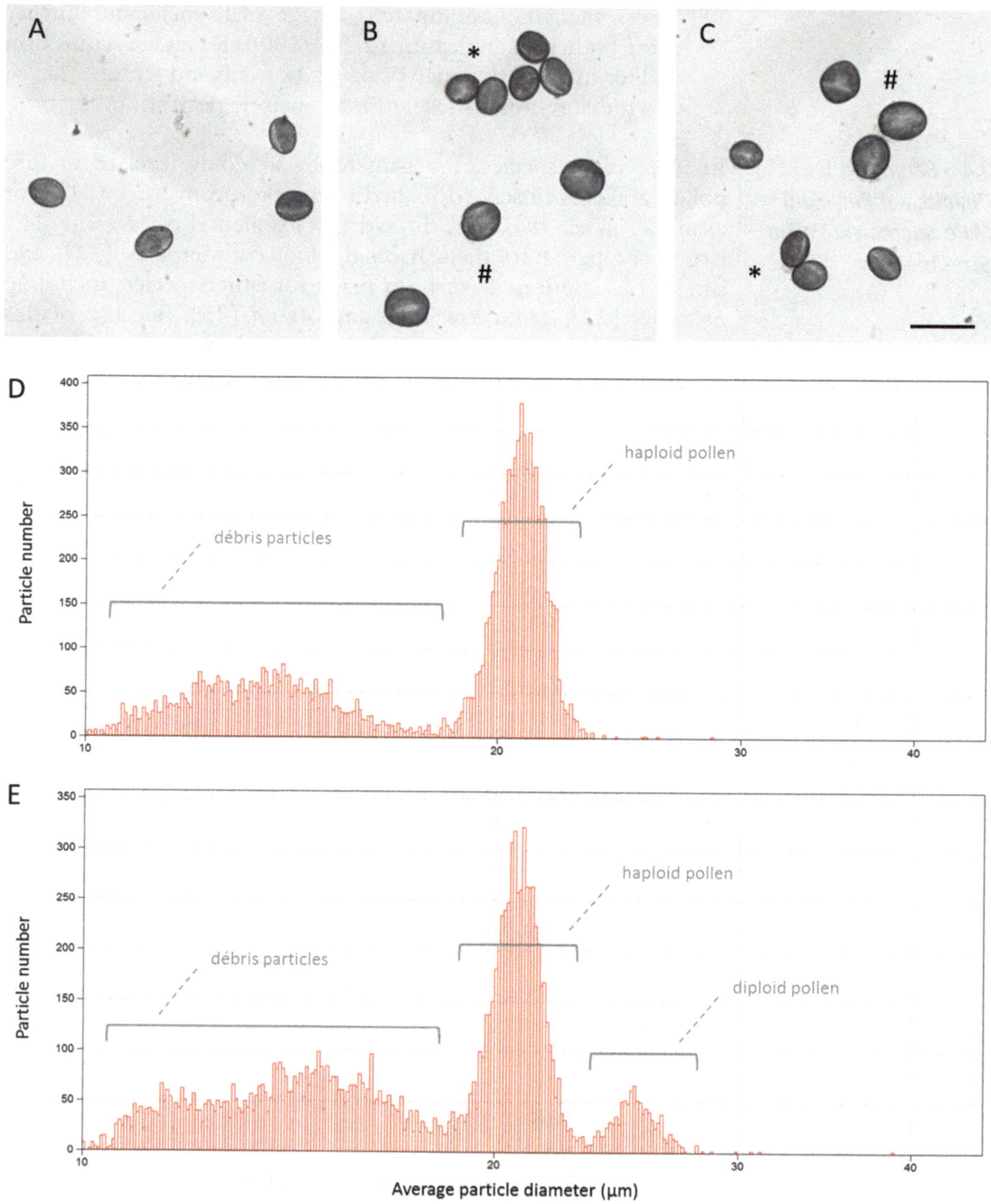

Fig. 3 Representative images of pollen isolated from *A. thaliana* Col-0 wild type (**a**) and *jason* mutant (**b**, **c**) plants; (asterisk) Haploid pollen grains, (hash) diploid pollen grains. Scale bar: 50 µm. Histograms showing spherical-based diameter values of pollen isolated from *Arabidopsis thaliana* Col-0 wild type (**d**) and *jason* mutant (**e**), as obtained via automated particle size analysis using the Multisizer III Coulter Counter device (Beckman Coulter). Wild type plants produce one single population of uniformly sized haploid pollen grains, whereas the *jason* mutant produces an extra population of significantly larger pollen grains, that is, representing diploid (2*n*) pollen grains formed via male meiotic restitution

4. Perform image-based pollen size analysis and distribution acquisition manually via diameter measurements (*see* **Note 10**) or in a highly automated manner using 2D particle size analysis software programs, such as Fiji or ImageJ [26].

In a hydrated state, haploid *Arabidopsis thaliana* pollen grains have an ellipsoid shape with three apertures symmetrically positioned along the length of the pollen grain. With increasing ploidy levels, pollen grains are not only bigger but also show a more spherical shape with an increasing number of apertures [23]. Overall, the mean minor diameter of haploid, diploid and tetraploid pollen in *Arabidopsis thaliana* amounts to 20.3, 25.6, and 32.3 μm with an averaged eccentricity of 0.50, 0.45, and 0.36, respectively (the eccentricity is a parameter that represents the deviation of an ellipsoid from a perfect circle and ranges from 0 to 1 with a value of 0 indicating for a perfect circle and a value of 1 representing an extreme ellipse) [15].

3.4.3 Volumetric Pollen Size Analysis Using Multisizer III

1. Fill a standard Multisizer analysis cup with 10 mL of Isoton II and add the resuspended pollen pellet into this cup by pipetting (*see* **Note 11**).
2. Place the Multisizer cup with the pollen solution at the designated place into the Beckman Coulter Multisizer III device. Make sure that the analysis tube installed in the device has the appropriate aperture size (*see* **Note 12**).
3. Start automated assessment of pollen grain size. Repeatedly check for putative blockages of the tube aperture using the built-in camera display. In case of a physical blockage, pause the analysis and manually remove the debris by carefully flushing or rubbing the tube aperture.
4. Stop automated analysis once an appropriate number of pollen grains have been monitored.

Via an integrated impedance measurement, the Multisizer III devices accurately measures the individual volume of a large number of isolated pollen, irrespective of their shape, size or morphology, and hence allows representation of pollen size either directly as a volume (μm^3) or alternatively as an average diameter (μm) [23]. In the latter case, however, the particular shape of the pollen grain is not taken into account, and the derived "average diameter" is calculated from the pollen volume under the assumption that they are perfect spherical.

In diploid, tetraploid, and octaploid *Arabidopsis thaliana* plants, volume-based pollen size always exhibits a Gaussian distribution, indicating that the size of corresponding pollen grains is highly uniform with only a minor biological variability [23].

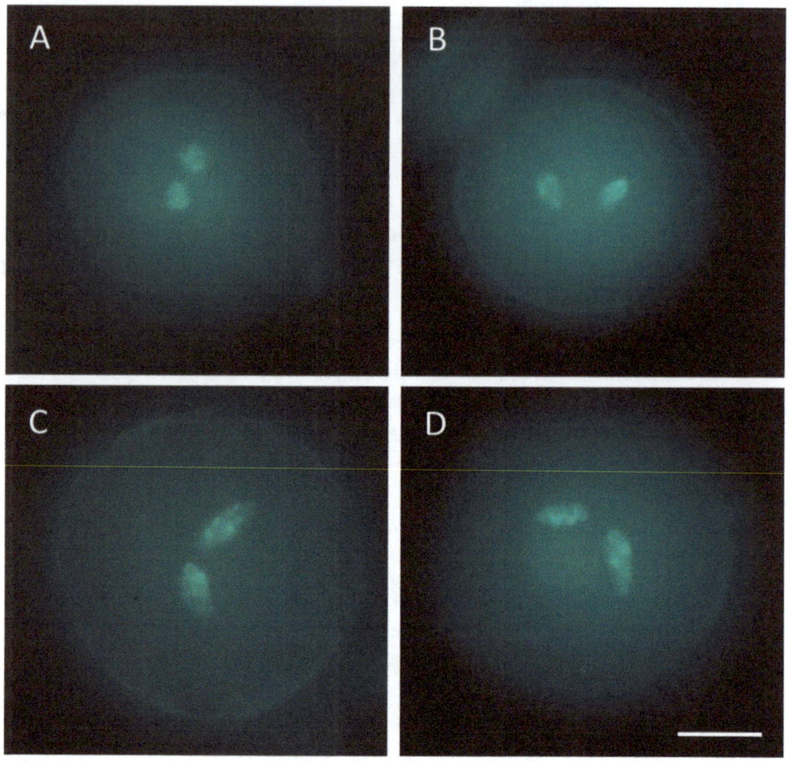

Fig. 4 DAPI-stained mature pollen grains isolated from *A. thaliana* Col-0 plants consistently showing two sperm nuclei in haploid (**a**, **b**) and diploid (**c**, **d**) pollen grains. Note that both the sperm nuclei in diploid pollen grains are significantly larger compared to their corresponding counterparts in haploid pollen grains. Scale bar: 5 μm

3.5 Male Gametophytic Ploidy Analysis via DAPI Staining of Mature Pollen Grains

Validation and quantification of diploid and polyploid pollen grain formation in plants can be performed by quantitative ploidy analysis of the sperm cells. Visualization of the sperm nuclei enables detailed analysis of the number and size of sperm, and hence allows quantitative analysis of the gametophytic ploidy level and nuclear configuration of mature pollen grains. A detailed procedure for DAPI staining of sperm nuclei in mature *A. thaliana* pollen grains is outlined (Fig. 4).

1. Harvest 5–10 Arabidopsis flowers at anthesis and collect them in a 2 mL Eppendorf tube. Only select flowers that are fully open and from healthy plants free of pests or infections.
2. Add 350 Ml of 0.5 M EDTA to the tube and vortex this solution at low speed for approximately 5 min to stimulate dissolution of pollen in the buffer.
3. Remove flowers carefully from the Eppendorf tube using tweezers and centrifuge the solution for 5 min at 18,000 × *g* using a microfuge.

4. Carefully remove the supernatant, leaving the yellowish pellet for further treatment.
5. Add 200 μL of fixative to the pellet, briefly vortex and incubate for 20 min at RT to fixate the pollen grains.
6. Centrifuge the sample for 5 min at 18,000 × *g* using a microfuge and remove the supernatant.
7. Wash the pollen pellet by adding 350 μL of distilled H_2O to the tube.
8. Incubate for 5 min at RT, centrifuge for 5 min at 18,000 × *g* using a microfuge, and remove the supernatant.
9. Add 10 μL of DAPI staining solution to the pollen pellet to label DNA and to visualize sperm nuclei.
10. Incubate for 30 min at RT, add 200 μL of distilled H_2O, centrifuge the sample for 5 min at 18,000 × *g* using a microfuge and remove the supernatant.
11. Wash the pollen pellet with distilled H_2O to reduce background fluorescence. To do so, add 350 μL of distilled H_2O to the pollen pellet, centrifuge samples for 5 min at 18,000 × *g* using a microfuge and carefully remove supernatant.
12. Add 15 μL of distilled H_2O to the washed pollen pellet and pipet the DAPI-stained pollen solution onto a microscopy slide.
13. Carefully cover the sample with an 8 × 8 mm cover slip.
14. In order to eliminate dehydration of the sample, seal cover slip edges with nail polish.
15. Visualize DAPI stained sperm nuclei using (epi)fluorescence microscopy; optimally using excitation light of 365 nm at a magnification of ×200–1000.

At anthesis, *Arabidopsis thaliana* pollen grains are trinucleate with one large, less condensed vegetative nucleus and two smaller, highly condensed sperm nuclei. The size of both types of gametophytic nuclei increases with increasing ploidy level, and thus allows for an estimation of the gametophytic ploidy level on a single cell basis.

4 Notes

1. The optimal duration of cold treatment can vary between species and generally depends on the duration of the meiotic cell cycle. In *A. thaliana*, male meiosis takes approximately 24 h from prophase up till the end of meiosis II. Using a cold shock treatment of 24 h or longer hence guarantees that all male meiocytes have undergone cold stress during the meiotic

stage(s) that are relevant for induction of meiotic restitution. For plants that show a shorter or a prolonged meiotic cell division, the duration of cold shock treatment thus may need to be optimized to guarantee and optimize the level of $2n$ pollen grain formation.

2. The frequency of $2n$ pollen grain formation via cold shock or via exogenous GA treatment in *A. thaliana* amounts to 10–15%. Remaining pollen grains are haploid and do not show any implications by the cold treatment.
3. Male meiocytes of plant species or variants with an aneuploid genome also generally show tetrad figures containing four microspores; however, resulting spores are variable in size due to unbalanced segregation of the chromosomes in MI or MII. These meiotic figures are often referred to as "unbalanced tetrads." Also, due to the occasional occurrence of lagging chromosomes or chromosome missegregation in MI or MII, male meiosis in this type of plants may also yield tetrad-stage figures that harbor more than four spores, that is, often referred to as polyads.
4. Alternatively, fixation of immature flower buds can also be done in Carnoy's fixative solution, which is composed of 60% (v/v) ethanol, 30% (v/v) chloroform, and 10% (v/v) glacial acetic acid.
5. Preserving LPO slide preparations for 1 to 3 h at room temperature (instead of incubation on a hot plate) also reduces cytoplasm staining and improves staining of male meiocytes and chromosomes. In this case, however, it is essential to carefully seal the coverslip edges to reduce sample dehydration and to maintain integrity of the slide preparation.
6. LPO slide preparations cannot be stored longer than 1 week.
7. Pollen size-based estimation of the male gametophytic ploidy level, either via 2D or 3D size measurements outlined here, can be applied for a broad range of plant species. Prior knowledge about putative size differences between haploid (1×) and diploid (2×) pollen is highly relevant, as this forms the basis for size-based detection and quantification of $2n$ pollen grains.
8. The number of mature flowers largely depends on the pollen production per flower. In case pollen production per flower is low, then it is advisable to pool a large number of flowers to obtain sufficient pollen material for size distribution analysis.
9. The optimal volume of pollen solution pipetted on the microscopy slide largely depends on the size of the pollen grains. When too little volume is applied for larger size pollen grains, capillary forces between microscopy slide and cover slip will squeeze the pollen grains, thereby artificially increasing their

transectional surface area. This causes biases in pollen size acquisition. In contrast, when too much volume is applied for smaller size pollen grains, pollen are not stationary but instead move around in the solution, thereby strongly impairing proper image acquisition.

10. Importantly, depending on the specific shape of pollen grains under investigation, one should take into account the three-dimensional position of the pollen grain on the microscopy slide when calculating pollen size (2D transsectional surface area or 3D volume) via 2D microscopy image analysis. For example, due to the prolate ellipsoid shape of *Arabidopsis thaliana* pollen, pollen volume measurement based on 2D data, such as microscopy images, requires accurate quantification of both the major and minor axis and calculation of the volume using the following formula:

$$4/3 \times \pi \times a \times b^2 \text{ with } a = \text{major axis length and } b = \text{minor axis length.}$$

11. Avoid air bubbles as these can interfere with impedance-based pollen size analysis.
12. For the analysis of Arabidopsis pollen grains, the Multisizer III analysis tube with an aperture diameter of 100 μm is used (i.e., accurate particle size analysis between 2 and 60 μm diameter).

Acknowledgments

We would like to thank C. Petit and P. Vandenbroucke for practical assistance with the growing and maintenance of Arabidopsis plants. Also many thanks to Prof. M. Höfte (Phytopathology Lab, UGent) for the use of the light microscopy platform.

References

1. Jiao YN, Wickett NJ, Ayyampalayam S et al (2011) Ancestral polyploidy in seed plants and angiosperms. Nature 473:97–100
2. Landis JB, Soltis DE, Li Z et al (2018) Impact of whole-genome duplication events on diversification rates in angiosperms. Am J Bot 105:348–363
3. Soltis PS, Soltis DE (2016) Ancient WGD events as drivers of key innovations in angiosperms. Curr Opin Plant Biol 30:159–165
4. Leitch AR, Leitch IJ (2008) Genomic plasticity and the diversity of polyploid plants. Science 320:481–483
5. Cheniclet C, Rong WY, Causse M et al (2005) Cell expansion and endoreduplication show a large genetic variability in pericarp and contribute strongly to tomato fruit growth. Plant Physiol 139:1984–1994
6. De Storme N, Geelen D (2013) Sexual polyploidization in plants cytological mechanisms and molecular regulation. New Phytol 198:670–684
7. Wijnker E, Schnittger A (2013) Control of the meiotic cell division program in plants. Plant Reprod 26:143–158
8. Pecrix Y, Rallo G, Folzer H et al (2011) Polyploidization mechanisms: temperature environment can induce diploid gamete formation in *Rosa* sp. J Exp Bot 62:3587–3597

9. De Storme N, Mason A (2014) Plant speciation through chromosome instability and ploidy change: cellular mechanisms, molecular factors and evolutionary relevance. Curr Plant Biol 1:10–33
10. Lu M, Zhang PD, Kang XY (2013) Induction of 2n female gametes in *Populus adenopoda* maxim by high temperature exposure during female gametophyte development. Breed Sci 63:96–103
11. Mason AS, Nelson MN, Yan GJ et al (2011) Production of viable male unreduced gametes in *Brassica* interspecific hybrids is genotype specific and stimulated by cold temperatures. BMC Plant Biol 11:103
12. Negri V, Lemmi G (1998) Effect of selection and temperature stress on the production of 2n gametes in *Lotus tenuis.* Plant Breed 117:345–349
13. Rezaei M, Arzani A, Sayed-Tabatabaei BE (2010) Meiotic behaviour of tetraploid wheats (*Triticum turgidum* L.) and their synthetic hexaploid wheat derivates influenced by meiotic restitution and heat stress. J Genet 89:401–407
14. Wang J, Li DL, Shang FN et al (2017) High temperature-induced production of unreduced pollen and its cytological effects in *Populus.* Sci Rep 7:5281
15. Younis A, Hwang YJ, Lim KB (2014) Exploitation of induced 2n-gametes for plant breeding. Plant Cell Rep 33:215–223
16. Engel ML, Davis RH, Mccormick S (2005) Green sperm. Identification of male gamete promoters in Arabidopsis. Plant Physiol 138:2124–2133
17. De Storme N, Copenhaver GP, Geelen D (2012) Production of diploid male gametes in arabidopsis by cold-induced destabilization of postmeiotic radial microtubule arrays. Plant Physiol 160:1808–1826
18. De Storme N, Geelen D (2011) The Arabidopsis mutant *jason* produces unreduced first division restitution male gametes through a parallel/fused spindle mechanism in meiosis II. Plant Physiol 155:1403–1415
19. Liu B, De Storme N, Geelen D (2018) Cold-induced male meiotic restitution in *Arabidopsis thaliana* is not mediated by GA-DELLA signaling. Front Plant Sci 9:91
20. Liu B, De Storme N, Geelen D (2017) Gibberellin induces diploid pollen formation by interfering with meiotic cytokinesis. Plant Physiol 173:338–353
21. Altmann T, Damm B, Frommer WB et al (1994) Easy determination of ploidy level in *Arabidopsis thaliana* plants by means of pollen size measurement. Plant Cell Rep 13:652–656
22. De Storme N, Zamariola L, Mau M et al (2013) Volume-based pollen size analysis: an advanced method to assess somatic and gametophytic ploidy in flowering plants. Plant Reprod 26:65–81
23. Sakhanokho HF, Islam-Faridi MN (2014) Spontaneous autotetraploidy and its impact on morphological traits and pollen viability in *Solanum aethiopicum.* HortScience 49:997–1002
24. Johansen B, Vonbothmer R (1994) Pollen size in *Hordeum* L.: correlation between size, ploidy level, and breeding system. Sex Plant Reprod 7:259–263
25. Katsiotis A, Forsberg RA (1995) Pollen grain size in four ploidy levels of genus *Avena.* Euphytica 83:103–108
26. Schindelin J, Rueden CT, Hiner MC et al (2015) The ImageJ ecosystem: an open platform for biomedical image analysis. Mol Reprod Dev 82:518–529

Chapter 24

Analysis of Pollen Grains by Immunostaining and FISH in Triticeae Species

Dan D. Wu, Alevtina Ruban, Twan Rutten, Yong H. Zhou, and Andreas Houben

Abstract

In this chapter we describe protocols for immunolabeling and FISH of pollen grains undergoing post-meiotic mitosis using *Aegilops speltoides*, *Secale cereale*, and *Hordeum vulgare* as models. Tissue sectioning of pollen overcomes the problem of the pollen grain wall impermeability that interferes with immunolocalization and in situ hybridization. The crucial element of the protocol is the generation and immobilization of tissue sections. Our method facilitates the investigation of the microspore cell divisions and pollen grain maturation.

Key words *Aegilops speltoides*, *Hordeum vulgare*, *Secale cereale*, Pollen mitosis, Tissue section, Immunostaining, FISH, Meiosis

1 Introduction

Microgametogenesis comprises events which lead to the development of the unicellular microspores into mature microgametophytes containing the two male gametes. The first mitosis in microspores of angiosperms results in two structurally and functionally different nuclei, one generative and one vegetative. This process is accompanied by an asymmetric cell division, which partitions the mother cell into two daughter cells of unequal size [1, 2]. The smaller generative cell undergoes a second mitotic division that results in the formation of two reproductive sperm cells, which will fuse during fertilization with the egg and central cell, respectively. The functional differentiation of pollen nuclei is accompanied by contrasting nuclear chromatin configurations. The generative nucleus is highly condensed, whereas the vegetative nucleus is characterized by diffuse chromatin and higher transcriptional activity [3]. The nuclear differentiation occurs during the first pollen mitosis or in early interphase bicellular pollen [4]. Soon

Mónica Pradillo and Stefan Heckmann (eds.), *Plant Meiosis: Methods and Protocols*, Methods in Molecular Biology, vol. 2061, https://doi.org/10.1007/978-1-4939-9818-0_24,

after the asymmetric division of the microspore, the generative cell enters S phase and then passes into G2, while the vegetative cell is arrested at G1 [1]. Such fundamental differences are correlated with differential histone modification patterns in corresponding nuclei. To elucidate the chromatin composition of generative and vegetative nuclei and to trace individual chromosomes, we developed a method suitable to combine immunostaining and FISH [5–8]. To overcome the problem of the impermeable pollen grain wall that interferes with immunolocalization and in situ hybridization studies, we employed a tissue sections method which we describe below.

2 Materials

2.1 Plant Material and Plant Cultivation

Aegilops speltoides, *Secale cereale*, and *Hordeum vulgare* plants are cultivated under greenhouse conditions (16 h light, 22 °C day/ 16 °C night).

2.2 Tools and Equipment

1. Parafilm M.
2. Moist chamber for slide incubation.
3. Glass and plastic staining jars, Hellendahl-type.
4. Histological microtome (Leica Biosystems, Leica RM 2265).
5. Low profile microtome blades (Leica Biosystems, 14035838925).
6. Coverslips for microscope slides 24 × 50 mm, #1 (Menzel-Gläser) (Fig. 1a2).
7. Standard microscopic slides (Roth).
8. Disposable base molds (Fig. 1a4).
9. Disposable base molds with PEG-wax (Fig. 1a5).
10. Silicone embedding mold (Plano, G3532) (Fig. 1a6).
11. Nitrocellulose membrane (Serva, 42516.01).
12. Sterile disposable filter, PES membrane, diameter 50 mm, pore size 0.45 μm (Thermofisher, 124-0045).
13. Diamond pencil (Fig. 1a7).
14. Micro spatula (Lab unlimited) (Fig. 1a10).
15. Flat synthetic brush (Fig. 1a11).
16. Fuzz-free tissue.
17. Glue (Marabu, fixo gum).

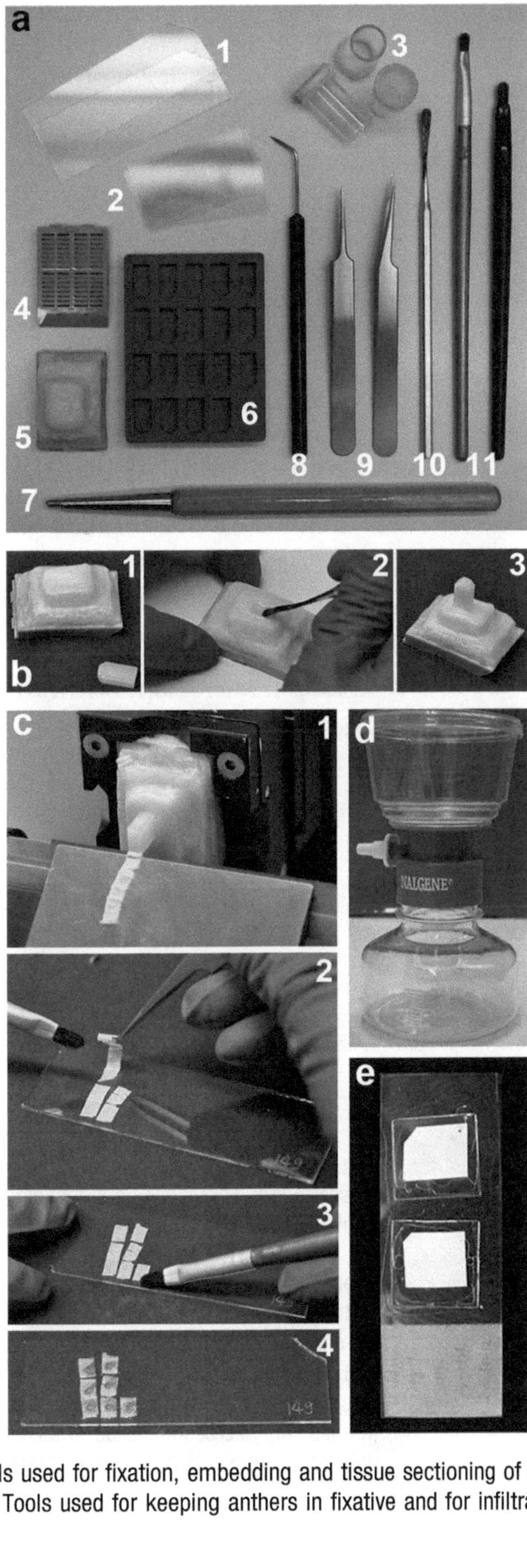

Fig. 1 Tools used for fixation, embedding and tissue sectioning of anthers and pollen. (**a**) Tools used for keeping anthers in fixative and for infiltration. Slides

2.3 Solutions

2.3.1 Chemicals and General Solutions

1. Phosphate buffered saline (PBS; 10×): 1.37 M NaCl, 0.026 M KCl, 0.1 M Na_2HPO_4, 0.018 M KH_2PO_4, pH 7.4. Dissolve 80 g of NaCl, 2 g of KCl, 14.4 g of Na_2HPO_4, and 2.4 g of KH_2PO_4 in 800 ml of dH_2O. Mix, adjust the pH to 7.4 with 1 M HCl and add dH_2O up to 1 l. Autoclave and store at 4 °C (*see* **Note 1**).
2. Microtubules stabilizing buffer (MTSB; 10×): 0.89 M KOH, 0.53 M PIPES, 1 M $MgSO_4$, 0.05 M EGTA, pH 7.2. Dissolve 50 g of KOH completely in 800 ml of dH_2O and add 160 g of PIPES, 12.3 g of $MgSO_4$, and 19.02 g of EGTA. Adjust the pH with 5 M KOH to 7.2 and add dH_2O up to 1 l. Autoclave and store at 4 °C (*see* **Note 1**).
3. Saline-sodium citrate (SSC; 20×): 3 M NaCl, 0.3 M $Na_3C_6H_5O_7{\cdot}2H_2O$, pH 7.2. Dissolve 175.3 g of NaCl and 88.2 g of $Na_3C_6H_50_7{\cdot}2H_2O$ in ~900 ml of dH_2O. Adjust pH to 7.2 with 1 M HCl and add H_2O to total volume of 1 l. Autoclave and store at 4 °C.
4. DAPI solution: 1 μg/ml DAPI (4,6-diamidino-2-phenylindole, dihydrochloride) in Vectashield (Linaris). Store at 4 °C.

2.3.2 Coating Slides

Poly-L-lysine 0.01% (Sigma-Aldrich) or Biobond Tissue Adhesive (Electron Microscopy Sciences).

2.3.3 Material Fixation and Embedding

1. Paraformaldehyde fixative (PFA; 3% freshly prepared): 0.05 M paraformaldehyde, 1 M NaOH, 0.05% Triton X-100, 10× PBS, or 10× MTSB. Dissolve 1.5 g of paraformaldehyde completely in 40 ml of deionized water with additional 5 μl of 1 M NaOH while heating to 60 °C. After cooling down to RT, add 5 ml either of 10× PBS and adjust the solution to pH 7.0 with 1 M HCl or 10× MTSB and adjust pH to 7.2 with 5 M NaOH. Add 250 μl of 10% Triton X-100 and adjust the total volume to 50 ml.

Fig. 1 (continued) and coverslips (**a**1, 2), container with filter (**a**3), embedding cassette (**a**4), PEG block on embedding cassette after removal of the mold (**a**5), mold (**a**6), diamond pencil to mark slides (**a**7), needle, forceps, brushes (**a**8–11). (**b**) Before tissue sectioning, block with anthers is fixed on disposable base molds with PEG-wax (**b**1–3). (**c**) Tissue sectioning with a microtome. Cut blocks with a microtome as 8–10 μm slices (**c**1), transfer slices using forceps and flat synthases brush (**c**2), stick slices on poly-L-lysine–coated slides with a brush (**c**3) and drop 1–1.5 μl of distilled water on each slice. (**d**) Disposable 45 μm filter system for immobilizing and washing pollen sections in PEG onto NC membranes. (**e**) Microscope slide with NC membranes carrying immobilized and labeled pollen sections ready for examination

2. PEG-wax: Heat 90 g of PAG (polyethylene glycol) distearate (Sigma-Aldrich, 305413) to 60 °C and add 10 g of cetyl alcohol. Keep liquid at 37 °C.
3. Infiltration solution series: PEG-wax in absolute ethanol. Three solutions of following ratios: one part of PEG-wax with two parts of ethanol; one part of PEG-wax with one part of ethanol; two parts of PEG-wax with one part of ethanol (v/v). Keep liquid at 37 °C.
4. Alginate (Roth, 7317).
5. $CaCl_2$ solution: 0.05 M $CaCl_2$ in dH_2O.
6. PEG 1500: polyethylene glycol 1500 (Roth, 0152.1).

2.3.4 Immunostaining

1. Sodium citrate buffer: 0.01 M $C_6N_5Na_3O_7{\cdot}2H_2O$. Dissolve 2.94 g of $C_6N_5Na_3O_7{\cdot}2H_2O$ in 1 l of deionized water and adjust pH to 6.0 with 1 M HCl. Store at 4 °C.
2. Blocking buffer: 5% BSA (bovine serum albumin, AppliChem Panreac), 0.3% Triton X-100 in 1× PBS or 1× MTSB. Store at −20 °C.
3. Primary antibody solution (*see* **Note 2**): rabbit anti-grass CENH3 [9] and mouse anti-α-tubulin (Sigma-Aldrich, T9026-.2ML) [2] were diluted 1:3000 and 1:300 respectively in 1× PBS or 1× MTSB buffer containing 1% BSA and 0.3% Triton X-100.
4. Secondary antibody solution: prepare a relevant dilution of, for example, anti-rabbit conjugated to Cy3 (Jackson ImmunoResearch, 111-165-003) and anti-mouse conjugated to Alexa Fluor 488 (Thermo Fisher scientific, #A11001). Dilution depends on the antibody used.

2.3.5 Fluorescence In Situ Hybridization

1. FISH probes are generated using the Atto550 NT labeling kit (Jena Bioscience) and Atto488 NT labeling kit (Jena Bioscience). Store probes according the manufacturer's instructions.
2. Fixative solution: three parts of absolute ethanol to one part of acetic acid (v/v).
3. Formaldehyde solution (4%): dilute 37% formaldehyde (Roth) with 2× SSC. Prepare fresh.
4. Hybridization mixture: dissolve 1 g of dextran sulfate in 1 ml of ddH_2O at 60 °C and adjust volume up to 2 ml. Add 5 ml of deionized formamide, 1 ml of 20× SSC, and 1 ml of fish sperm DNA (Sigma-Aldrich, 1146714001, working concentration 100 μg/ml). Leave for 1 h at RT, aliquot, and store at −20 °C.

3 Methods

In this section we describe how to analyze pollen grains within anthers or in tissue culture suspension. In most experiments pollen can be retained within the anthers facilitating their concentration (Subheading 3.1). In cases of pollen suspensions it is necessary to contain the pollen in an artificial matrix such as alginate (Subheading 3.2).

3.1 Analysis of Pollen Grains in Anthers

3.1.1 Coating Glass Slides with Poly-L-Lysine (Poly-L-Slides, Fig. 1a1)

1. Cut one edge and mark the working side of the glass slides with a diamond pencil.
2. Keep marked slides in 100% ethanol overnight at RT.
3. Dry and clean slides with fuzz-free tissue.
4. Apply 60 μl of poly-L-lysine on the working side of one slide and cover slide face to face with another slide.
5. Keep slides in a moist chamber for 2 h at RT.
6. Separate slides under water in a 20 cm diameter petri dish filled with distilled water and wash slides in ddH_2O for 2 min twice.
7. Dry slides with a flow of pressurized air. Keep slides at 4 °C.

3.1.2 Fixation and Embedding of Material

1. Isolate post-meiotic anthers (*see* **Note 3**) from spikelets and cut-open both ends of the anthers using a blade razor.
2. Transfer anthers in freshly prepared 3% paraformaldehyde solution in 1× PBS or 1× MTSB, apply vacuum for 5 min, and fix for 2 h. All steps with 1× PBS are conducted on ice, with 1× MTSB are conducted at RT.
3. Wash material twice in 1× PBS on ice or 1× MTSB at RT for 15 min.
4. Dehydrate materials in ethanol series: 30%, 50%, 70%, 90% and 100% for 30 min each at RT, and in 100% ethanol for 30 min at 37 °C.
5. Incubate anthers in an infiltration solution series of 1:2, 1:1, 2:1 (PEG-wax: ethanol, v/v) and pure PEG-wax up to 24 h at 37 °C in each solution.
6. Keep container with pure dissolved PEG-wax in warm water bath (50–75 °C) to avoid solidification during embedding. The temperature should be above 37 °C to prevent premature solidification.
7. Using a Pasteur pipette fill half of the embedding silicone mold with pure PEG-wax. Let the surface solidify.
8. Per block place three fixed anthers on the PEG-wax surface and fill them up with pure PEG-wax (*see* **Note 4**).
9. Let blocks harden overnight at RT, take the blocks out of the mold and store in a closed plastic bag at 4 °C for up to 1 year.

3.1.3 Tissue Sectioning

1. Before tissue sectioning, keep poly-L-slides and the PEG-wax blocks containing the anthers at RT overnight.
2. Micro spatula is heated using an alcohol lamp and employed to melt disposable base molds with PEG-wax. Fix block with anthers on disposable base molds with PEG-wax (Fig. 1b1–3).
3. Cut blocks into 8–10 μm thick slices using a microtome (Fig. 1c1).
4. Transfer slices onto poly-L-slides using forceps and a synthetic brush. Spread slices well on the surface of a slide with a flat synthetic brush (Fig. 1c2–3).
5. Add a 1–1.5 μl drop of distilled water on top of each section and let them dry at RT overnight (Fig. 1c4).

3.1.4 Immunostaining

1. Remove traces of PEG-wax by washing the slides twice for 10 min in 96% ethanol and one time for 10 min in 90% ethanol. Air dry slides until there are no traces of ethanol on the surface detectable. At this stage the quality of sections can be briefly checked under a phase contrast microscope (*see* **Note 5**). Wash the slides for 10 min in 90% ethanol and twice in ddH_2O for 5 min.
2. Transfer wet slides into a plastic Hellendahl jar containing 70 ml of 10 mM sodium citrate buffer and microwave the sample for 60 s at 800 W (solution begins to boil).
3. Immediately after the microwave treatment, transfer the slides into 1× PBS or 1× MTSB for 5 min at RT using forceps.
4. Perform blocking with 200 μl of blocking buffer per slide in a moist chamber for 1 h at RT. Ensure that the blocking buffer covers all sections. Do not cover the sections with Parafilm or coverslips.
5. Apply 100 μl of primary antibody solution per slide. Ensure that each section is covered. Shield with Parafilm and incubate in a moist chamber for 16–24 h at 4 °C.
6. Gently remove the Parafilm and wash the slides in 1× PBS or 1× MTSB three times for 5 min.
7. Apply 200 μl of secondary antibody solution per slide. Ensure that each section is covered. Do not shield the sections with Parafilm or coverslips. Incubate slides in a moist chamber for 1 h at 37 °C in darkness. Avoid strong light after incubation with secondary antibodies.
8. Gently wash slides in 1× PBS or 1× MTSB twice for 10 min.
9. Dehydrate slides in an ethanol series of 70%, 80% and 90% for 1 min each.

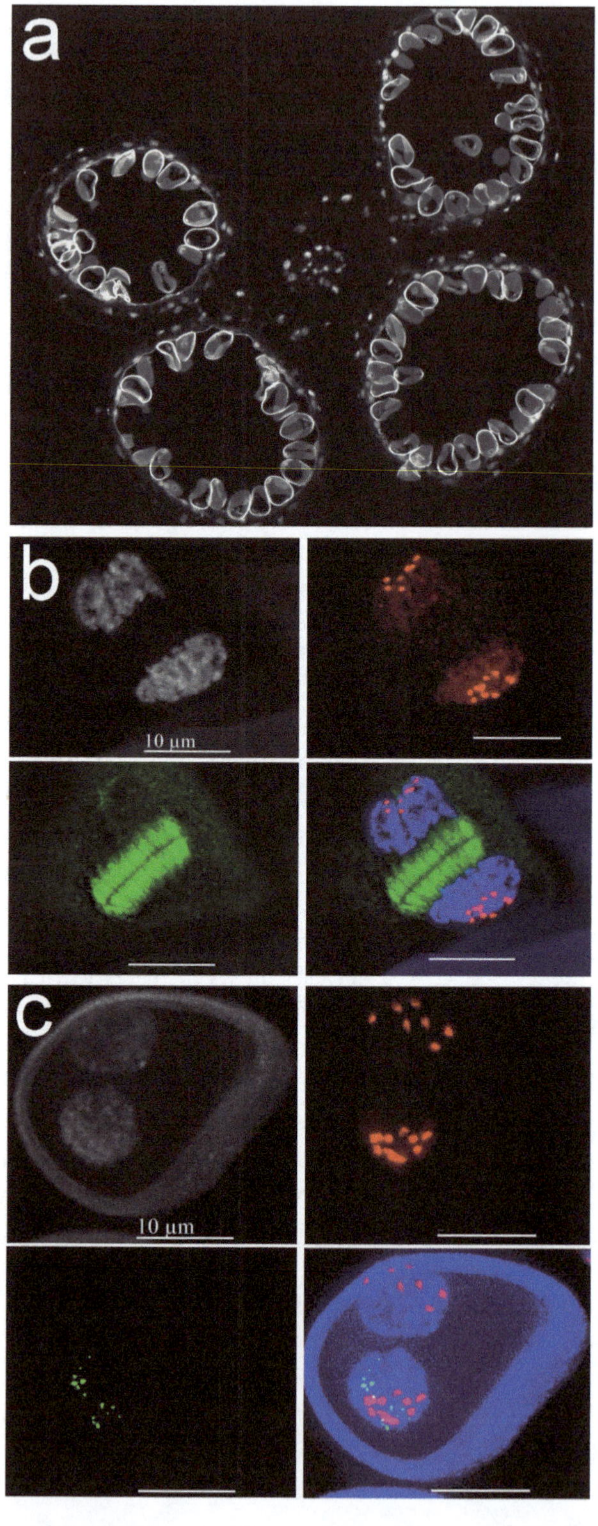

Fig. 2 Anther and pollen tissue section analysis. (**a**) Cross section of an entire rye (*S. cereale*) anther after DAPI staining. (**b**) Pollen grain section of *Ae. speltoides*

10. Air-dry the slides until no droplets on the surface of slides are visible and counterstain with 35 μl of DAPI solution in antifade and cover the sections with a coverslip. Avoid intense light.
11. Analyze the slides using an epifluorescence microscope. The selection of fluorescence filter depends on the fluorochrome used for probe labeling. Slides may be stored at 4 °C in darkness up to 1 year (Fig. 2a, b).

3.1.5 FISH Procedure After Immunostaining

1. To remove the coverslips preheat slides on a 38 °C warm hot plate for 10 min. Move carefully the coverslips aside and do not lift the coverslips. Transfer the slides in a jar and incubate jar containing slides under slow running tap water for 10 s.
2. Wash slides in 2× SSC for 5 min at RT.
3. Treat with freshly prepared 3:1 (ethanol: acetic acid, v/v) fixative solution for 10 min.
4. Wash slides in 2× SSC twice for 5 min.
5. Treat samples with freshly prepared 4% formaldehyde solution for 10 min.
6. Wash slides in 2× SSC three times for 5 min.
7. Dehydrate slides in an ethanol series of 70%, 85% and 90% for 1 min in each.
8. Air-dry slides until all drops of liquid visible on the surface of slides are disappeared.
9. Meanwhile add 3 μl of FISH probe (40–60 ng/μl of labeled DNA, *see* Subheading 2.3.5) to 57 μl of hybridization mixture (per slide). Denature mixture at 95 °C for 10 min and keep on ice until use.
10. Apply hybridization mixture onto each slide and cover with a coverslip. Ensure that each section is covered and there are no air bubbles under the coverslip.
11. Fix the coverslip with glue and denature the samples at 80 °C for 5 min on a hot plate.
12. Incubate in a moist chamber at 37 °C for 14–24 h.
13. Short wash in 2× SSC at RT for removing the coverslip.
14. Post wash in 2× SSC at 58 °C for 20 min.

Fig. 2 (continued) undergoing the first mitosis after immunostaining with anti-CENH3 (in red) and anti-α-tubulin (in green). DAPI-stained chromatin is shown in blue. (**c**) Pollen grain section of *Ae. speltoides* at binucleated stage after immunostaining with anti-CENH3 (in red) and subsequent FISH with a B chromosome-specific probe (in green). Supernumerary B chromosomes (Bs) are dispensable chromosomes, which do not confer advantages on the organisms that harbor them. DAPI-stained chromatin is shown in blue

15. Dehydrate slides in an ethanol series of 70%, 85% and 90% for 1 min in each.
16. Air-dry the slides until no obvious droplets are visible on the slide surface, counterstain with 35 μl of DAPI solution in antifade and cover with coverslips. Avoid intense light.
17. Analyze the slides using an epifluorescence microscope. The selection of fluorescence filter depends on the fluorochrome used for probe labeling. If necessary, store slides at 4 °C under dark conditions up to a year (Fig. 2c).

3.2 Analysis of Isolated and Cultivated Pollen

1. Barley pollen was cultured according to [10].
2. Collect cultured pollen by gentle centrifugation at 200 × *g* for 3 min at RT.
3. Short wash in 1× PBS.
4. Fix for 60 min in freshly prepared 3% paraformaldehyde at RT.
5. Short wash twice in 1× PBS and once in distilled water.
6. Resuspend pollen pellet in 3% aqueous alginate (1:1, v/v) at RT.
7. Pipette the suspension into a 0.05 M $CaCl_2$ solution using a 3 ml plastic pipette from 20 cm height resulting in a droplet size of approx. 25–30 μl.
8. Dehydrate in an ascending ethanol series of 30%, 50%, 70%, 90%, and 2× 100% each for 15 min. Subsequent infiltration in 24 h steps of PEG 1500–ethanol (w/v) 1:2, 1:1, and 2:1 at 46, 48, and 50 °C, respectively, followed by another 24 h in pure PEG 1500 at 52 °C.
9. After hardening of the PEG 1500, cut semithin sections (12 μm) on a microtome. These sections are used immediately or stored in a dry place at RT until use for at least 1 month.
10. Prepare 8 × 12 mm pads of nitrocellulose (NC) membrane and cut off part of the right-upper corner for marking.
11. Place the NC membrane pads on a filtration unit (Fig. 1d).
12. Distribute PEG sections on the NC membrane and carefully pipet 2–3 ml of 1× PBS under mild vacuum filtration, first to moisten the filter, subsequently to wash the sections on the NC membrane. This removes PEG and alginate, flattens pollen sections and immobilizes them onto the NC membrane.
13. Use freshly immobilized pollen sections on NC membranes immediately for labeling experiments.
14. For immunolabeling place NC pads carrying pollen sections face-up on a piece of Parafilm in a moist chamber. For blocking and antibody incubation *see* Subheading 3.1, **step 4**.

15. Place labeled NC membranes face up on a microscope slide, add 20 μl of DAPI solution in antifade, cover with a coverslip, and store in a fridge (Fig. 1e). Since fluorescence on NC membranes deteriorates over time, labeled sections have an exploitable shelf life of about 2 weeks and should therefore be analyzed as soon as possible with an epifluorescence microscope.

4 Notes

1. In case tubulin is visualized use MTSB buffer for fixation and immunostaining, otherwise PBS buffer is used.
2. Store antibody stocks at −20 °C as aliquots. Diluted antibodies are stored at 4 °C. When a new antibody is tested, try different dilutions, starting with 1:100, 1:500, and 1:1000.
3. One flower contains three anthers at the same developmental stage. Use one anther to determine the stage of pollen development and fix remaining anthers for subsequent tissue sections preparation. Pollen undergoing the first pollen mitosis is found in *Ae. speltoides* anthers of roughly 5 to 5.5 mm in length when the distance between the end of the spike and flag leaf is roughly 1.5 to 1 cm.
4. Place anthers in the same direction as cutting.
5. During immunostaining and FISH, it is always important to avoid shaking of the jar in order to keep more section slices on the surface of slides. Over-dried slides will impair the immunostaining results. A slide with high-quality sections is characterized by section slices firmly stuck to the glass surface and pollen are visible under the phase contrast microscope. Glass slides with wavy slices and without pollen should be discarded. Drying of slides at this step ensures better adhesion of sections to the slide surface. However, it is important not to overdry the slides, which may influence further immunostaining results.

Acknowledgments

We thank the Deutsche Forschungsgemeinschaft DFG (Ho1779/26-1, 30-1) and the China Scholarship Council (CSC, 201606910015) for financial support.

References

1. Tanaka I (1997) Differentiation of generative and vegetative cells in angiosperm pollen. Sex Plant Reprod 10:1–7
2. Borg M, Brownfield L, Twell D (2009) Male gametophyte development: a molecular perspective. J Exp Bot 60:1465–1478
3. Honys D, Twell D (2003) Comparative analysis of the Arabidopsis pollen transcriptome. Plant Physiol 132:640–652
4. Terasaka O, Tanaka R (1974) Cytological studies on the nuclear differentiation in microspore division of some angiosperms. Bot Mag Tokyo 87:209–217
5. Houben A, Kumke K, Nagaki K, Hause G (2011) CENH3 distribution and differential chromatin modifications during pollen development in rye (Secale cereale L.). Chromosome Res 19:471–480
6. Banaei-Moghaddam AM, Schubert V, Kumke K, Weibeta O, Klemme S, Nagaki K, Macas J, Gonzalez-Sanchez M, Heredia V, Gomez-Revilla D, Gonzalez-Garcia M, Vega JM, Puertas MJ, Houben A (2012) Nondisjunction in favor of a chromosome: the mechanism of rye B chromosome drive during pollen mitosis. Plant Cell 24:4124–4134
7. Pandey P, Daghma DS, Houben A, Kumlehn J, Melzer M, Rutten T (2017) Dynamics of post-translationally modified histones during barley pollen embryogenesis in the presence or absence of the epi-drug trichostatin A. Plant Reprod 30:95–105
8. Pandey P, Houben A, Kumlehn J, Melzer M, Rutten T (2013) Chromatin alterations during pollen development in *Hordeum vulgare*. Cytogenet Genome Res 141:50–57
9. Sanei M, Pickering R, Kumke K, Nasuda S, Houben A (2011) Loss of centromeric histone H3 (CENH3) from centromeres precedes uniparental chromosome elimination in interspecific barley hybrids. Proc Natl Acad Sci 108: E498–E505
10. Kumlehn J, Serazetdinova L, Hensel G, Becker D, Loerz H (2006) Genetic transformation of barley (*Hordeum vulgare* L.) via infection of androgenetic pollen cultures with *Agrobacterium tumefaciens*. Plant Biotechnol J 4:251–261

Chapter 25

Analyzing Somatic DNA Repair in Arabidopsis Meiotic Mutants

Annika Dorn and Holger Puchta

Abstract

Meiotic and somatic recombination share a common set of factors. Thus, the analysis of somatic DNA repair in meiotic mutant lines should be of special interest. Growth defects of mutant plants induced by specific genotoxins can thereby hint to DNA repair functions of the affected proteins. Here, we describe two kinds of approaches to characterize deficiencies in DNA repair in mutant lines of *Arabidopsis thaliana*, after genotoxin treatment.

Key words *Arabidopsis thaliana*, Cisplatin, DNA repair, Genotoxin, Methyl methanesulfonate, Mitomycin C, Root meristem

1 Introduction

Homologous recombination is essential for the correct course of meiosis. As meiotic recombination is initiated by the introduction of double strand breaks and shares a multitude of mechanisms with somatic double strand break repair, many factors involved in meiosis also possess an important role in somatic DNA repair [1, 2]. Therefore, it is highly recommended to test whether Arabidopsis mutants affected in meiosis also have somatic DNA repair defects. A prominent example is the proteins of the RTR (RECQ4A/TOP3α/RMI1/2) complex [3]. The complex is involved in the dissolution of recombination intermediates like double Holliday junctions, thus playing an important role in DNA repair via homologous recombination. RTR-mutant lines exhibit characteristic hypersensitivities against a number of genotoxins [4–6]. Furthermore, *top3α* and *rmi1* mutants are characterized by a unique meiotic phenotype, with extended chromosome fragmentation leading to an arrest after meiosis I [5, 7]. Similarly, the nuclease MUS81 is involved in resolving class II crossovers in meiosis, whilst also being important for somatic DNA cross-link repair [8]. Another example is the cohesion cofactor PDS5

Mónica Pradillo and Stefan Heckmann (eds.), *Plant Meiosis: Methods and Protocols*, Methods in Molecular Biology, vol. 2061, https://doi.org/10.1007/978-1-4939-9818-0_25,

(characterized as meiotic factor SPO76 in a number of organisms), shown to be involved not only in meiosis in Arabidopsis but also exhibiting an involvement in DNA double-strand break repair [9]. This demonstrates the close link between meiotic recombination and DNA repair. For the analysis of DNA repair, mutant analysis testing the hypersensitivity of plants against certain genotoxins is a well-established and reliable method. By the selection of specific genotoxins, different kinds of DNA damage can be induced. Commonly used genotoxins for the induction of genotoxic stress are the methylating agent methyl methanesulfonate (MMS) and the two DNA cross-linking agents cisplatin and mitomycin C. DNA cross-links pose a dangerous threat for genome stability by blocking essential processes like replication and transcription. While cisplatin induces mostly intrastrand cross-links that affect only one DNA strand and leave the complementary strand as a repair template, mitomycin C mainly causes interstrand cross-links that are highly toxic by connecting both DNA strands [10, 11]. Mutants defective in the repair of specific DNA damages, exhibit a slower growth in the presence of the causal genotoxin, thus resulting in a reduced fresh weight compared to wild-type plants treated with the same substance [4, 12]. This allows for the conclusion to be made whether or not the affected protein is involved in the repair of this kind of DNA damage. As the repair of specific DNA damages require specialized pathways, depending on the involvement of the affected factor, mutant lines can vary in their response to different genotoxins. A second approach for the characterization of genotoxin sensitivity is based on the analysis of cell death in the root meristem. Meristematic cells are characterized by a high proliferation rate, but at the same time, an accumulation of mutations has to be prevented. As plant stem cells form precursors of all tissues during the whole lifetime of a plant, DNA damage protection is especially important here. Thus, cells of the root and apical meristem are highly sensitive against genotoxic stress. This results in a selective appearance of programmed cell death in meristematic tissue, even at a low level of DNA damage [13]. Due to its transparency and ease of accessibility for microscopy, the root meristem is an ideal system to analyze the effect of DNA damage. Because of their high proliferative rate, root cells are particularly suitable indicators of replication-associated DNA damage [14]. Propidium iodide (PI) is a fluorescent dye that intercalates into nucleic acids, but is only able to penetrate the perforated cell membrane of dead cells, thus enabling the differentiation between living and dead cells. The application of PI staining enables the analysis of cell death in root cells for the quantification of replication-associated DNA damage in mutant lines. Depending on the strength and the nature of the defect to process aberrant DNA replication intermediates, enhanced cell death is detectable under normal plant growth conditions [15, 16]. An additional genotoxin-treatment might be advisable in cases where no clear defect is visible under standard conditions [17, 18].

In the following, we describe the analysis of Arabidopsis mutant lines in response to genotoxic stress. Thereby, two different kinds of protocols are described. The sensitivity assay relies on the reduction in total fresh weight in response to the treatment with genotoxic agents. The second method is based on the analysis of cell death in the root meristem, using microscopic analysis of PI stained roots that can be genotoxin treated.

2 Materials

2.1 Reagents

Prepare all solutions with ultrapure water. Prepare and store all reagents at room temperature (unless indicated otherwise).

1. Germination medium (GM) for cultivation of *Arabidopsis thaliana*: 4.9 g/l Murashige and Skoog medium including vitamins and MS buffer, 10 g/l sucrose, adjust pH to 5.7 with KOH and sterilize by autoclaving. Add 7.6 g/l (10 g/l for vertical cultivation) Plant Agar before sterilization for solid media (*see* **Note 1**). Pour solid media, cooled to 50 °C, after sterilization into usual round petri dishes for horizontal cultivation or rectangular petri dishes for vertical cultivation.
2. 5 μg/ml propidium iodide (PI) solution: Dissolve in water and store in a lightproof container at 4 °C.
3. 1 mM *Cis*-diamminedichloridoplatinum(II) (cisplatin): Dissolve in water and store in aliquots at −20 °C.
4. Methyl methanesulfonate (MMS): Prepare aliquots at room temperature.
5. 1 mg/ml mitomycin C (MMC): Dissolve in water immediately before use.

3 Methods

Surface-sterilize your Arabidopsis seeds with your favorite method and incubate the seeds over night at 4 °C in water for stratification. All steps until weighing/slide preparation should be performed under sterile conditions in a laminar flow hood to prevent microbial contamination. Pay special attention to the handling and disposal of genotoxins and all material contaminated with genotoxins. Prevent aerosol formation especially during the disposal of liquid genotoxin waste by using a pump.

3.1 Sensitivity Assay

1. Sow 100 seeds per line on solid GM, evenly spaced. Incubate the plates horizontally in a plant growth chamber for 7 days (*see* **Notes 2** and **3**).

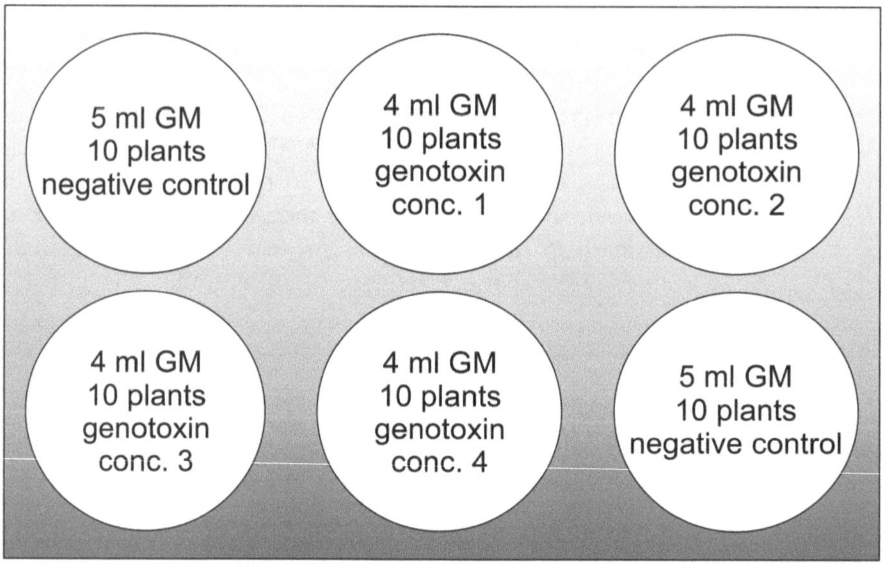

Fig. 1 Six-well plate setup for sensitivity assays. The two outermost wells are used for negative controls; the other wells can be used for treatment with four different genotoxin concentrations. The wells for negative controls are filled with 5 ml liquid GM medium; the wells for genotoxin treatments are filled with 4 ml GM, and the appropriate genotoxin can be added diluted in 1 ml GM. Every well is equipped with 10 plants

2. Use an individual six-well plate for each line and fill two wells with 5 ml liquid GM for negative controls; fill the other wells with 4 ml GM (*see* **Note 4**). Using forceps, transfer ten seedlings into each well (Fig. 1, *see* **Note 5**). Incubate the six-well plates in a plant growth chamber overnight.
3. Dilute your genotoxin stock in liquid GM medium. Add 1 ml of appropriately diluted genotoxin to the wells with 4 ml GM for a total of 5 ml (*see* **Note 6**). Thus, four different genotoxin concentrations can be tested per six-well plate. Incubate the six-well plates in a plant growth chamber for 13 days.
4. Transfer the plantlets onto tissue paper, place another tissue on top and press with the lid of the six-well plate to remove excess liquid. Determine the fresh weight of the plants in each well with a fine scale balance (*see* **Note 7**).
5. For data analysis, form the mean weight from both negative controls and calculate the fresh weight of the treated samples in relation to the control. This enables the elimination of the influence of basic growth differences from different lines.

3.2 Analysis of Cell Death in the Root Meristem

1. Sow three lines of 13 seeds on rectangular petri dishes with solid GM. Seal the plates with Parafilm and incubate them vertically in a plant growth chamber for 4 days.

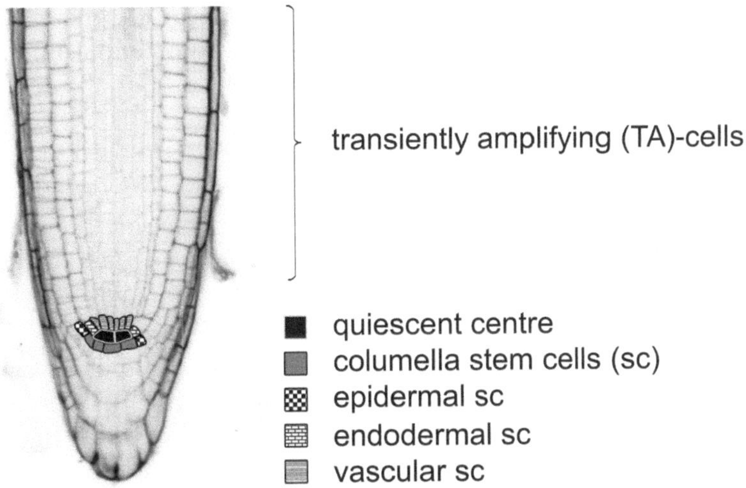

Fig. 2 Different cell types in the root meristem. The root meristem contains the stem cell niche, consisting of the quiescent center surrounded by four different stem cell types. These stem cells give rise to the different organs of the root. At the apex of the stem cell niche, the transiently amplifying cells are located, that are characterized by a high division activity

2. Transfer ten seedlings into a well of a six-well plate filled with 5 ml liquid GM for untreated approaches or 4 ml liquid GM for genotoxin treated approaches (*see* **Note 8**).
3. Dilute your genotoxin stock in liquid GM medium. Add 1 ml of appropriately diluted genotoxin to the wells containing 4 ml liquid GM (*see* **Note 9**). Incubate the six-well plates in a plant growth chamber for at least 24 h.
4. The next day, carefully wash the seedlings by dipping them shortly into water. Then remove the leaves with forceps and transfer the root onto a microscope slide with 180 μl PI solution. Place 7–10 roots on one slide and add a coverslip (24 × 40 mm, *see* **Note 10**).
5. Analyze the roots with a confocal laser scanning microscope using an appropriate filter set and count the number of dead cells per root in the anterior 200 μm (*see* **Notes 11–13**). By taking the special arrangement of the cells around the quiescent center into account, different cell types can be distinguished (Fig. 2).

4 Notes

1. Make sure to use media with at least 10 g/l agar for vertical cultivation, as medium with less agar tends to sag in petri dishes.

2. For plant growth in a growth chamber, we use a day–night rhythm of 16 h illumination at 22 °C and 8 h darkness at 20 °C. The number of days for all incubation steps is optimized for these growth conditions. You can alter the incubation time according to your needs, but make sure to stay consistent with time periods once decided upon.
3. It is possible to elongate incubation time to 2 weeks, for example for segregating lines that enable phenotypic identification of homozygous plants. Then transfer only five plants per well in the next step.
4. We always use two wells, as far away from each other as possible, for negative controls, as the six well plates are contamination prone and the experiment relies on at least one suitable negative control for each line.
5. Be careful not to tear the plant roots off during transfer. If this is a recurring problem in your experiment, consider using less agar for your GM plates.
6. As a starting point for final genotoxin concentrations we use 40–100 ppm MMS, 5–20 μg/ml MMC and 5–20 μM cisplatin. Try different ranges of concentrations for your lines to find a suitable sensitivity window. Of course the assays described here are not limited to these genotoxins only, there is a wide selection of different genotoxins available. If your genotoxin is not soluble in water, make sure to add the maximal amount of solvent contained in your treated wells also to the negative control.
7. This assay also allows the determination of plant dry weight. For this purpose, write down the empty weight of each weighing dish before weighing plant fresh weight. After fresh weight determination, dry the plants in the individual weighing dishes at 60 °C at least overnight. Then weigh the weighing dishes including the dry plants and subtract the empty weight of the weighing dish.
8. Be extremely careful with plant roots as the lightest touch can damage root cells, which can affect results.
9. As a starting point for final genotoxin concentration, we use 40 μM cisplatin, 3 μg/ml MMC or 75 ppm MMS. Try different concentrations for your lines to find a suitable sensitivity window (*see* **Note 14**).
10. Prepare the slides immediately before analysis and do not press on the coverslip as roots are highly sensitive.
11. Dead cells are apparent by a red fluorescence. As PI also penetrates the intercellular space, this allows for the visualization of the different cells in the root.
12. Make sure to check all focal planes for dead cells.

13. For an alternative evaluation method, you can count the number of roots exhibiting at least one dead cell per cell type.
14. In the case that a DNA repair defect can be revealed by the described protocols, it might be advisable to also test the efficiency of intrachromosomal homologous recombination (HR) in somatic cells. Several powerful transgenic assay systems based on the restoration of the ß-glucuronidase marker by HR are available for Arabidopsis, and a detailed protocol of their application has already been published in this series [19, 20].

Acknowledgments

This work was supported by the European Research Council (Advanced Grant "COMREC" and "CRISBREED").

References

1. Osman K, Higgins JD, Sanchez-Moran E et al (2011) Pathways to meiotic recombination in *Arabidopsis thaliana*. New Phytol 190 (3):523–544. https://doi.org/10.1111/j.1469-8137.2011.03665.x
2. Sanchez-Moran E, Armstrong SJ (2014) Meiotic chromosome synapsis and recombination in *Arabidopsis thaliana*: new ways of integrating cytological and molecular approaches. Chromosom Res 22(2):179–190. https://doi.org/10.1007/s10577-014-9426-8
3. Knoll A, Schröpfer S, Puchta H (2014) The RTR complex as caretaker of genome stability and its unique meiotic function in plants. Front Plant Sci 5:33. https://doi.org/10.3389/fpls.2014.00033
4. Hartung F, Suer S, Puchta H (2007) Two closely related RecQ helicases have antagonistic roles in homologous recombination and DNA repair in *Arabidopsis thaliana*. Proc Natl Acad Sci U S A 104(47):18836–18841. https://doi.org/10.1073/pnas.0705998104
5. Hartung F, Suer S, Knoll A et al (2008) Topoisomerase 3alpha and RMI1 suppress somatic crossovers and are essential for resolution of meiotic recombination intermediates in *Arabidopsis thaliana*. PLoS Genet 4(12):e1000285. https://doi.org/10.1371/journal.pgen.1000285
6. Dorn A, Röhrig S, Papp K et al (2018) The topoisomerase 3α zinc-finger domain T1 of *Arabidopsis thaliana* is required for targeting the enzyme activity to Holliday junction-like DNA repair intermediates. PLoS Genet 14 (9):e1007674. https://doi.org/10.1371/journal.pgen.1007674
7. Chelysheva L, Vezon D, Belcram K et al (2008) The Arabidopsis BLAP75/Rmi1 homologue plays crucial roles in meiotic double-strand break repair. PLoS Genet 4(12):e1000309. https://doi.org/10.1371/journal.pgen.1000309
8. Berchowitz LE, Francis KE, Bey AL et al (2007) The role of AtMUS81 in interference-insensitive crossovers in *A. thaliana*. PLoS Genet 3(8):e132. https://doi.org/10.1371/journal.pgen.0030132
9. Pradillo M, Knoll A, Oliver C et al (2015) Involvement of the Cohesin Cofactor PDS5 (SPO76) During Meiosis and DNA Repair in *Arabidopsis thaliana*. Front Plant Sci 6:1034. https://doi.org/10.3389/fpls.2015.01034
10. Rink SM, Lipman R, Alley SC et al (1996) Bending of DNA by the mitomycin C-induced, GpG intrastrand cross-link. Chem Res Toxicol 9(2):382–389. https://doi.org/10.1021/tx950156q
11. Eastman A (1985) Interstrand cross-links and sequence specificity in the reaction of cis-dichloro(ethylenediamine)platinum(II) with DNA. Biochemistry 24(19):5027–5032
12. Hartung F, Suer S, Bergmann T et al (2006) The role of AtMUS81 in DNA repair and its genetic interaction with the helicase AtRecQ4A. Nucleic Acids Res 34 (16):4438–4448. https://doi.org/10.1093/nar/gkl576

13. Fulcher N, Sablowski R (2009) Hypersensitivity to DNA damage in plant stem cell niches. Proc Natl Acad Sci U S A 106 (49):20984–20988. https://doi.org/10.1073/pnas.0909218106

14. Curtis MJ, Hays JB (2007) Tolerance of dividing cells to replication stress in UVB-irradiated Arabidopsis roots: requirements for DNA translesion polymerases η and ζ. DNA Repair (Amst) 6(9):1341–1358. https://doi.org/10.1016/j.dnarep.2007.03.004

15. Röhrig S, Schröpfer S, Knoll A et al (2016) The RTR complex partner RMI2 and the DNA helicase RTEL1 are both independently involved in preserving the stability of 45S rDNA repeats in *Arabidopsis thaliana*. PLoS Genet 12(10):e1006394. https://doi.org/10.1371/journal.pgen.1006394

16. Klemm T, Mannuß A, Kobbe D et al (2017) The DNA translocase RAD5A acts independently of the other main DNA repair pathways, and requires both its ATPase and RING domain for activity in *Arabidopsis thaliana*. Plant J 91(4):725–740. https://doi.org/10.1111/tpj.13602

17. Recker J, Knoll A, Puchta H (2014) The *Arabidopsis thaliana* homolog of the helicase RTEL1 plays multiple roles in preserving genome stability. Plant Cell 26 (12):4889–4902. https://doi.org/10.1105/tpc.114.132472

18. Herrmann NJ, Knoll A, Puchta H (2015) The nuclease FAN1 is involved in DNA crosslink repair in *Arabidopsis thaliana* independently of the nuclease MUS81. Nucleic Acids Res 43 (7):3653–3666. https://doi.org/10.1093/nar/gkv208

19. Puchta H, Hohn B (2012) In planta somatic homologous recombination assay revisited: a successful and versatile, but delicate tool. Plant Cell 24(11):4324–4331. https://doi.org/10.1105/tpc.112.101824

20. Schmidt-Puchta W, Orel N, Kyryk A et al (2004) Intrachromosomal homologous recombination in *Arabidopsis thaliana*. Methods Mol Biol 262:25–34. https://doi.org/10.1385/1-59259-761-0:025

Chapter 26

A Modular Tray Growth System for Barley

Mikel Arrieta, Isabelle Colas, Malcolm Macaulay, Robbie Waugh, and Luke Ramsay

Abstract

Determining when a barley plant starts and finishes meiosis is not trivial as when the spikelets undergo meiosis, the spike is not visible as it is still well within the leaf sheath on the developing tiller. This is a general constraint for any experiment involving meiosis, such as cytology, RNA extractions, or abiotic stress treatments aiming to target such a developmental stage. The lack of synchronicity between barley tillers within the same plant exacerbates the difficulty to determine the overall meiotic stage of a plant at a certain time.

Given the lack of a nondestructive staging system for predicting the entry into meiosis and the problems of working with large pot plant systems, a modular plant growing is proposed. This system enables the growth of a high number of plants in a small surface, each producing a single tiller. The modular tray system was used to generate a nondestructive prediction tool for meiosis by using external morphological features. As an example, the system is used here for heat treating F_1 plants in early meiosis stages to modify recombination.

Key words Barley, Meiosis, Staging, Nondestructive, Prediction, High-throughput, Temperature, Modular tray system, Heat shock

1 Introduction

The study of meiosis and recombination in plants requires a precise understanding of when and where a plant initiates floral development and undergoes meiosis. This is important because many experiments such as cytological, transcriptomics, and protein studies, involve collecting tissue, usually anthers at certain meiotic stages from plants. In addition, other experiments may not require tissue collection, but still rely on estimating the meiotic stage of a plant for deciding when to apply treatments such as temperature stress to target certain stages of meiosis (Fig. 1), etc.

Electronic supplementary material: The online version of this chapter (https://doi.org/10.1007/978-1-4939-9818-0_26) contains supplementary material, which is available to authorized users.

Mónica Pradillo and Stefan Heckmann (eds.), *Plant Meiosis: Methods and Protocols*, Methods in Molecular Biology, vol. 2061, https://doi.org/10.1007/978-1-4939-9818-0_26, © Springer Science+Business Media, LLC, part of Springer Nature 2020

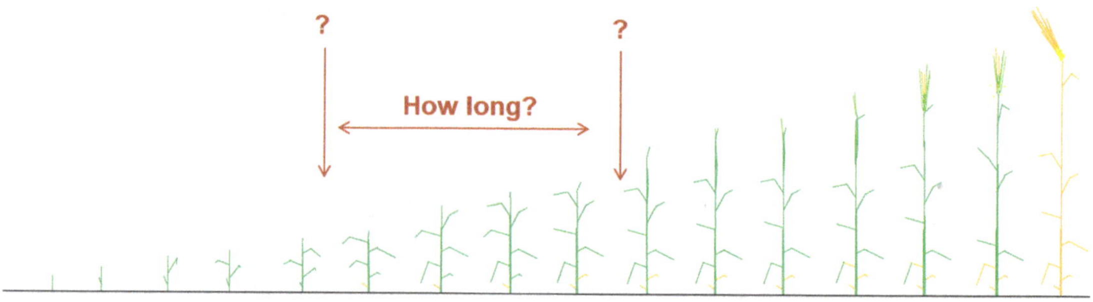

Fig. 1 A prediction of the internal floral development stage is required for example when applying a short temperature stress treatment targeting meiosis, in order to decide when to start and finish the treatment

In barley, when the meiotic cells undergo differentiation, the spike is still inside the leaf sheath on the expanding tiller (barley stem), so it is not possible to assess the meiotic stage of the cells in a nondestructive way. Furthermore, the lack of synchronicity of meiosis between the different spikelets and tillers within the same plant complicates the assessment of the meiotic stage of a plant.

The different cereal developmental scales available, such as the widely used Zadoks scale [1, 2] do not make a link between external developmental features and meiotic stage and do not make a link to spike growth at all until the ear starts to emerge. The Waddington scale [3] does measure the developmental growth of the sexual organs, but does not give any insight of meiotic progress at the cellular level, and requires the dissection and destruction of the material. Gómez and Wilson published a nondestructive method [4] that related spike size using an adaptation of the Zadoks scale to predict pollen stages and the anther development. Unfortunately, the earlier stages of meiotic progression were not differentiated. Another nondestructive system was published by Tracy et al. [5] using X-ray micro-computed tomography (μCT) for floral stage phenotyping of *Arabidopsis thaliana* and barley cultivar Optic. This system provides a detailed internal 3D phenotype of flowers of living plants but is of limited use for larger scale application and did not differentiate meiotic stages. The meiotic entry and progress of barley is thus not covered by any published nondestructive staging system.

In this chapter, a new protocol for predicting meiotic stages that is more amenable to higher throughput is introduced. It uses modular trays widely used in some breeding applications, where plants are grown in small modules restricting their size and development, constraining them to develop a single tiller per plant, and to grow faster. Using staging data, a protocol has been devised that provides a nondestructive meiosis reference table for barley experiments, and its use is demonstrated in an experiment that uses temperature treatments to modulate meiotic recombination.

2 Materials

2.1 Modular Tray Growing Platform

The modular tray growing platform is prepared before sowing the plants.

1. Capillary matting: This matting is widely used in gardening and is made of a mixture of fabrics capable of holding and transporting large volumes of water. The watering matting is often sold in rolls and is cut when mounting the system (e.g., AquaMat® from LBS).
2. Outer plastic tray: A large enough tray is required to contain the matting and the size of modular plant tray used inside. Plastic gardening trays, often known as propagator trays, can be used. These trays require a system to allow the egress of water, so some holes are drilled in the corners of the tray if necessary (*see* **Note 1**, Fig. 2a) (e.g., G16B Garden tray, Garland Products Ltd)
3. Modular tray: These are plastic trays containing a large number of cells to grow plants. They are widely used in gardening, plant nurseries, and for some breeding applications such as speed breeding or single seed descent [6, 7]. The number and cell size can be variable, as an example (QPD104 trays, PG Horticulture) the following sizes can be found (*see* **Note 1**).
   - 104 round cell trays (13 × 8; approx. Diameter: 3.5 cm/5 cm depth).
   - 96 square cell trays (12 × 8; approx. 3.7 × 3.7 cm/8 cm depth).
4. Containers for watering: Buckets or plastic containers can be used to keep the water. The size must be suitable to fit in the space between the plastic tray and the bench.

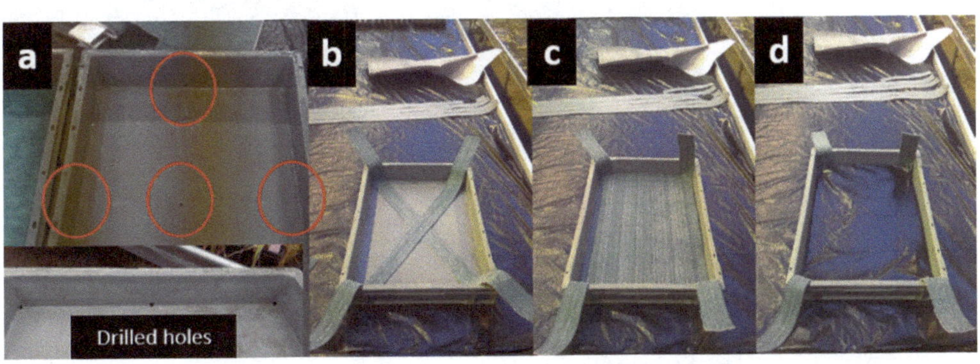

Fig. 2 Examples of different stages of mounting the watering system

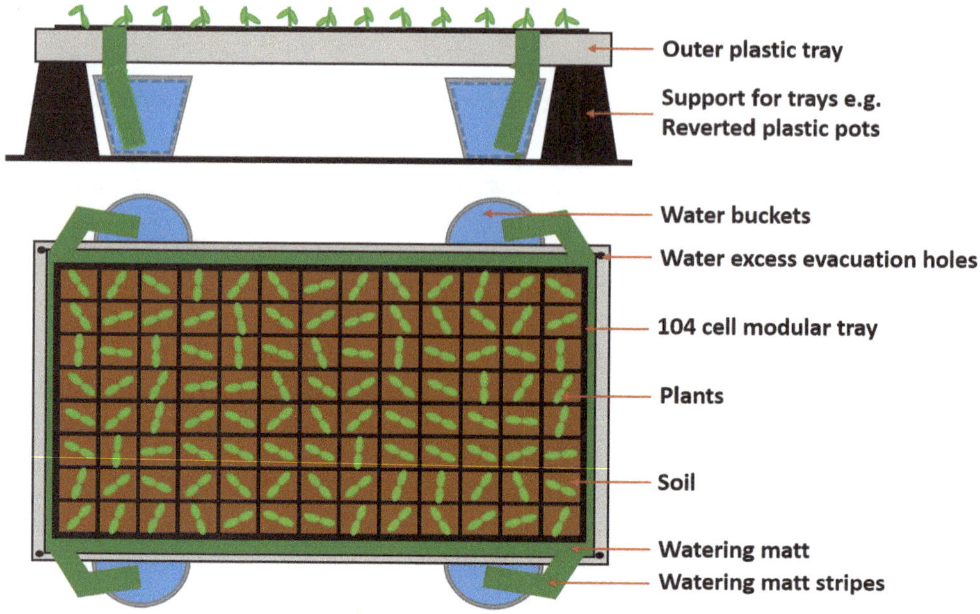

Fig. 3 Diagram of the modular tray system

Fig. 4 Example of support system to hold the help the plants grow and avoid falling on each other

5. Support for the trays (Fig. 3): The tray system needs to be supported to a height so that the watering buckets do fit underneath. As an example, inverted plant pots can be used for this task.
6. String/Twine as a support for the plants (e.g., Gardening jute or polypropylene twine) (Fig. 4).

2.2 Growing Material

1. A controlled environment for growing the plants (*see* **Note 2**).
2. Soil (*see* **Note 3**).
3. Barley seeds. In the case of the present example cultivar "Morex" (six-rowed spring barley), cultivar "Bowman" (two-rowed spring barley), and F_1 seed produced from a cross between them are used.
4. Petri dishes. Variable size depending on the number of seeds. In the present example 140 mm ø Petri dishes are used.
5. Sterile Distilled Water (SDW).
6. Filter paper to fit petri dishes (Whatman™). Variable size depending on the number of seeds. In the present example 150 mm ø filter papers are used.
7. Parafilm™.

2.3 Staging Material

1. Ruler.
2. Tweezers.
3. Optical microscope with graticule (e.g., Olympus CK2® light microscope).
4. Microscope Slides. No coating required.
5. Coverslips (VWR).
6. Acetocarmine (as Chapter 15 of this book and [8]).

3 Methods

3.1 Modular Tray Platform Preparation

1. Cut the capillary matting to the size and shape of the surface of the outer plastic tray (Fig. 2b, c).
2. Cut strips of capillary matting. They need to be long enough to cross the area of the plastic trays and hang below long enough to reach the depth of the water containers (Fig. 2b, c).
3. Place the matting strips on the tray. The strips can be positioned in different ways, such as a cross (Fig. 2b) or crossing the trays horizontally or vertically (*see* **Note 4**). The tips of the strips need to hang down the sides of the tray to where the watering system will be located.
4. Cover the strips by the square matting cut in **step 1** (*see* **Note 5**).
5. Water the matting until is completely wet (*see* **Note 6**).
6. Fill the modular tray with the soil (*see* **Note 7**) and position it on the watering matting in the tray.
7. Mount the whole system on the supporting pots and place the water containers (filled with water, *see* **Note 8**) underneath with the matting strips in the water (Figs. 3 and 4).

3.2 Sowing and Plant Growing

1. Place the seeds in petri dishes (around 30 seeds per plate) with filter paper and add SDW so that the paper is wet but does not drip if turned on the side. Prepare more seeds than needed, as not all of them may germinate (*see* **Note 9**). In the present example, 104 cell trays are used, and around 30% more seeds than needed are germinated.
2. Seal the petri dish with Parafilm™, and place the plate in the dark at RT.
3. Once the first root and coleoptile apices emerge from the seeds (Fig. 5b), sow them in the cells (coleoptile facing the top) (*see* **Note 10**), choosing healthy seeds with as little evident variation in terms of coleoptile emergence as possible.
4. Keep the capillary matting always wet. To achieve this in the initial stages, watering over the capillarity matting through the sides of the tray is required to ensure that the whole matting system including the strips is totally wet. Additional watering might be required on the matting before the root system is developed. Once the plants are 2–3 weeks old the watering should work through capillary action from the watering containers, so only keeping them full of water is required.
5. Grow the plants (*see* **Note 11**) in the most controlled and constant conditions as possible. In the present example growth conditions in the cabinets is 18 °C day and 15 °C night for the control conditions before the treatment, and 30 °C day and 25 °C night for the heat shock conditions.

Fig. 5 Different stages of the modular tray system, temperature treatments, and harvesting

3.3 Collecting Staging Data

While the population grows, some plants are sampled, dissected, and staged for building the reference table that is used to predict meiosis. Ideally, a whole tray is used for this to ensure a reasonably sized dataset. It is important to sample only healthy plants, and discard all the plants that show problems, such as late or bad emergence, or sickness, like yellow leaves.

1. After the third week, dissect some plants at regular intervals to detect the first stages of spike morphogenesis (Waddington stages 1–2) [3].
2. Once the spike detected is around 3–4 mm, start sampling more frequently, by picking (ideally) three random plants every 2 or 3 days (*see* **Note 12**).
3. Measure the following morphological features (*see* **Note 13**):
   - Plant height (cm): Measure starting from the ground to the last unfolded leaf's ligule (*see* **Note 14**).
   - Number of unfolded leaves: Count the number of unfolded leaves in the main shoot (*see* **Note 15**).
   - Spike size (mm): Measure the total length of the spike.
   - Anther size in the middle of the spike (mm): Measure the length of the anther of a floret in the middle of the spike by using the graticule of the light microscope. It is taken from the middle of the spike as the most advanced meiotic stage in the spike is usually found here.
   - Meiotic stage by acetocarmine spread (when possible): Place the same anther used to measure the length on a slide with a drop of acetocarmine as described in Chapter 15 of this book and cover by a reusable plastic coverslip. Push the coverslip gently with a rod glass to break the anther. Assess the meiotic stage under the light microscope using a magnification of ×20 and ×40 (Fig. 6). For this work four categories of meiotic stages are classified: Premeiosis ("Premeio"), Leptotene to Zygotene ("Lept.-Zyg."), Pachytene to Tetrad ("Pach.-Tet."), and Pollen (all stages after young pollen grains).
4. Once the three plants dissected show pollen, the sampling can be stopped, unless further stages are needed in the reference table.

3.4 Building a Reference Table

The measured variables, linked to the meiotic stage can be summarized in a table (like Table 1), which is used as reference to link nondestructive morphological features with the meiotic stage and the spike size or the anther size. This type of table can be produced by using the R script supplied, which uses the package "psych" (https://cran.r-project.org/web/packages/psych/psych.pdf) on

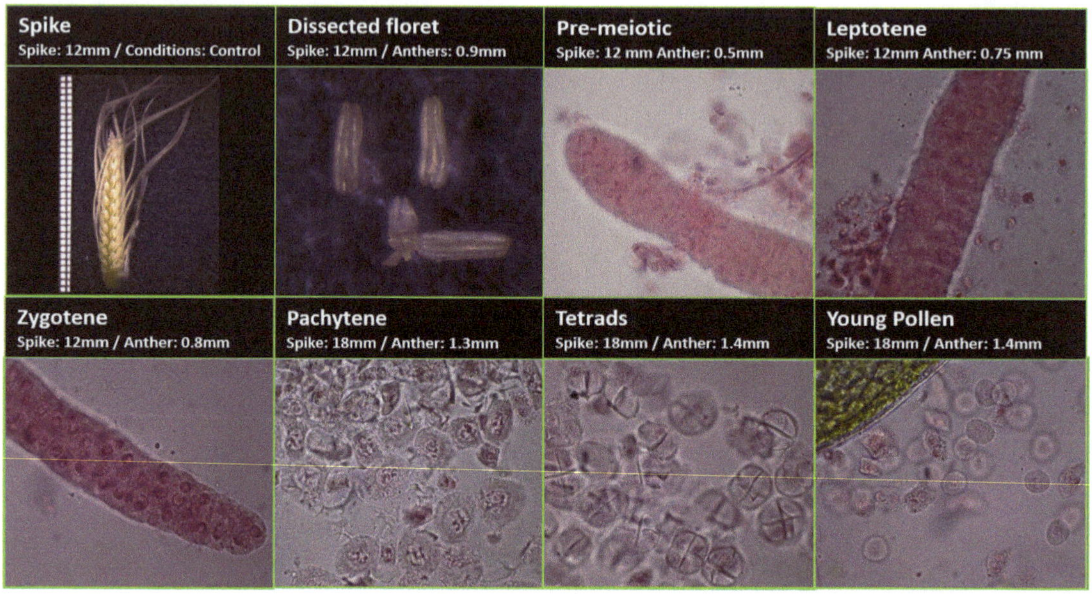

Fig. 6 Acetocarmine spreads and meiotic stages of barley anthers

Table 1
Reference summary table with the means of different developmental features for each meiotic stage per treatment ("Treat."), (H.S., Heat Shock, Ctr., Control). All genotypes were used together to create this table.

Genotype	Meiotic stage (middle of the spike)	Treat.	*n*	Mean height (cm)	S. E.	Mean *n*, unfolded leaves	S. E.	Mean spike size (mm)	S. E.	Mean anther size (mm)	S. E.
All	0_Pre-mei	Ctr.	8	19.6	1.3	4.7	0.3	4	0.7	0.3	0.1
All	1_Lept.Zyg.	Ctr.	7	26.9	1.6	5.6	0.3	9.3	0.8	0.7	0.1
All	2_Pach.-Tetr	Ctr.	7	29.9	0.5	6.1	0.2	13.6	1.1	1	0.1
All	3_Pollen	Ctr.	7	36.1	2.1	7.6	0.5	27.1	2.9	1.6	0.2
All	0_Pre-mei	H.S.	5	20.4	1.3	5.9	0.5	5.6	0.7	0.4	0
All	1_Lept.Zyg.	H.S.	7	23.1	1.0	6.1	0.3	11.1	0.8	0.6	0
All	2_Pach.-Tetr	H.S.	7	28.5	0.9	6.8	0.2	13.7	1.4	0.9	0.1
All	3_Pollen	H.S.	8	33.2	1.5	6.9	0.2	25.2	3.4	1.7	0.2

the example dataset. In the example, samples from Bowman, Morex, and the F_1s from the cross between them are collected.

1. Open R Studio and import the file where the staging data is included (example dataset provided).

2. Follow the instructions of the example R script (Supplementary Material Data 1) included to generate the reference summary table. The outcome is a *csv* file table that can be then edited manually with a spreadheet software to select only the desired information, and tidy up the table for visualization (Table 1).

3.5 Using the Reference Table in Temperature Experiments

The following example shows how the nondestructive reference table for spike size and meiosis is used in a heat stress temperature experiment to study the effect of heat on meiotic recombination (*see* **Note 16**, Fig. 5):

1. Prepare a modular tray as in Subheading 3.1 with barley seeds (in this case with F_1 seeds).
2. When the population of plants is getting close to the stage of treatment initiation (in number of leaves and height of plants), do whole tray staging but only for the plant height and number of unfolded leaves. In the present example, the temperature treatment aimed to cover the premeiotic and early meiotic stage, so the initiation of temperature stress is premeiosis.
3. When the mean of the tray for the number of leaves and height reaches the level associated with the desired meiotic stage to start the treatment, the batch is moved to the cabinet for heat treatment (30 °C day, 25 °C night). In this example this was done with a mean of 3.9 (±0.04) unfolded leaves and 15 cm (±0.43) height, based on the data presented in the reference table (*see* **Note 17**, Table 1).
4. As it is assumed the plants are in premeiosis, the control tray is moved to the glasshouse at (20 °C day, 16 °C night).
5. The plants are kept in the heat treatment until it is considered the premeiotic stage and entry in meiosis is completed. In this case, this is when plants have a mean of 6.5 (±0.2) unfolded leaves and 25.6 cm (±1.34) height (Table 1).
6. To assess recombination, F_2 seeds from both control and heat treatment are sown, and plants are genotyped. In this case using the BeadXpress® (Illumina) 96-SNP genotyping technology that uses a subset of the markers on the 9K Illumina SNP chip [9].
7. The comparison of the resultant genetic maps (calculated with MapDisto [10]) shows an increase of recombination in the heat-treated population compared to the control between the pericentromeric markers 11_10461 and 11_20969, from 9.5 cM under control conditions to 20.4 cM under heat stress (Fig. 7). These results indicate that the distribution of crossovers in barley can be modified by temperature stress, as it has been observed in previous studies [11].

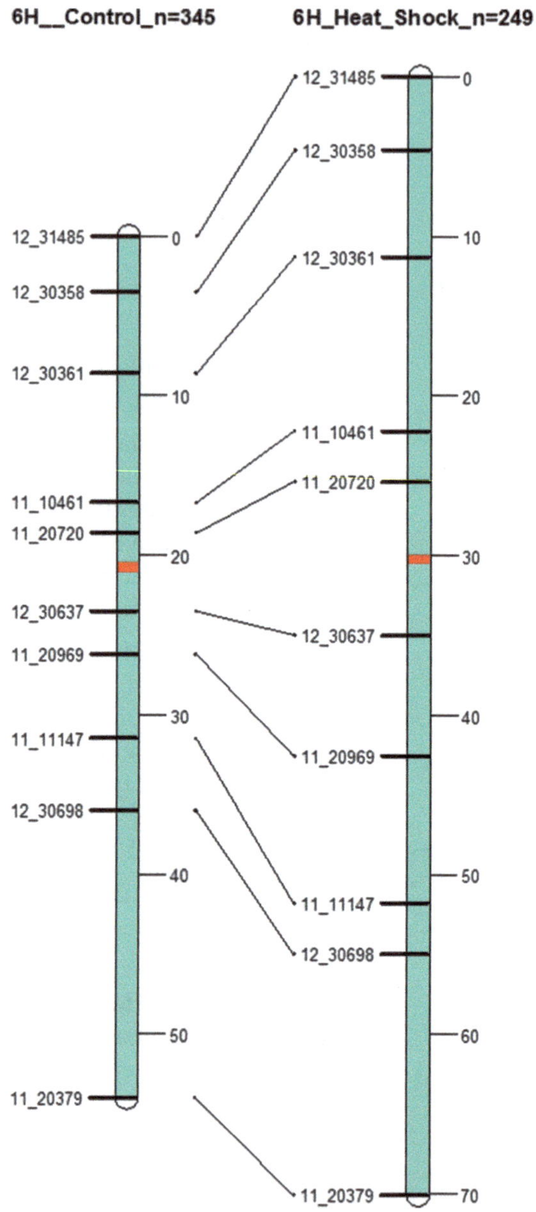

Fig. 7 Comparison of the genetic maps of chromosome 6H for control (left) and heat-treated (right) conditions. The barley genomic zone based on [12] is colored Blue (as "Zone 2"). The red bar indicates the approximate position of the centromere

4 Notes

1. The size of the outer tray or the modular cell tray does not need to be the same as in this protocol, smaller or bigger systems could be prepared by cutting or merging together modular trays.

2. The control of the environmental variables during the growth of the plants will depend on the place of the experiment and the resources available. Ideally plant growth cabinets present the best control of the environment, this being the crucial factor to ensure synchronicity between the plants and homogeneity between batches of trays.
3. The soil and substrate used will depend in the experiment. In the present example the soil used was general purpose mixture compost including peat, sand, limestone, perlite, Celcote®, Osmocote®, and Exemptor®.
4. The position of the strips and the capillary matting need to ensure a flat surface in the tray so that all the cells of the modular tray make contact with it.
5. The capillary matting system could be covered with a black microperforated plastic film (LBS) to maintain moisture better and avoid the formation of algae on the matting (Fig. 2d).
6. It is important to keep the matting wet all the times as if it gets dry the water transport can be compromised.
7. The soil in the cells of the modular trays must be filled equally. It can be compacted with the fingers to make sure all the cells are full of soil to the same level.
8. The watering system through capillarity could be used also to apply fertilizers, which could be mixed in the water container.
9. The seeds should be chosen with aim of ensuring as good quality of seeds and homogeneity between seeds as possible, in order to improve synchronicity in germination and growth.
10. To facilitate sowing, a hole can be formed in the soil (around 1 cm depth) of each of the cells using the back side of a thick pen, or the tip of a Falcon™ tube, for example.
11. The plants can be held by a support system to direct their growth, and to hold them erect and separated to avoid falling over each other (Fig. 4).
12. The aim of the sampling is to cover all the stages from the premeiotic stages to the end of meiosis. For this reason, the sampling frequency and intensity could be changed at early sampling to ensure that there are sufficient plants at later stages of particular interest.
13. Additional variables can be measured and included in the reference summary table to see if it can be improved or simplified. Some examples of more nondestructive variables include the number of days after sowing when the sample was collected, the Accumulated Growing Degree Days (AGDD) when the sample was collected, or the number of nodes detected in the stem.

14. The last height is taken to the flag leaf's ligule, even when the inflorescence has emerged completely, as the collar of the ear sometimes surpasses the flag leaf, and sometimes does not.
15. As [2] describes, "A leaf can be described as unfolded when its ligule has emerged from the sheath of the preceding leaf." In this study, a decimal number was added after a comma, referring to the elongation of the next leaf to the one already unfolded from 1 to 10. This elongation was referred to the previous unfolded leaf; for example, if the sixth leaf was the latest leaf emerging (still folded) and had reached a length of the last unfolded leaf (the fifth), the stage would be 5.5.
16. The temperature experiment was carried out on F_1 plants of a cross between cultivars Morex and Bowman. The control conditions were in glasshouse with around 20 °C day and 16 °C night (Although ideally, it would be done also in a control growth cabinet), and the heat treatment was done in a growth cabinet (SANYO 1700) with 30 °C day and 25 °C night.
17. Most of the data in the reference summary table (Table 1) was generated with Morex and Bowman data; however, the decisions were made on the F_1s. Whilst such fine scale genotypic differences should be taken into account if seed quantities allow it, it was assumed that such genetic variation was less than the within tray environmental variation shown by the staging data and thus ignored in this case.

Acknowledgments

The author would like to acknowledge Jim McNicol and BioSS for their advice and helpful discussions on the staging data and Ed Byrne and KWS-UK for their advice and collaboration in the COMREC network. This research was funded by the European Community's Seventh Framework Programme FP7-PEOPLE-2013-ITN COMREC- 606956. I.C. was funded by the European Community's Seventh Framework Programme *FP7/2007–2013* under grant agreement n° 222883 MeioSys and by ERC advanced grant "Shuffle" (Project ID: 669182). L.R. and R.W. were funded from the Scottish Government's Rural and Environment Science and Analytical Services Division Theme 2 Work Program 2.1.

References

1. Zadoks J, Chang T, Konzak C (1974) A decimal growth code for the growth stages of cereals. Weed Res 14:415–421
2. Tottman DR, Makepeace RJ, Broad H (1987) An explanation of the decimal code for the growth stages of cereals, with illustrations. Ann Appl Biol 110:441–454

3. Waddington SR, Cartwright PM, Wall PC (1983) A quantitative scale of spike initial and pistil development in barley and wheat. Ann Bot 51:119–130
4. Gómez JF, Z a W (2012) Non-destructive staging of barley reproductive development for molecular analysis based upon external morphology. J Exp Bot 63:4085–4094
5. Tracy SR, Gómez JF, Sturrock CJ et al (2017) Non-destructive determination of floral staging in cereals using X-ray micro computed tomography (μCT). Plant Methods 13:9
6. Watson A, Ghosh S, Williams MJ et al (2018) Speed breeding is a powerful tool to accelerate crop research and breeding. Nat Plants 4:23–29
7. Ghosh S, Watson A, Gonzalez-Navarro OE et al (2018) Speed breeding in growth chambers and glasshouses for crop breeding and model plant research. bioRxiv 2018:369512
8. Chaudhari AK, Chaudhary BR (2012) Meiotic chromosome behaviour and karyomorphology of Aloe vera (L.) Burm. f. Chromosom Bot 7:23–29
9. Comadran J, Kilian B, Russell J et al (2012) Natural variation in a homolog of antirrhinum CENTRORADIALIS contributed to spring growth habit and environmental adaptation in cultivated barley. Nat Genet 44:1388–1392
10. Lorieux M (2012) MapDisto: fast and efficient computation of genetic linkage maps. Mol Breed 30:1231–1235
11. Phillips D, Jenkins G, Macaulay M, et al (2015) The effect of temperature on the male and female recombination landscape of barley. New Phytol 208:421–429
12. Mascher M, Gundlach H, Himmelbach A et al (2017) A chromosome conformation capture ordered sequence of the barley genome. Nat Publ Gr 544:1–43

Chapter 27

In Planta Delivery of Chemical Compounds into Barley Meiocytes: EdU as Compound Example

Yun-Jae Ahn, Maria Cuacos, Mohammad A. Ayoub, Judith Kappermann, Andreas Houben, and Stefan Heckmann

Abstract

Here, we describe a protocol for in planta delivery of chemical compounds into meiocytes of different barley genotypes not impacting plant fertility allowing to harvest seeds from treated plants. Compound uptake into meiocytes is assessed by determining 5-ethynyl-2′-deoxyuridine (EdU) incorporation. Similar to EdU, other compounds being soluble in an aqueous solution can be delivered in planta before/during meiosis to decipher their impact on meiosis and meiotic recombination.

We give practical advice on how to deliver EdU as compound example (delivery via injection or *needle and thread*, addition of detergents or surfactants to increase compound uptake), how in planta compound delivery can be established for your plant material under specific growing conditions, how to generate and characterize barley hybrid plants, and how to conduct a meiotic cytological study of (treated) barley plants.

Key words Meiotic recombination, In planta compound delivery, Meiosis, EdU (5-ethynyl-2′-deoxyuridine), Cytology, *Hordeum vulgare*, Kompetitive allele specific PCR (KASP) analysis

1 Introduction

Meiosis generates genetic variation harnessed in plant breeding by genetic exchange between homologous chromosomes (i.e., crossovers (COs)) through homologous recombination (HR) [1, 2]. However, in cereal crops such as barley or wheat CO events are mainly restricted to distal chromosome regions, hindering exploitation of genetic resources from interstitial chromosome regions in breeding programs [3, 4]. Hence, manipulating CO frequency and/or distribution is of interest for crop plant breeding.

Meiotic recombination is tightly regulated [1, 2] but it is influenced by several factors including genetic, epigenetic and environmental factors; for example, (1) either disrupting genes encoding for anti-crossover factors or overexpressing *HEI10* increase meiotic recombination [5, 6], (2) recombination hotspots are associated with epigenetic (histone and DNA) modifications and

Mónica Pradillo and Stefan Heckmann (eds.), *Plant Meiosis: Methods and Protocols*, Methods in Molecular Biology, vol. 2061, https://doi.org/10.1007/978-1-4939-9818-0_27, © Springer Science+Business Media, LLC, part of Springer Nature 2020

mutations altering the chromatin landscape can also alter the meiotic recombination landscape [7–9], (3) posttranslational modifications (PTMs), such as SUMOylation, ubiquitinylation, or neddylation/rubylation, of proteins involved in different meiotic processes impact CO frequency and distribution by regulating various aspects of HR [10], and (4) temperature, pathogen infection, nutrient availability, salinity and drought stress, heavy metals, insecticides, herbicides, and fungicides can increase, redistribute, or decrease recombination events [11]. Thus, meiotic recombination landscapes are to some extent "plastic".

One way to evaluate the role of, for example, epigenetic modifications in different aspects of plant development is employing chemical compounds impacting DNA and histone modifications [12, 13]. To study their impact on meiotic progression or recombination frequency, compounds need to be delivered before/during meiosis in planta. Various approaches were used; for example incubating cut Arabidopsis stems with the histone deacetylase inhibitor Trichostatin A causes a redistribution of CO without changing the overall CO number [14], foliar spraying of various chemicals on tomato plants decreases recombination frequency between two linked markers [15], or incubating detached tillers of wheat–rye interspecific hybrid plants in okadaic acid positively affects chromosome pairing in a dose-dependent manner [16]. In barley, chemical compounds injected into stems can be incorporated into meiocytes [17–19].

Based on these reports, we developed a reliable protocol for treating barley meiocytes in planta with the thymidine analog EdU (5-ethynyl-2′deoxyuridine). This synthetic nucleoside can be substituted with any compound soluble in aqueous solution to study its impact on meiosis, by delivering it in the tiller allowing to acquire offspring of treated plants. In barley, as in other cereals, during the time meiosis occurs, the spike cannot be seen (being within the leaf sheath on the developing tiller), making it difficult to know the exact developmental stage of the spike. Scales correlating different aspects of plant morphology with the spike developmental stage do not consider differences between primary, secondary, or later tillers, and differences between cultivars or growth conditions or do not cover meiosis [20–23]. Alternatively, floral staging based on tomography and a modular tray system to synchronize spike development were developed [24, 25]. We describe here how to correlate plant morphology with developmental stage of the spike under your growing conditions, and subsequently how to perform pulse-chase EdU experiments to establish and monitor a reliable EdU uptake into barley meiocytes. We give advice on how to efficiently deliver (injection or *needle and thread*) chemical compounds in planta enabling seed production for subsequent analysis. Injecting simultaneously Tween 20 or Silwet L-77 increases the uptake efficiency. Consecutive injections or constant delivery of compounds by *needle and thread* enable to increase the total number of treated meiotic cells per spike during spike development. We also give practical advice on how to generate and analyze (Kompetitive allele specific PCR (KASP)-marker based SNP genotyping) barley F_1 hybrids (Barke × Morex) enabling to compare recombination landscapes between treated and untreated plants and on how to prepare chromosome spreads including EdU-detection for cytological evaluation.

2 Materials

2.1 Plant Material and Cultivation

1. Seeds of *Hordeum vulgare* cultivars Barke and Morex as well as F_1 Barke × Morex hybrids.
2. Precultivation soil: substrate 1.
3. Greenhouse soil: two parts self-made compost, two parts substrate 2, one part sand.
4. Fertilizer: Plantacote depot 4 m granulate (added once to the soil at the ~3 leaf stage) and 1% Hakaphos red (applied to the soil 4–5 weeks after germination).
5. Pots: 6 and 18 cm Ø pots.
6. Filter paper: 125 mm Ø.
7. Petri dish: 145 mm Ø and 20 mm height.
8. Parafilm.

2.2 Barke × Morex F_1 Hybrids: Plant Crossing and Hybrid Confirmation

1. Forceps: fine.
2. Pair of scissors.
3. Glassine bags (e.g., 16.5 × 40 cm).
4. Labels.
5. 2 and 1.5 ml centrifuge tubes.
6. Metal beads (e.g., 2.4 mm Ø).
7. Liquid nitrogen.
8. Homogenizer (e.g., mixer mill).
9. gDNA extraction buffer: 1% *N*-lauryl-sacrosine, 100 mM Tris–HCl (pH 8), 10 mM EDTA (pH 8) and 100 mM NaCl.
10. Phenol–chloroform–isoamyl alcohol: mix 25 parts of phenol and 24 parts of chloroform to 1 part of isoamyl alcohol (v/v/v).
11. Sodium acetate: 3 M sodium acetate in dH_2O, adjust pH to 5.2 with glacial acetic acid.
12. Isopropanol.
13. Fume hood.
14. 70% ethanol

15. 384-microwell plate (e.g., MicroAmp Optical 384-well reaction plate).
16. KASP SNP genotyping chemistry: enzyme mix and assay probes.
17. Assay16 (Polymorphism [Morex/Barke]): GATATGAGGAG AGGATCAGGCTGGTCATGTTCAACACCCATTTTGGAG CCCCCCACAAGC**[A/G]**CATCCATCCAGAGTCTTTCTA CACAAACTCATGGCCAGAGTGCTTCTGCCAGATGAA TGC.
18. Assay28 (Polymorphism [Morex/Barke]): TACAACCGAGA TCCTGCAAGCAGGATGCCTCTAGCTAGCACAACCGGA TTATTGATCATCATACAGACATCCATAACTTAAGAAAC TTTACAGAACCAAATATCAGCAGCTAATAATATA**[A/G]** TTAAGCGCAACCACGACGTTATTATTTTATACAGTAAT AAAATACAAAAAAGGAACGCAGCGCACGAACTTAGC AAGAGCTACATGCATGTAGTAGAAGCAGAAACCCGA CTTCTACTTC.
19. Adhesive films for microtiter plates.
20. Microcentrifuge.
21. Real-time PCR system including plate reader capable of detecting FAM (520 nm) and HEX (556 nm) (e.g., QuantStudio 6 Flex Real-Time PCR system).

2.3 Plant Selection

1. Forceps: fine.
2. Pair of scissors.
3. Fixative solution: three parts of absolute ethanol to one part of acetic acid fixative (v/v). Prepare fresh.
4. Ruler.
5. Glass vial (e.g., 2.8 cm Ø and 5 cm height).

2.4 In Planta Delivery of Compounds

1. 1× Phosphate Buffered Saline (PBS): 137 mM NaCl, 2.7 mM KCl, 10 mM Na_2HPO_4, 1.8 mM KH_2PO_4, pH 7.
2. EdU solution: 10 mM 5-ethynyl-2′deoxyuridine (EdU) in 1× PBS (*see* **Note 1**).
3. Syringe: 1 ml single-use syringe.
4. Injection needle: 25 G × 1″, size 17/23, Ø 0.50 × 25 mm.
5. Vac-In-Stuff (Silwet L-77).
6. Tween 20.
7. Labeling tape.
8. Fixative solution: three parts of absolute ethanol to one part of acetic acid fixative (v/v). Prepare fresh.
9. Commercial needles and threads.

10. 1.5 ml microcentrifuge tube.
11. 15 ml reaction tube.
12. Permanent marker.
13. Parafilm.
14. Crystal violet solution: ~100 μg/ml crystal violet in 1× PBS.

2.5 Cytological Analysis

1. Fixative solution: three parts of absolute ethanol to one part of acetic acid fixative (v/v). Prepare fresh.
2. Pair of scissors.
3. Forceps: fine.
4. Glass vial (e.g., 2.8 cm Ø and 5 cm height).
5. Petri dish: 94 mm Ø and 16 mm height.
6. Watch glass.
7. Commercial acetocarmine solution.
8. Alcohol burner.
9. Filter paper: 90 mm Ø.
10. 1.5 ml microcentrifuge tubes.
11. Enzyme digestion mix: 1% cellulase and 1% pectolyase in 0.01 M citrate buffer, pH 4.5 (stock solution). Prepare 333 μl aliquots which can be frozen at −20 °C for >1 year. Working solution: add 667 μl of 0.01 M citrate buffer, pH 4.5, to a stock aliquot.
12. Citrate buffer: 0.1 M citric acid monohydrate and 0.1 M Tri-sodium citrate dehydrate each in 1× PBS (10× stock solutions). Working solution: mix 444 μl of 0.1 M citric acid monohydrate and 556 μl of 0.1 M trisodium citrate dehydrate with 9 ml of dH_2O, prepare fresh.
13. 60% acetic acid.
14. Brass rod.
15. Hot block.
16. Microscope slides.
17. Coplin jar.
18. Coverslip: 22 × 22 mm.
19. Diamond pen.
20. Commercial blow-dryer.
21. EdU Click-488 kit.
22. 1× PBS: 137 mM NaCl, 2.7 mM KCl, 10 mM Na_2HPO_4, 1.8 mM KH_2PO_4, pH 7.
23. PBS-T: 0.1% Triton X-100 in 1× PBS.
24. Parafilm: pieces cut to approximately the size of a coverslip.

25. Humid chamber (e.g., plastic slide storage box with a layer of damp paper towels in the bottom).
26. VECTASHIELD mounting medium with 1.5 μg/ml of DAPI.
27. Stereo microscope.
28. Light microscope with a magnification of 20× or 40×.
29. Fluorescence microscope: equipped with a filter for DAPI and for the fluorochrome used for EdU detection (e.g., FITC filter) as well as with an image capture and analysis system.

3 Methods

3.1 Plant Growth and Cultivation

1. Place two filter papers in a petri dish and add 7.5 ml of distilled water.
2. Place 6–8 seeds (Barke, Morex, or F_1 hybrid) per petri dish on the wet filter paper and seal the petri dish with Parafilm (*see* **Note 2**).
3. Keep petri dish on the window bench for 5–6 days at RT (*see* **Note 3**).
4. Select similar-sized seedlings, transfer them to 6 cm Ø pots filled with precultivation soil, and precultivate them in a growth chamber [12 h light (~200 μmol/m^2 s), 80% relative humidity, 12 °C day/14 °C night] for 2 weeks.
5. Transfer plants to 18 cm Ø pots filled with greenhouse soil and relocate to a greenhouse [16 h light (~20 klx), 60–70% relative humidity, 17 °C day/13 °C night] (*see* **Note 4**).
6. To support plant growth and development apply fertilizer.
7. Eight to nine (Barke) or seven to eight (Morex) weeks after germination plants are ready for crossing to produce F_1 hybrid seeds (*see* Subheading 3.2). Five to six (Barke) or 4–5 (Morex, F_1 hybrids) weeks after germination plants are ready for plant selection (*see* Subheading 3.3) or in planta delivery of EdU/-compounds (*see* Subheading 3.4).

3.2 Barke × Morex F_1 Hybrids: Plant Crossing and Hybrid Confirmation

To produce hybrids seeds perform crosses of cultivars Barke (2-rowed) and Morex (6-rowed).

1. Use for emasculation 8–9-week-old Barke plants (Fig. 1a), when the spike is within the leaf sheath and ~1 cm of the first emerging awns are visible (Fig. 1b; *see* **Note 5**).
2. Remove the flag leaf, open up the outer leaf sheath carefully with your hand exposing the young spike (Fig. 1c) not breaking the fragile material (*see* **Note 6**).
3. Cut the awns with scissors enabling access to florets during emasculation.

Fig. 1 Barke × Morex F_1 hybrid barley production and confirmation via SNP genotyping. (**a**) Eight-week-old Barke (left) and Morex (right) barley plant; (**b**) Spike of an 8-week-old Barke plant with the first emerging visible awns being ~1 cm ready for emasculation and crossing; (**c**) Dissected Barke spike (left top) versus emasculated Barke spike (right top); left bottom panel: close-up of intact (left) and opened (right) spikelet with three immature anthers surrounding the female gynoecium, right bottom panel: close-up of dissected and emasculated spikelet, a ~2 mm vertical incision was made and green (immature) anthers were removed with forceps (see close-up); (**d**) Pale yellow (mature) Morex anthers collected in a petri dish (top) for pollination of emasculated Barke spikelets, Morex anthers inside of emasculated Barke spike for pollination (bottom); (**e**) 9-week-old Barke (left), Barke × Morex F_1 hybrid (middle), and Morex (right) plants; inset: close-up of spikes of parental genotypes and of their F_1 hybrids resembling either of their two parental genotypes (2-rowed Barke or 6-rowed Morex). (**f**) KASP marker based SNP genotyping: Plot showing Barke (red dots), Morex (blue dots) and Barke × Morex F_1 hybrid (green dots) allele calls based on SNP assay 16 and assay 28. Black squares represent negative controls

4. Trim away all lateral spikelets (two sets of three spikelets form at each node along the rachis, however in 2-rowed Barke only the central spikelets from each set are fertile, while the two laterals are not) as well as the most distal spikelets (typically too young and/or asynchronously developed) to exclude self-pollinated offspring.

5. Doing one spikelet at a time, make a ~2 mm vertical incision in each spikelet along the whole spike using forceps.
6. Remove all three green (immature) anthers from each remaining spikelet along the whole spike while not breaking them (Fig. 1c). Also, try not to touch stigma or style which might damage it.
7. Put the emasculated spike back into the leaf sheath carefully not breaking it, cover it with a glassine bag, close it with paper clip, and label it indicating the date and cross.
8. Roughly 2–3 days later, remove glassine bag, and take out the emasculated spike from the leaf sheath. Note, a female spike ready for pollination is characterized by swollen spikelets and elongated hairy stigma.
9. Cut off the tiller a few centimeters below the base of the spike from 6- to 7 -week-old Morex plants, remove the leaf sheath exposing the spike, hold the awns and pull them toward the base of the spike, open up each spikelet and collect pale yellow anthers in a clean petri dish (Fig. 1d).
10. Carefully using forceps transfer 2–3 mature Morex anthers into each of the emasculated Barke spikelets through the incision (Fig. 1d).
11. After pollination, transfer the spike back into the leaf sheath, cover it with a glassine bag and label it indicating date and cross info. Developing seeds will be seen within ~10 days in case of a successful cross.
12. Remove the bag 2 weeks after pollination to prevent the spike from becoming moldy due to excessive humidity.
13. ~30–40 days later, when the pollinated spike has set seeds and become dry and yellow, cut the spike from the base, store in a glassine bag at RT for ~1 week, harvest seeds and store at 4 °C.

    Germinate hybrid seeds, extract genomic DNA, and perform KASP-marker based SNP genotyping to unambiguously confirm hybrid genotype of plants in order to exclude self-pollinated offspring (morphology of hybrid plants varies, that is, they can resemble Barke or Morex with spikes being 2- or 6-rowed, respectively; Fig. 1e).
14. Germinate hybrid seeds as well as Barke and Morex seeds as controls, as described in Subheading 3.1.
15. Three to four weeks after germination, isolate 200–400 mg of young leaf material from each plant and freeze material in 2 ml microcentrifuge tubes with metal beads in liquid nitrogen.
16. Grind frozen samples using a homogenizer at 28 Hz for 45 s, open microcentrifuge tubes, add 800 μl of gDNA extraction buffer at RT and mix by vortexing.

17. Working under a fume hood, add 800 µl of phenol–chloroform–isoamyl alcohol, vortex samples for 2 min, centrifuge at 16,000 x *g* in a microfuge for 3 min at RT and transfer 700 µl of the supernatant to a new 1.5 ml microcentrifuge tube.
18. Add 70 µl of 3 M sodium acetate and 700 µl of isopropanol, invert the tubes gently but thoroughly, and centrifuge samples at 16,000 x *g* for 10 min at 4 °C in a microfuge.
19. Discard the supernatant, wash the pellet with 500 µl of 70% ethanol, and centrifuge at 16,000 x *g* for 5 min at 4 °C in a microfuge.
20. Remove the supernatant and air-dry the pellet at RT until no ethanol remains and a dry white pellet is seen.
21. Add 20 µl of dH_2O and vortex for a few seconds to resuspend the pellet.
22. Use these DNA samples to perform SNP genotyping with two independent KASP markers, assay 16 and assay 28, distinguishing Barke and Morex alleles to confirm hybrid status.
23. Prepare the following for each KASP marker reaction: 2.5 µl of 6 ng/µl template DNA, 2.5 µl of enzyme mix and 0.07 µl of assay mix. Include genomic DNA of Morex and Barke (positive control) and no DNA but dH_2O (negative control).
24. Load samples into a 384-well plate, cover with an adhesive film, and centrifuge to collect all samples at the bottom of each well.
25. Perform KASP reactions according to the following thermal cycling conditions: 30 s at 72 °C; 6 s at 26 °C; 15 min at 94 °C; 20 s at 94 °C and 1 min at 65 °C (step-wise decrease by 0.8 °C per cycle) for 10 cycles; 20 s at 94 °C and 1 min at 57 °C for 40 cycles.
26. Measure fluorescence at 520 nm (FAM) and 556 nm (HEX) by performing end-point fluorescence read-out at 30 °C for 1 min.
27. Plot FAM fluorescence values against HEX fluorescence values for each sample (including controls) (Fig. 1f) and also export into Microsoft Excel raw data.
28. Assign manually the allelic state of each sample—use as reference positive controls.
29. Once hybrid status of plants is confirmed via SNP genotyping, proceed to deliver in planta compounds/EdU (*see* Subheading 3.4).

3.3 Plant Selection

To study the effect of chemical compounds in meiosis, compounds need to be in planta delivered before/during the time a barley plant undergoes meiosis. During this time the spike cannot be seen being within the leaf sheath on the developing tiller. Thus, correlating plant morphology with the spike developmental stage going along with meiosis progression is crucial. In barley meiosis takes place in a gradient within the spike: most mature flowers are found toward the center of the spike, and younger stages are found both toward the top and the bottom. Spikes ~1.8–2.5 cm in length are suitable to find most meiotic stages, although variation exists. Depending on the cultivar and the growing conditions, the first tillers containing cells undergoing meiosis are found roughly 6–8 weeks after sowing. However, variation exists between and within plants. Therefore, below we explain a recommended procedure to get familiar with the material.

1. Start monitoring plants at regular intervals roughly 5 weeks after germination (*see* **Note 7**).
2. Find out where the spike is located along the stem by pressing the tiller softly with the thumb and the first finger, and move hand up and down along the tiller. Some nodes are found along the tiller and the spike sits above the uppermost one (*see* **Note 8**).
3. To identify a spike containing the first anthers with cells undergoing meiosis, find a node which has a hollow stem right below it, or in other words, two nodes close to each other with a small gap in between (usually <1 cm), and one leaf auricle 4–5 cm above this double node (Fig. 2a). The spike is within this region, located right above the second inner node which is felt very close to the bottom node (Fig. 2b, c). This is also the region where one should inject the compound solution to target the first anthers with cells undergoing meiosis (Fig. 2d; *see* Subheading 3.4). A rough indication to find where the spike is located is that below this double node the tiller feels empty when pressing whereas above this double node it feels hard and compact.

These indications are helpful, but one needs to acquire experience with the specific material under given growth conditions to identify plants with spikes around the time meiosis occurs. To do so, we suggest the following procedure:

1. Identify tillers containing spikes with the first anthers with cells undergoing meiosis according to the criteria described above.
2. Open the tiller in the greenhouse by cutting the stem with a scissor and carefully removing the leaf sheath on the developing tiller (*see* **Note 9**).

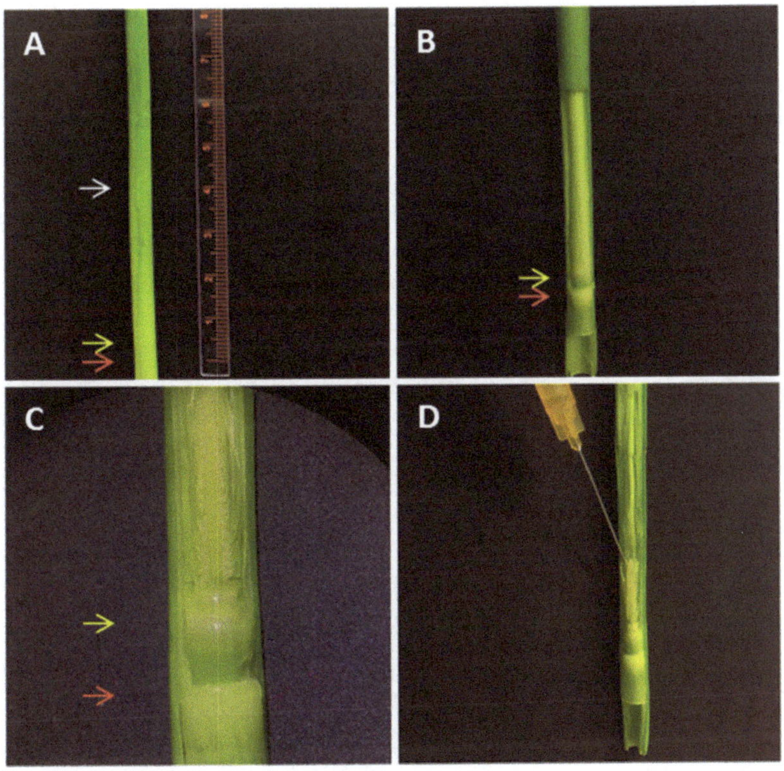

Fig. 2 Tiller and spike morphology of barley cultivar Barke during the time of in planta EdU/compound injection. (**a**) Spike containing meiocytes (in early meiosis) inside of the tiller. Yellow and red arrows indicate the position of the most upper nodes separated by a hollow part. ~4–5 cm above this thick node a leaf auricle (white arrow) is found. One can feel this leaf auricle by touching the outmost leaf layer as a small lump. Note, most meiotic stages within a spike are found when the distance between the leaf auricle and the upper node is ~5–7 cm. (**b**, **c**) Tiller with surrounding leaves removed showing the position of the two nodes and the spike. At the time of injection, the distance between the two nodes is less than 1 cm. (**d**) Representation of the injection procedure (surrounding leaves are removed for visualization). The needle is inserted diagonally and as vertical as possible into the target area not pierced into the spike to release the syringe content within the open space surrounding the spike

3. Measure the size of the spike and note in addition to the size also any morphological clue which may assist you in identifying similar-sized spikes (*see* **Note 10**).
4. Fix spikes within a size range of 1.0–3.5 cm in ice-cold fixative solution and check under the microscope the meiotic stages found in flowers at different positions along the spike (*see* Subheading 3.5).
5. Continue doing this process, ideally for 3–4 weeks every second or third day. Try to guess the spike size and then open the tillers to see if you were right (*see* **Note 10**).

Once you are familiar with your material under your growth conditions, perform pulse-chase EdU injections (*see* Subheading 3.4.1) or *needle and thread* EdU deliveries (*see* Subheading 3.4.3) to evaluate your success of identifying and treating plants containing meiocytes.

3.4 In Planta Delivery of Compounds

3.4.1 Pulse–Chase Injection: EdU as Compound Example to Evaluate Uptake into Meiocytes

Use initially EdU, that is incorporated into newly replicated DNA during premeiotic S-phase allowing to assess its uptake into meiotic cells cytological, for single pulse–chase injections to evaluate your success of identifying and treating plants containing meiocytes with a compound solution; however, any chemical compound soluble in an aqueous solution can be delivered in a similar way replacing EdU within 1× PBS. The duration of barley meiosis has been analyzed in different cultivars and growing conditions estimated to last from 39 h [26–28] to 43 h [18]. In our materials under our growth conditions, the earliest time point after a single EdU injection when we found tetrads labeled was 72 h. Thus, to assess and optimize compound uptake into meiocytes use a single injection of EdU and fix spikes 43 and 72 h after the injection.

1. Prepare 0.5 ml of a 10 mM EdU solution in 1× PBS per spike to be treated (*see* **Notes 1** and **11**).
2. Load the syringe with 0.5 ml of EdU solution per treatment.
3. Carefully (wear protective gloves) insert needle within the 4–5 cm space between the upper leaf auricle and the node where the spike is located, slightly above the spike (Fig. 2d).
4. Hold the stem with your other hand above the injection position, and firmly thrust the needle 1.5 cm diagonally (almost vertical) through the outer leaves (*see* **Note 12**).
5. Empty the content of the syringe. Successful delivery inside the tiller is characterized by no visible liquid dropping along the tiller.
6. Keep plants in the greenhouse.
7. To assess delivery of EdU into meiocytes fix treated spikes 43 and 72 h after injection (*see* Subheading 3.5).

3.4.2 Compound Delivery via Consecutive Injections

To increase the number of meiocytes exposed to a compound over the course of spike development, consecutive injections (this section) are performed or the *needle and thread* method (*see* Subheading 3.4.3) is used to deliver compounds for a defined period of time.

If the impact of a certain compound will be studied, spikes are either fixed after defined periods of time after the delivery to study meiocytes cytological or treated plants are kept in the greenhouse until seeds are formed used for further analysis.

1. Prepare 0.5 ml of EdU/compound solution per spike to be treated and 0.5 ml of 1× PBS to treat control plants (*see* **Note 11**). *Optional*: Adding surfactants or detergents to the compound solution further increases uptake rates not impacting plant fertility (*see* **Note 13**).
2. Load the syringe with 0.5 ml of solution per treatment and perform **steps 3–5** described in Subheading 3.4.1.
3. Mark treated tiller using labeling tape.
4. Keep plants in the greenhouse.
5. Repeat **steps 1–4** every 24 h for up to five times keeping plants in the greenhouse to increase the number of meiocytes exposed to the compound during spike development (*see* **Note 14**).
6. To assess delivery of EdU into meiocytes fix treated spikes 43 and 72 h after the last injection (*see* Subheading 3.5). To study the impact of a given compound on meiotic progression and/or recombination fix spikes after, for example, 12, 24, 36, 48, 60, and 72 h for a detailed cytological analysis or keep plants in the greenhouse to acquire seeds in order to, for example, analyze meiotic recombination rates in the F_2 offspring of untreated versus treated F_1 hybrid plants.

3.4.3 Compound Delivery via Needle and Thread

As alternative method for compound delivery use the *needle and thread* method (Fig. 3a). By inserting a thread with a needle through the node immediately below the spike and immersing both sides of the thread into a 1.5 ml microcentrifuge tube containing the compound solution, the compound is transmitted via the thread in planta. This enables a constant compound delivery that can be stopped at any time by removing the thread. However, the

Fig. 3 Delivery of compound solution via *needle and thread*. (**a**) Delivery of compounds in tiller of barley cultivar Barke via *needle and thread*. (**b**, **c**) Evaluation of the suitability of threads and the thread length to transmit solutions: Threads are passed with the help of a needle through a paper tissue and are immersed into a staining solution (e.g., 1× PBS with crystal violet). If the thread is transmitting the stained solution, the paper tissue gets stained. (**b**) Successful transmission dependent on thread used and (**c**) unsuccessful transmission due to too long thread used

frequency of treated plants showing EdU uptake into meiocytes is lower compared to injections likely due to not consistently inserting the thread into the vascular system delivering substances to the spike. By inserting two threads per tiller the frequency of treated plants showing EdU uptake can be increased. However, not all treated plants develop seeds, but instead early spike abortion can occur. Thus, a higher number of plants need to be treated when compared to the injection.

1. Thread needles (one or two needles per spike) with a ~40 cm long thread and join together both ends of the thread with a knot (*see* **Note 15**).
2. Fill 1.5 ml microcentrifuge tubes (one per spike) with 1 ml of the desired compound solution.
3. Prepare an additional 15 ml reaction tube filled with 1× PBS which will be used to wet the threads before inserting them into the plants.
4. Identify the uppermost node where the spike is located and make a small mark with a permanent marker.
5. Open one 1.5 ml microcentrifuge tube, cover it with Parafilm, and attach it to the spike with tape closely below the uppermost node where the spike is located (*see* **Note 16**).
6. Repeat **steps 1–5** for all tillers.
7. Take a threaded needle and wet using the 15 ml reaction tube filled with compound solution.
8. Thrust the needle firmly, horizontally through the node, and pull until roughly half the length of the (double) thread has been through.
9. Immediately insert both ends of the thread (one of them carrying the needle) into the attached 1.5 ml microcentrifuge tube through the Parafilm. To minimize evaporation keep the hole in the Parafilm as small as possible. Use the top part of the needle (where the hole or eye is) to push all the thread into the tube (Fig. 3a).
10. *Optional*: Insert a second thread with a needle in the same way, this time perpendicular to the first one. Immerse this second needle and thread into the same 1.5 ml microcentrifuge tube as in **step 9**.
11. Repeat **steps 7–10** for each spike.
12. Keep this set-up leaving the plants in the greenhouse (*see* **Note 17**).
13. After treatment fix material in fixative solution for cytological evaluation. Alternatively, either replace compound solution within attached 1.5 ml microcentrifuge tube with 1× PBS or carefully remove threads and keep plants in the greenhouse

until later cytological evaluation or to allow the plants to set seeds. *Optional*: Close the hole in the node with Parafilm or tape.

3.5 Cytological Analysis

3.5.1 Fixation of Barley Inflorescence

Barley spikes (~1.8–2.5 cm spike) containing anthers with cells undergoing meiosis are found in tillers characterized by a distance of 5–7 cm between the upper leaf auricle and the lower node where the spike is located.

1. Cut the tiller below the node where the spike is located.
2. Dissect leaves surrounding the spike with scissor and forceps.
3. Transfer isolated spike(s) in glass vials filled with ice cold fixative solution and keep on ice.
4. Replace fixative solution after three hours with fresh ice-cold fixative solution and store overnight at RT.
5. The following day replace fixative solution with fresh fixative solution and store overnight a RT.
6. Repeat **step 5** until the solution remains clear. Proceed with slide preparation (*see* Subheading 3.5.2) or store material at 4 °C. Material can be stored for at least six months without impact on sample quality.

3.5.2 Chromosome Spread Preparation

One barley flower has three anthers in which meiosis progresses fairly synchronized. To preselect anthers at desired meiotic stage for subsequent analysis use acetocarmine chromosome staining for one of the three anthers per flower. Using a light microscope different categories of meiotic stages can be easily identified: (1) Early prophase I (leptotene → early pachytene): Meiocytes are found as columns of connected cells (Fig. 4a), (2) late prophase I (pachytene → diplotene): Associated meiocytes lose contacts forming discrete cells (Fig. 4b), (3) meiotic divisions (diakinesis → telophase II): Condensed meiotic chromosomes are visible undergoing both meiosis I and II division (Fig. 4c, d), and (4) tetrad stage: four haploid nuclei forming tetrads (Fig. 4e); later stages represent various intermediates of pollen development (Fig. 4f). After preselection, the remaining two anthers per flower are used for chromosome spread preparations and subsequent analysis.

1. Place fixed spike(s) in a petri dish filled with fixative solution, and dissect the flowers using forceps.
2. Dissect one of the three anthers per flower and keep the remaining two anthers in the petri dish.
3. Carefully transfer one anther onto a microscopic slide.
4. Add 5-10 μl of acetocarmine solution to the anther on the slide and heat specimen by passing it briefly above the flame of an alcohol burner two to three times to increase staining intensity.

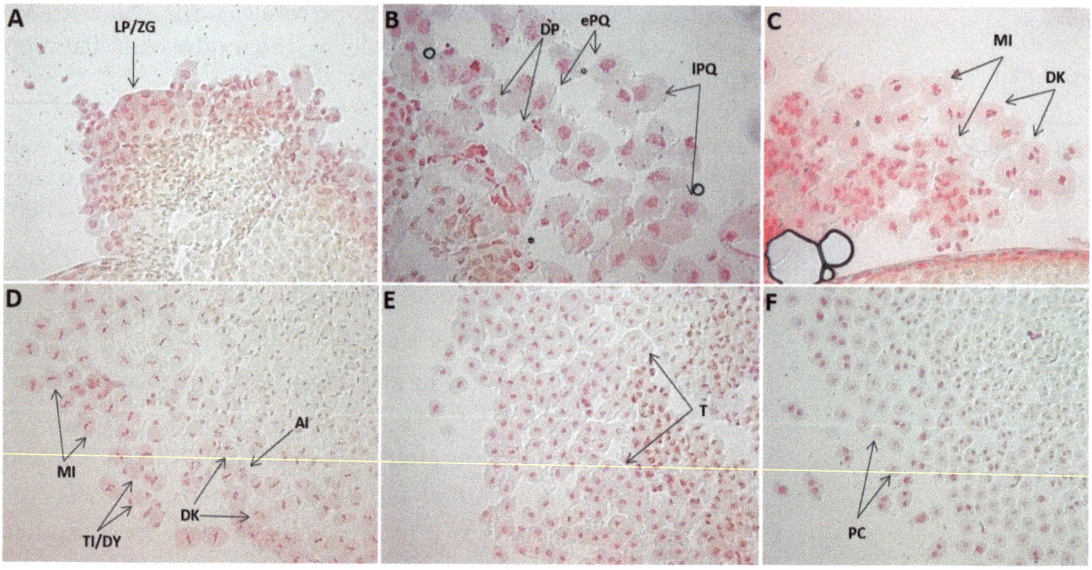

Fig. 4 Analyzing male meiocytes of barley cultivar Barke using acetocarmine. (**a**) Leptotene/zygotene: meiocytes as columns of connected cells. (**b**) Pachytene/diplotene: meiocytes as discrete cells. (**c**, **d**) Condensed meiotic chromosomes undergoing meiosis I division forming dyads. (**e**) Tetrad stage: four haploid nuclei forming tetrads. (**f**) Different stages of pollen development. *LP* leptotene, *ZG* zygotene, *ePQ* early pachytene, *lPQ* late pachytene, *DP* diplotene, *DK* diakinesis, *MI* metaphase I, *TI* telophase I, *DY* dyad, *T* tetrad, *PC* pollen cell

5. Place a coverslip on top and press gently between half-folded filter paper with the thumb to release the meiocytes and to remove excess of staining solution.
6. Observe under a light microscope with a magnification of 20× or 40× and determine the meiotic stage(s) (*see* **Note 18**).
7. Dissect the remaining two anthers of the flower(s) at desired stage and proceed with slide preparation or store anthers in 1.5 ml microcentrifuge tubes filled with 800 μl of fixative solution at 4 °C (material can be stored for at least six months without impact on sample quality).
8. To prepare meiotic chromosome spreads, transfer anthers, for example, according to meiotic stage to independent watch glasses containing 300 μl of citrate buffer at RT for 5 min.
9. Remove citrate buffer, add 300 μl of fresh citrate buffer, and incubate for 5 min at RT.
10. Repeat **step 9**.
11. Replace the citrate buffer with at least 200 μl of enzyme digestion mix covering completely the anthers and incubate at 37 °C for 45 min in a humid chamber.
12. Remove enzyme digestion mix carefully and add 500 μl of ice-cold citrate buffer to stop enzymatic digestion reaction. Place the watch glasses on ice meanwhile preparing the slides.

13. Transfer one or two anthers (*see* **Note 19**) onto a slide with a 3 μl drop of citrate buffer by scooping it up softly with forceps.
14. Squash the anther(s) thoroughly with a brass rod.
15. Confirm under a stereo microscope that all material is crushed. If large pieces remain intact continue squashing the anther (s) thoroughly with a brass rod.
16. Add 10 μl of 60% acetic acid to the drop and mix with a needle.
17. Place the slide on a hot block at 52 °C and move the drop for 60 s using a needle horizontal to the slide not scratching the slide surface.
18. After chromosome spreading, mark the spreading area with a diamond pen (slightly smaller than the size of a coverslip) and remove the slide from the hot block.
19. Immediately add 100 μl of ice-cold fixative solution in a circle surrounding the drop and invert the slide at a 45–90° angle over a waste container to remove excessive fluid.
20. Dry the preparation by applying hot air toward the back of the slide with a blow-dryer.
21. Proceed with EdU click reaction and DAPI staining (*see* Subheading 3.5.3; Fig. 5) or continue with the preparation for chromosome spread analysis.

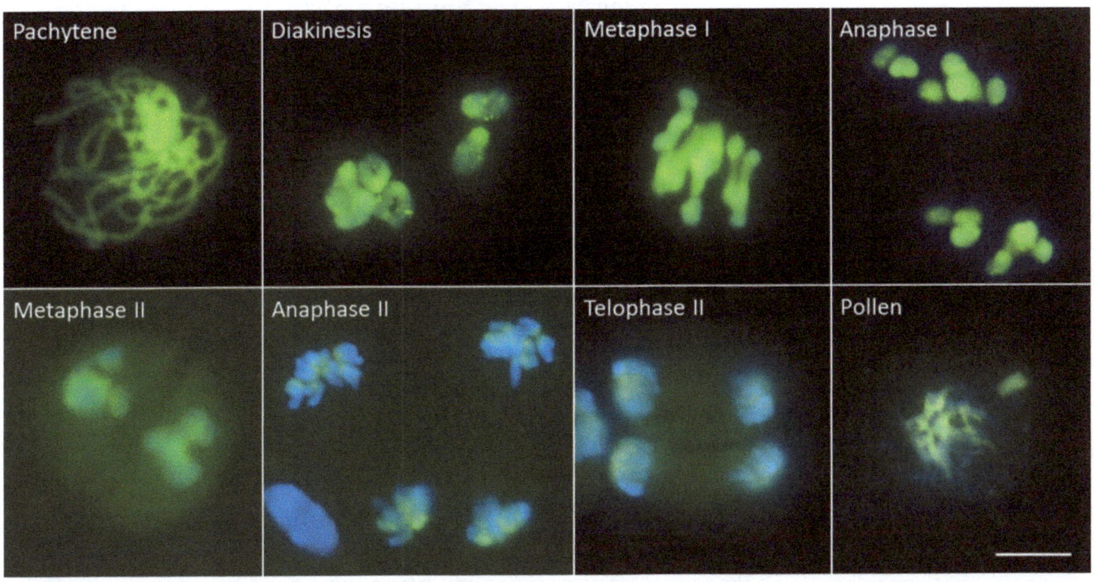

Fig. 5 EdU-positive male meiocytes of barley cultivar Barke. Chromosome spreads of EdU-positive male meiocytes at pachytene, diakinesis, metaphase I, anaphase I, metaphase II, anaphase II, telophase II, and early pollen after single injection of a 10 mM EdU solution in planta. DNA-counterstained with DAPI. Bar represents 10 μm

22. Add 18 μl of VECTASHIELD mounting medium with DAPI on top of the spreading area, cover with a coverslip, and press gently between half-folded filter paper with the thumb to remove excess of mounting medium.
23. View chromosome spread preparations under a fluorescence microscope.

3.5.3 EdU Click Reaction and DAPI Staining

1. Incubate slides (*see* **Note 18**; *see* Subheading 3.5.2, **step 21**) in a Coplin jar filled with 1× PBST for 15 min at RT.
2. Remove 1× PBST by carefully inverting the Coplin jar holding the slides, fill the Coplin jar with 1× PBS, and incubate for 5 min.
3. Remove 1× PBS by carefully inverting the Coplin jar holding the slides, fill the Coplin jar with 1× PBS, and incubate for 5 min.
4. Prepare EdU detection cocktail according to the manufacturer's instruction. Take care of adding components in the exact given order and protect cocktail from direct light.
5. Remove slides from the Coplin jar, add 50 μl of EdU detection cocktail per slide on top of the spreading area, and cover with Parafilm.
6. Incubate the slides in a humid chamber at 37 °C for 30 min in darkness.
7. Carefully remove Parafilm using forceps and wash slides in a Coplin jar filled with 1× PBS for 2 min.
8. Remove slides from Coplin jar and allow air-drying in darkness.
9. Add 18 μl of VECTASHIELD mounting medium with DAPI on top of the spreading area, cover with a coverslip, and press gently between half-folded filter paper with the thumb to remove excess of mounting medium.
10. Analyze slides under a fluorescence microscope.

4 Notes

1. 10 μM [17] and 10 mM [18] EdU were used to label meiotic cells in barley. We tested 100 μM, 1 mM, and 10 mM EdU concentrations, none of them impacting meiotic progression. 1 and 10 mM work reliable for us, whereas the percentage of labeled cells with 100 μM typically decreased.
2. Place 50% more seeds than plants you need for your experiment to germinate on filter paper to assure a sufficient number of plants, since not all seeds will (synchronously) germinate.

3. Monitor plates every second day. When the filter paper is getting dry add more distilled water.
4. In our experience, under these conditions barley plants (Barke, Morex, and F_1 Barke × Morex hybrids) grow vigorous and uniform as well as they produce a high number of tillers per plant.
5. Timing and morphology suitable for emasculation differ from cultivar to cultivar. For example, in cultivar Golden Promise if the awns just emerged outside the flag leaf, anthers are already yellow and mature, meaning that it is too late for emasculation. We recommend sowing Barke plants ~2 weeks earlier than Morex plants, since under our growth conditions Morex plants grow faster and are ready earlier for crossing than Barke.
6. The spike is too old for emasculation if anthesis (pollen shed) occurred, that is, if anthers particularly toward the center of spike (typically most developed spikelets) are pale yellow. To exclude self-pollination in this case, remove those spikelets or do not use the spike.
7. We recommend sowing initially at regular intervals plants (e.g., every second or third day) to have sufficient plant material available as well as once a proper plant is identified the next batch of plants will be ready foreseeable.
8. Initially, available scales [20–23] are of great help to get familiar with your plant material and to identify plants undergoing meiosis.
9. We tried an alternative method (similar as for maize [29]) to determine the spike size in alive barley plants: making a small incision in the tiller to carefully measure the spike size through the incision and/or to dissect a few flowers in order to determine meiotic stages, afterward suturing the incision with labeling tape or Parafilm, keeping the plants in the greenhouse and repeating this procedure at defined intervals to follow spike development in planta. Unfortunately, barley stems are relatively thin and soft making it very difficult to make an incision without damaging the tiller. In addition, after closing the incision using labeling tape or Parafilm, we could not assure undisturbed spike development.
10. Note down morphological features (e.g., we use a number of phenotypic parameters to estimate the spike size, particularly the position of nodes, the internode distance, and position of the spike relative to the nodes to estimate the spike size) for consecutive days before opening the material and when opening the material, to know how the plant looks like before and during the time meiosis occurs within the spikes.

11. Initially varying amounts of fluid (typically between 300 and 800 μl) were injected until a drop of liquid appeared at the top of the tiller. To deliver consistent amounts of fluids across plants, we changed to delivering a fixed amount of 500 μl which is sufficient to flood the space surrounding the spike within the tiller and to obtain EdU incorporation into meiotic cells.
12. By inserting the needle with a vertical angle thrusting the needle through the tiller is avoided. Always place your hand above and never behind the injection site to avoid piercing yourself in case the needle goes through the tiller.
13. Different detergents and surfactants were tested in order to increase EdU uptake rates into meiocytes. Addition of 0.5%, 1%, and 5% of Tween 20 increased consistently the percentage of EdU-positive meiotic cells across independent experiments without causing any detrimental effects in the plants, that is, meiosis progression was unaltered and seeds were harvested. Similar results were obtained by adding 0.025%, 0.01%, and 0.05% of Silwet L-77. Adding 0.1%, 0.5%, or 1% of Triton X-100 was detrimental to the plants turning them yellow and dying. Thus, we did not further evaluate addition of Triton X-100 at, for example, lower concentrations.
14. We performed up to five consecutive injections every 24 h not impacting plant fertility, that is, seeds of treated plants were harvested.
15. Commercial needles and threads were employed. However, not all threads transmit solutions with the same efficiency and successful transmission of solutions depends on the length of the thread. An easy test for the suitability of the thread and the thread length is to pass them with the help of a needle through a paper tissue and immersing the thread subsequently into a staining solution (e.g., 1× PBS with Crystal violet). If the thread is transmitting the liquid, the paper gets stained (Fig. 3b).
16. Keeping the distance between the 1.5 ml microcentrifuge tube and the insertion point short, minimizes evaporation of the solution along the thread and increases efficiency of transmission to the plant. We recommend performing **steps 1–5** for all scheduled treatments before inserting the needles into the plants, to minimize the time lapse between different spikes, especially if many plants are to be treated simultaneously.
17. It is important to monitor the 1.5 ml microcentrifuge tubes regularly, that is, at least once but better twice per day, to refill the absorbed/evaporated content.
18. Slides prepared from acetocarmine-squashed materials can be used for DAPI-stained chromosome spread analysis and/or

EdU-detection. To do so, freeze the slide for 10 s in liquid nitrogen, remove coverslip using a razor blade, incubate slides in 1× PBS at RT for 2 min, air-dry slides, and continue with DAPI staining (*see* Subheading 3.5.2, **step 22**) or EdU detection (*see* Subheading 3.5.3).

19. Using two anthers per slide results in a high number of meiocytes per preparation. To monitor consistent EdU uptake across a spike, prepare independent slides from the two remaining anthers from individual flowers across the spike. In our materials we found consistent EdU uptake in individual anthers of each flower independent of the position along the spike, that is, located at the top, middle, or bottom of the spike.

Acknowledgments

We would like to thank the gardeners team from the IPK Gatersleben for their excellent support with plant cultivation, Jana Lorenz for excellent technical assistance, all members of the Meiosis team at the IPK for fruitful discussions, as well as the German Federal Ministry of Education and Research (BMBF—031B0188), Bayer CropScience AG (Grants4Traits 2016-1-18), the Marie-Curie "MEICOM" network H2020 ITN-765212, and the IPK Gatersleben for financial support.

References

1. Mercier R, Mézard C, Jenczewski E, Macaisne N, Grelon M (2015) The molecular biology of meiosis in plants. Annu Rev Plant Biol 66:297–327. https://doi.org/10.1146/annurev-arplant-050213-035923
2. Wang Y, Copenhaver GP (2018) Meiotic recombination: mixing it up in plants. Annu Rev Plant Biol 69:577–609. https://doi.org/10.1146/annurev-arplant-042817-040431
3. Choulet F, Alberti A, Theil S, Glover N, Barbe V, Daron J et al (2014) Structural and functional partitioning of bread wheat chromosome 3B. Science 345:1249721. https://doi.org/10.1126/science.1249721
4. Phillips D, Wnetrzak J, Nibau C, Barakate A, Ramsay L, Wright F et al (2013) Quantitative high resolution mapping of HvMLH3 foci in barley pachytene nuclei reveals a strong distal bias and weak interference. J Exp Bot 64:2139–2154. https://doi.org/10.1093/jxb/ert079
5. Fernandes JB, Séguéla-Arnaud M, Larchevêque C, Lloyd AH, Mercier R (2018) Unleashing meiotic crossovers in hybrid plants. Proc Natl Acad Sci U S A 115:2431–2436. https://doi.org/10.1073/pnas.1713078114
6. Ziolkowski PA, Underwood CJ, Lambing C, Martinez-Garcia M, Lawrence EJ, Ziolkowska L et al (2017) Natural variation and dosage of the HEI10 meiotic E3 ligase control Arabidopsis crossover recombination. Genes Dev 31:306–317. https://doi.org/10.1101/gad.295501.116
7. Choi K, Zhao X, Tock AJ, Lambing C, Underwood CJ, Hardcastle TJ et al (2018) Nucleosomes and DNA methylation shape meiotic DSB frequency in Arabidopsis thaliana transposons and gene regulatory regions. Genome Res 28:532–546. https://doi.org/10.1101/gr.225599.117
8. Underwood CJ, Choi K, Lambing C, Zhao X, Serra H, Borges F et al (2018) Epigenetic activation of meiotic recombination near *Arabidopsis thaliana* centromeres via loss of H3K9me2 and non-CG DNA methylation. Genome Res 28:519–531. https://doi.org/10.1101/gr.227116.117

9. Yelina N, Diaz P, Lambing C, Henderson IR (2015) Epigenetic control of meiotic recombination in plants. Sci China Life Sci 58:223–231. https://doi.org/10.1007/s11427-015-4811-x
10. Lambing C, Heckmann S (2018) Tackling plant meiosis: from model research to crop improvement. Front Plant Sci 9:829. https://doi.org/10.3389/fpls.2018.00829
11. Fuchs LK, Jenkins G, Phillips DW (2018) Anthropogenic impacts on meiosis in plants. Front Plant Sci 9:1429. https://doi.org/10.3389/fpls.2018.01429
12. Pecinka A, Liu CH (2014) Drugs for plant chromosome and chromatin research. Cytogenet Genome Res 143:51–59. https://doi.org/10.1159/000360774
13. Zhang H, Wang B, Duan CG, Zhu JK (2013) Chemical probes in plant epigenetics studies. Plant Signal Behav 8:e25364. https://doi.org/10.4161/psb.25364
14. Perrella G, Consiglio MF, Aiese-Cigliano R, Cremona G, Sanchez-Moran E, Barra L et al (2010) Histone hyperacetylation affects meiotic recombination and chromosome segregation in Arabidopsis. Plant J 62:796–806. https://doi.org/10.1111/j.1365-313X.2010.04191.x
15. Griffing B, Langbridge J (1963) Factors affecting crossing over in the tomato. Aust J Biol Sci 16:826–837
16. Knight E, Greer E, Draeger T, Thole V, Reader S, Shaw P et al (2010) Inducing chromosome pairing through premature condensation: analysis of wheat interspecific hybrids. Funct Integr Genomics 10:603–608. https://doi.org/10.1007/s10142-010-0185-0
17. Colas I, Macaulay M, Higgins JD, Phillips D, Barakate A, Posch M et al (2016) A spontaneous mutation in MutL-Homolog 3 (HvMLH3) affects synapsis and crossover resolution in the barley desynaptic mutant *des10*. New Phytol 212:693–707. https://doi.org/10.1111/nph.14061
18. Higgins JD, Perry RM, Barakate A, Ramsay L, Waugh R, Halpin C et al (2012) Spatiotemporal asymmetry of the meiotic program underlies the predominantly distal distribution of meiotic crossovers in barley. Plant Cell 24:4096–4109. https://doi.org/10.1105/tpc.112.102483
19. Sinha RP, Helgason SB (1969) The action of actinomycin D and diepoxybutane on recombination of two closely linked loci in *Hordeum*. Can J Genet Cytol 11:745–751
20. Zadoks JC, Chang TT, Konzak CF (1974) A decimal code for the growth stages of cereals. Weed Res 14:415–421
21. Gómez JF, Wilson ZA (2012) Non-destructive staging of barley reproductive development for molecular analysis based upon external morphology. J Exp Bot 63:4085–4094. https://doi.org/10.1093/jxb/ers092
22. Tottman D, Makepeace R, Borad H (1979) An explanation of the decimal code for the growth stages of cereals, with illustrations. Ann Appl Biol 93:221–234
23. Waddington S, Cartwright P, Wall P (1983) A quantitative scale of spike initial and pistil development in barley and wheat. Ann Bot 51:119–130
24. Tracy SR, Gómez JF, Sturrock CJ, Wilson ZA, Ferguson AC (2017) Non-destructive determination of floral staging in cereals using X-ray micro computed tomography (μCT). Plant Methods 13:9. https://doi.org/10.1186/s13007-017-0162-x
25. Arrieta M, Colas I, Macaulay M, Robbie Waugh R, Luke Ramsay L (2019) A modular tray growth system for barley. In: Plant meiosis: methods and protocols. Humana Press, Totowa, New Jersey
26. Bennett MD, Finch RA (1971) Duration of meiosis in barley. Genet Res 17:209–214
27. Bennett MD, Finch RA, Smith JB, Rao MK (1973) The time and duration of female meiosis in wheat, rye and barley. Proc R Soc Lond B 183:301–319
28. Finch RA, Bennett MD (1972) The duration of meiosis in diploid and autotetraploid barley. Can J Genet Cytol 14:507–515
29. Dukowic-Schulze S, Garcia N, Shunmugam ASK, Kagale S, Chen C (2019) Isolating male meiocytes from maize and wheat for "-omics" analyses. In: Plant meiosis: methods and protocols. Humana Press, Totowa, New Jersey

INDEX

Mónica Pradillo and Stefan Heckmann (eds.), *Plant Meiosis: Methods and Protocols*, Methods in Molecular Biology, vol. 2061, https://doi.org/10.1007/978-1-4939-9818-0, © Springer Science+Business Media, LLC, part of Springer Nature 2020

FSC
www.fsc.org
MIX
Papier | Fördert
gute Waldnutzung
FSC® C083411

Kolibri 360 GmbH
Ferdinand-Jühlke-Straße 7
99095 Erfurt, Deutschland
produktsicherheit@kolibri360.de